Thomist Realism
and the
Linguistic Turn

THOMIST REALISM
and the
LINGUISTIC TURN

Toward a More Perfect Form
of Existence

JOHN P. O'CALLAGHAN

University of Notre Dame Press
Notre Dame, Indiana

Copyright © 2003 by
University of Notre Dame
Notre Dame, Indiana 46556
http://www.undpress.nd.edu
All Rights Reserved

Manufactured in the United States of America

Library of Congress Cataloging-in-Publication Data
O'Callaghan, John (John P.)
   Thomist realism and the linguistic turn : toward a more perfect form of existence / John P. O'Callaghan.
      p.   cm.
   Includes bibliographical references and index.
   ISBN: 978-0-268-04218-9
   1. Knowledge, Theory of.   2. Mental representation.   3. Language and languages—Philosophy.   4. Thomists.   I. Title.
BD161 .O3   2002
149'.91—dc21
                        2001006430

∞ *This book is printed on acid-free paper.*

For Ralph McInerny

Christian, Scholar, Teacher, and Friend
Among the many things I have learned from you,
the greatest has been indirect,
that there is no word for gratitude.
So, to you my teacher, I dedicate these words.

Contents

	Acknowledgments	ix
	Introduction: Words, Thoughts, and Things	1
Chapter 1	Aristotle's Semantic Triangle in St. Thomas	15
Chapter 2	Three Rival Versions of Aristotle	41
Chapter 3	Language and Mental Representationalism: Historical Considerations	79
Chapter 4	The Language of Thought: A Revival of Mental Representationalism	113
Chapter 5	Hilary Putnam's Criticism of Aristotelian Accounts of Language and Mental Representationalism	135
Chapter 6	The Third Thing Thesis	159
Chapter 7	The Introspectibility Thesis	199
Chapter 8	The Internalist Thesis and St. Thomas's "Externalism"	237
Chapter 9	Conclusion: Toward a More Perfect Form of Existence	275
	Notes	299
	Bibliography	337
	Index	347

ACKNOWLEDGMENTS

Words, as conventional signs, cannot express the particular gratitude I feel for those who have helped me along the way to writing this work. Indeed, any effort to acknowledge the help I have received in producing it must inevitably fall short. But that is no reason not to try. Among the many I must thank are Ralph McInerny, Tom Hibbs, Scott Moore, Fred Freddoso, David Burrell, Stanley Hauerwas, Alasdair MacIntyre, David Solomon, Patrick Murray, John Safranek, Tim Smith, Michael Dauphinais, Michael Letteney, Chris Kaczor, and my colleagues at Creighton University. I must also thank the University of Notre Dame Press, and Jeff Gainey in particular, as well as Jude Dougherty and *The Review of Metaphysics*. I would also like to thank in a special way my editor Carole Roos, who has been tireless and patient in improving my text. The University of Notre Dame, Creighton University, and the Jacques Maritain Center have provided me with invaluable resources and financial support. My wife and children may someday understand how much their forbearance and patience amidst my efforts to finish this work have helped me to grasp its importance. After a mystical vision, St. Thomas compared his written work to straw. I do not need a mystical vision. I need only catch a momentary glimpse of them.

Finally, the gratitude I feel toward my father is without measure. There is no page of this work that was written without him in mind. Years before I ever heard the phrase *fides quaerens intellectum*, I knew in him the reality that it signifies. Though St. Thomas was his master, I think St. Augustine was his first love. It is only fitting that I recall how in the *Confessions* St. Augustine interrupts the train of his narrative to ask his readers to pray for his mother and father. I am no Augustine and he was no Monica. Still, I would ask the same of my readers. May he rest in peace.

Introduction

Words, Thoughts, and Things

> The frontier between explaining the meaning of words and describing the nature of things is easily violated.
> —C.S. Lewis

How do our words "attach" to objects in the world? Most assuredly they do. If I ask my five-year-old son to bring me an apple, unsurprisingly, he does; he brings me just what I wanted. If I ask him to bring me an orange, he brings me something quite different than in the first instance, but also just what I wanted. Not only does 'apple' attach in some way to certain objects in the world, but 'to bring' seems to attach to certain acts, even before those acts are performed and could be called existing things. My son performs the act at my request; he knows *what* to do even before the doing occurs. He doesn't turn on the television, or twiddle his thumbs, or needle his sister, unless of course he is disobedient. He gets up from his chair, walks to the bowl with apples in it, picks one, walks to me, and hands it to me. I count my request a success.

I do not mean to suggest that every word we use is like 'apple' or 'orange' which attach fairly straightforwardly to beings in the world. Words like 'oh' and 'every' and 'no' and 'but' do not seem to be as directly related to beings in the world, if at all. But given a moment's reflection even the connection of 'apple' or 'orange' to the world begins to take on the appearance of something quite odd. If my son randomly picks an apple from the bowl and brings it to me, I do not respond, "That's not the one I asked for," since there is in fact no one that I asked for. But if I didn't ask for a particular apple, what did I ask for, since the only apples that exist are particular apples? How does one ask for an apple without asking for a particular apple?

Even if it is unremarkable that some words attach, or connect, or signify, or refer to things in the world, philosophers have wondered just how this is possible. We all know the fact; what we want to know is the why of the fact. Some philosophers like Plato and Wittgenstein have considered the suggestion that our words should be compared to pictures or images of the things they attach to, though what Wittgenstein meant by a *picture* in this context is very complex. But as Plato realized in the *Cratylus,* and Wittgenstein later in the *Philosophical Investigations*, it is very difficult to make sense of this suggestion.

Other philosophers take particular note that my son does not originally succeed in bringing me an apple. When he is very young, he stands perplexed at my requests, then slowly over time fails less and less, and finally generally succeeds. There is a similarity between his behavior and that of other animals that can be trained to bring me an apple or an orange. Perhaps my son acquires certain dispositions to behave in certain overt ways in the appropriate circumstances and responds to certain vocal stimuli like 'apple' and 'bring', just like a chimp. The words attach to reality because of the ways in which the behavioral dispositions attach to reality. But this suggestion has also proven very difficult to sustain. The combinatorial possibilities of language use seem to be virtually infinite in their openness to future applications; trained dispositions to behave, on the other hand, do not seem to be so open. And it is not at all clear how this analysis is supposed to handle such statements as "Caesar crossed the Rubicon and started a civil war in Rome" or "the googol root of the googolplex is 10."

Another suggestion is that we are all born with an innate language of the mind, a *lingua mentis* that is the same for all human beings, a part of human nature. The innate language attaches to the world because it is causally related to beings in the world, which causal relations fall under certain natural laws. What in my son looks like learning a first language, a native language like English or Spanish, is really not learning how that language attaches to the world. He is born knowing a first language, the *lingua mentis*, the meaning of its terms and the possibilities for their syntactical arrangement. What he must learn are the translation rules into a native language like English. This suggestion, however, has its own problems with the nature and number of the innate terms of the *lingua mentis*.

Still others have suggested that spoken words are not translations of mental terms, but attach to reality because of something like a baptism. Someone originally conferred a name upon a being, or type of being, perhaps by pointing to it and designating it 'Y' or to be 'an X', and then that baptismal name was passed on from speaker to speaker. Perhaps these philosophers have found some literal truth in Genesis 2:19–20.[1] Once the baptism has been performed the term may well have a place in the expression of a very complex structure of relations among beliefs, desires, and actions within and among individuals, about and involving the being so designated; but the attachment to reality is nothing other than the causal-baptismal designation and subsequent community practice. St. Augustine in the *De magistro* and Wittgenstein again in the *Philosophical Investigations* had difficulties with the notion of pointing as a way of attaching words to things. Isn't pointing itself quasi-linguistic in the sense that it needs to be interpreted, much like linguistic symbols do?

The tradition with which I am mainly concerned in this work—the Thomistic-Aristotelian tradition—holds that words express what we understand of things. Words attach to reality because our cognitive faculties attach to reality in some way. There is something about us, over and above a *merely* causal relation to things, namely, our understanding of things that we seek to communicate with our words. Understanding is itself a kind of becoming identical with the being understood. The concepts that constitute and express understanding are not innate but are acquired developmentally, nor do they constitute an inner language spoken, as it were, to ourselves.[2] No understanding, no words. We may not understand everything there is to understand about an apple; indeed, such comprehensive understanding might be impossible for us.[3] We might be mistaken in many instances when we say of some A that 'A is B'. Still, unless we understand something of beings in the world, however inchoate, we cannot speak of any beings. We name as we know. Of that which one cannot know, one ought to remain silent; indeed one cannot but remain silent, however much one might babble on.

In this work, I hope to make some progress toward a better understanding of what the Thomistic-Aristotelian tradition does and does not claim about the relations that hold among words, thoughts, and things. This tradition has been criticized for trading in a flawed distinction between understanding and linguistic practice. It is not difficult to see why. A striking feature of St. Thomas's commentaries on the logical works of Aristotle is the assumption of unity embodied in them, a unity that depends upon a prior unity in the various acts of reason. He only wrote two commentaries on the logical works, an incomplete one on the *De interpretatione* (*Peri hermeneias* in Greek), and the other on the *Posterior Analytics*. Yet in both commentaries, he begins with very similar prefatory remarks, detailing the overall architectonic of Logic as a rational science. He then explains how Aristotle's treatises, including those he does not comment upon, exemplify this structure. In his preface to the *Commentary on the Posterior Analytics*, for example, he emphasizes the unity of Aristotle's logical treatises by relating them to the order and unity of human acts of reason.

> [T]his art is Logic, that is rational science. Which is not only rational from this, namely that it is according to reason (which is common to all arts); but also from this, that it is concerned with the very act of reason as concerning its proper matter.
>
> And so it seems that it is the art of arts, since it directs our reason in act, from which all arts proceed. Therefore it is necessary that the parts of Logic be taken according to the diversity of the acts of reason.
>
> But, there are three acts of reason, of which the first two are of reason considered as a certain act of understanding (*est intellectus quidam*).
>
> For one act of understanding is the understanding of indivisibles or incomplex things, according as it conceives *what* the thing is. And this operation is called by some the informing of the intellect or intellectual conception. And to this operation of intellect is ordered the teaching, which Aristotle treats in the book *Praedicamentorum* (*Categories*). The second operation of the intellect is the intellect's composition and division, in which the true or the false is then present. And the teaching devoted to this act of reason Aristotle treats of in the book *Peri hermeneias* (*De interpretatione*). The third act of reason concerns that which is proper to reason, namely to conclude one thing from another, as through that which is known, one may come to the cognition of the unknown. And the remaining books of Logic are devoted to this act.[4]

He goes on to include in these "remaining books of Logic" even Aristotle's *Rhetoric* and *Poetics*. As St. Thomas presents these comments, there is no air of discovery or wonder. His text reads like a calm, almost pedestrian analysis. Much the same is written in his preface to the *De interpretatione*.

These comments are striking to the perspective of a twenty-first–century reader. Nowadays the assumption of unity may seem somewhat naive and a little quaint, and possibly a step toward a serious misreading of Aristotle.

The linchpin of St. Thomas's claim for the unity of the logical works is human acts of reason. Why should human acts of reason provide the unity of the logical works? After all, those works seem to be about words. The *Categories* talk about "equivocation," "univocal predication," what is "said of but not in," what is "said in but not of." The *De interpretatione* treats of "spoken and written signs," "nouns and verbs," "enunciations," and so on. Finally, the *Posterior Analytics* speaks of taking the meaning of a term from the "*usus loquentium*" (the use of the ones who speak) and of the differences among "questions," "axioms," and "propositions." The logical works would seem to indicate that Logic concerns itself with words, statements, and their ordering to one another, not the order of human acts of reason, an order which seems more appropriate to Psychology than Logic.

Why then does St. Thomas say that it treats of the acts of reason? Aristotle himself provides part of the answer to that question in the opening passages of the *De interpretatione*.

> Thus, those which are in articulated sound are signs (*notae*) of those passions which are in the soul: and those which are written are of those things which are in articulated sound. And just as letters are not the same for all, so neither are articulated sounds the same: but the passions of the soul are the same for all, of which first (*primorum*) passions these [articulated sounds] are signs (*notae*), and the things of which these are likenesses are also the same. But these things were spoken of in those which were said of the soul, for that is another work.[5]

This very brief passage is described by Norman Kretzmann as the single most influential text in the history of semantics. It expresses what traditionally has been referred to as "Aristotle's semantic triangle." The significance of words is subordinated to the understanding of things via passions of the soul. Indeed, words have whatever significance they have only from their conventional relation to passions of the soul. Aristotle's suggestion that words are *conventionally* related to passions of the soul which are *likenesses* of things would seem to be a reply to the view considered in Plato's *Cratylus* that words have a natural relation to their referents. But passions of the soul are naturally related by likeness to things, and the same for all men. Whatever logical ordering of *words* to one another that there might be, it seems to be derivative upon the logical ordering of *passions of the soul* to one another. *Words, passions of the soul,* and *things* form the vertices of a "semantic triangle."

Still, to a twenty-first–century reader this partial answer in one of the logical works itself is no less striking than the original statement that gave rise to the question, and for some not a little bewildering. How, after all, can "passions" have a logical ordering? How can a "passion" be a likeness of things? How can the private contents of one's mind, one's thoughts, undergird the public meaning of vocal utterances?

Throughout the history of philosophy there are accounts that attempt to explain how words relate to things that can be said to share a family resemblance with Aristotle's, namely that words acquire their meanings by somehow being conventionally associated with mental states, in particular with what are often called "mental representations." Likewise, many of our contemporary philosophers make similar associations of words, thoughts, and things. For example, the *lingua mentis* view can be understood in that way. In our present idiom of intentionality, these philosophers will assert that the meaning or intentionality of words is derived from the intentionality of mental representations. Jerry Fodor writes:

> The idea, to put it in a nutshell, is that it might be possible to pull off a *double* reduction: First derive the semantic properties of linguistic symbols from the intentional properties of mental states; then postulate a population of *mental symbols*—mental representations ... and derive the intentional properties of beliefs and desires from these.[6]

Given these family resemblances, it is not at all odd for these philosophers to suggest the influence of Aristotle as their remote ancestor. On the other hand, philosophers following Wittgenstein find this way of approaching the meaning of terms deeply troubling and level severe criticism at it.

The thesis that language "hooks onto the world" via the mind has not faired well at all in the hands of these influential philosophers of mind and language. Aristotle is seen as originating a relatively continuous tradition of reflection on language that proceeds through the British Empiricists to recent accounts of language and mind referred to as "mental representationalism." In *Representation and Reality* Hilary Putnam explicitly refers to the theses from the *De interpretatione* as "a scheme that has proved remarkably robust" for characterizing meaning and reference, though he ultimately rejects it as fatally flawed. He refers to it broadly as "the Aristotelian view," and associates its constitutive theses with a number of others purportedly advocated by Mill, Frege, Russell, Carnap, and to a certain extent Chomsky, Fodor, and Searle. Putnam is not alone in this view. Michael Dummett writes:

> The vague conception, common, for instance, to both the British empiricists *and* Aristotle, whereby a word represents an "idea," and a phrase or sentence accordingly represents a complex of ideas, is

simply too crude to serve even as a starting point; it virtually forces us to adopt the conception whereby the meaning of a word is embodied in a mental image.[7]

The fundamental criticism is that our language never actually succeeds in attaining the world, but remains trapped in the murky internal depths of the mind. In our apparent talk about trees, we really succeed only in calling to mind our mental representations of trees.

This "Aristotelian" tradition posits inner objects of mind, concepts or ideas, that are supposed to mediate between the knower and the known. By possessing these inner objects, the knower understands the world. When these inner objects that constitute understanding are arbitrarily associated with sounds they constitute the meanings of those sounds, which then become words. To put it crudely, my son has an image of an apple come before his mind, and goes in search of something in the world like it. But no such inner objects can perform this function. Thus, the fundamental structure of the recent objections is that the "Aristotelian view" of language, thought, and world undermines a serious account of the semantic character of language, because it "throws up" mental representations before the mind. The mental representations obtrude themselves in one way or another, either as "primary objects" of signification, or "internal mental entities" serving as obstacles through which language must pass, in order to "hook on to the world." In principle this approach fails by yoking language to an untenable philosophy of mind.

Lurking in the background here is the problem of epistemological skepticism about the world and other minds. If these mental objects stand between my mind and the world, how can I know that they adequately represent the world? How can I discover the content of another's mind if it is just another part of the world that may be unknown to me? Are the contents of my mind fundamentally private, while the contents of others' minds are unknown?

Two Traditions

Philosophy today remains in many ways informed by the issues and difficulties raised by the philosophers anthologized in Richard Rorty's *The Linguistic Turn*.[8] The general methodological position of the *Linguistic Turn* is that philosophical problems are problems of language that can be eliminated either by reforming language to some ideal form or simply by paying greater attention to the facts of ordinary use. Hand in hand with this methodological claim went a substantive claim about the error of "traditional philosophy" which regularly misconstrued how it is that language is meaningful, often positing "in the head" a realm of objects called meanings. Putnam captures very well this aspect of the *Linguistic Turn:*

> [T]he traditional account suggests that finding out that someone has a concept is finding out that he has a particular mental presentation, and finding out that two people have the same concept is finding out that they have identical mental presentations....
>
> [W]e have followed one reason for upgrading the importance of language in philosophy. Concepts and ideas were always thought important; language was thought unimportant, because it was considered to be merely a system of conventional signs for concepts and ideas (considered as mental entities of some kind, and quite independent of the signs used to express them). But today it seems doubtful that concepts and ideas can be thought of as mental *events* or *objects* (as opposed to *abilities*) and even more doubtful that they are independent of *all* signs.[9]

The value of the *Linguistic Turn* is found in the extent to which it reconceives the problems of "traditional philosophy" as problems of language and how language is related to the world. It is not obvious, however, that those who have taken the linguistic turn are in any better position in this regard than the representative figures of "traditional philosophy" from whom they turned. One problem with "traditional philosophy" is thought to be the *privacy* of the inner mediating objects of mental attention—ideas or concepts, and so on. When we look for the meaning of a word we do not look for images in a speaker's mind. The problem is overcome by making the mediating objects public in language. According to Putnam,

> this has the advantage of being a public study, and more in the spirit of modern social science. If the way to find out what the concept Cause is . . . is to introspect one's own images, etc., then we would hardly expect to get very reliable reports, let alone agreement. If analyzing the concept Cause is rather a matter of studying the way in which we use the word . . . then the hope for reliable reports . . . is much greater.[10]

Now we are not trapped in the privacy of our own individual minds, and the epistemological skepticism on the horizon of the critical turn is presumably overcome by our recognition of the public and social character of knowledge embodied in linguistic forms. But is the problem of skepticism generated by mediating objects solved? What difference does it make whether the artifact is a private artifact in an individual mind, like an idea, or a public artifact in a social mind, like a word? Thus, Ian Hacking suggests:

> [W]hen mental discourse was taken for granted, ideas were the interface between the Cartesian *ego* and reality. We have displaced mental discourse by public discourse, and "ideas" have become unintelligible. Something in the domain of public discourse now serves as the interface between the knowing subject and the world.[11]

If Hacking is right, then isn't it possible that the problems of epistemological skepticism and anti-realism will simply be displaced by the problems of semantic skepticism and anti-realism? I might not be as lonely as when I was on the inside of the mental horizon, but the view from the inside of the social horizon looks as barren.

Still, an opening arises to see why a turn to language might be in a slightly better position with regard to semantic skepticism than "traditional philosophy" was with regard to epistemological skepticism. Words have this advantage over ideas—they are without question physical worldly "beings." They have, for example, a duration and volume that can be measured. But if they are worldly objects to which we have unmediated access, what is so special about them? If we can have unmediated cognitive access to these physical beings, why not to others? Of course, they are not simply physical beings like any others. They have a life that the sounds of cars crashing and water splashing do not have. Still, even as they have this life, they remain physical beings.

What gives life to our words, these physical worldly beings to which it seems we have unmediated access? We do. But, on pain of a regress back to "traditional philosophy," we had better not make an appeal to some special relation that our words bear to some inner mental objects to which we have unmediated cognitive access. Consider the vocal utterance of 'apple' and the vocal utterance of 'orange'. These sound different, and their difference of sound can be precisely measured with very sophisticated instruments. But we do not simply say that these are different sounds. Water dripping and cars crashing make different sounds, but they do not make different words. How is it that we manage to identify 'apple' as a different word, and not merely a different sound from 'orange'? To make the situation more difficult, consider two or more vocal utterances of 'the bat dropped on the cat on the mat'. Is 'bat' in these vocal utterances always a token of the same word-type, given that to a close approximation the occurrent utterances sound the same, that is, their physical sound-type characteristics are the same between the token instances? If not, how do they differ in word-type?

Words are supposed to mediate our cognitive relation to the non-human physical world, but on pain of regress we are supposed to have unmediated access to some physical beings in the world, namely, our words. But to recognize and identify our words as distinct from one another, indeed, as words at all, it would seem that we must have cognitive access to something other than spoken words (or mental images), a cognitive access unmediated by words or mental images. Here, we can turn once again to Plato and Wittgenstein. In the *Cratylus* Plato writes:

> But if this is a battle of names, some of them asserting that they are like the truth, others contending that *they* are, how or by what criterion are we to decide between them? For there are no other names to which appeal can be made, but obviously recourse must be had to another standard which, without employing names, will make clear which of the two are right, and this must be a standard which shows the truth of things.[12]

And Wittgenstein wrote about paradoxes of interpretation generated by taking the use of a word to be the application of a rule which must in turn be interpreted, that "what this shews is that there is a way of grasping a rule which is *not* an *interpretation*, but which is exhibited in what we call 'obeying the rule' and 'going against it' in actual cases."[13] Rorty, discussing Hacking's problem of the "interface," asks:

> "Given that we no longer take the 'idea' seriously, why need we assume that there is *any* 'interface' between the knowing subject and the world?" Why not say that the relation between the two is as unproblematic as that between the ball and the socket, the dove and the light air it cleaves? Why must there be something "in the domain of public discourse" for philosophers to vex themselves over as they once vexed themselves over "private associated ideas"?[14]

Here it might seem that the "semantic skepticism" generated by taking words as mediating objects in place of ideas is not solved, but rather dissolved. These comments come from one of the most metaphysical of philosophers and two of the most anti-metaphysical. So it seems that the point does not depend upon any particularly Platonic "metaphysical" aspirations to know the reality of "things in themselves." Comparing 'apples' and 'oranges' may indeed be very much like comparing apples and oranges, however we do that.

It would be premature, however, to suggest that the issue of how language is related to the world has been settled. John Searle writes:

> [S]ince Frege, reference has been regarded as the central problem in the philosophy of language; and by reference I mean not predication, or truth, or extension but *reference,* the relation between such expressions as definite descriptions and proper names on the one hand, and the things they are used to refer to on the other. I now think it was a mistake to take this as the central problem in the philosophy of language, because we will not get an adequate theory of linguistic reference until we can show how such a theory is part of a general theory of Intentionality, a theory of how the mind is related to objects in the world in general.
>
> On my view, the philosophy of language is a branch of the philosophy of mind; therefore no theory of language is complete without an account of the relations between mind and language and of how meaning—the derived intentionality of linguistic elements—is grounded in the more biologically basic intrinsic intentionality of the mind/brain.[15]

Searle believes that it is a mistake to separate questions of meaning from questions of how the mind is constituted, and the philosophy of language is accordingly subordinated to the philosophy of mind, but this need not be understood as a project of overcoming skepticism, epistemological or semantic. Different as Aristotle may be from Fodor and Searle, they seem to agree on how an account of language presupposes an account of how a human being understands the world. Indeed, one might wonder whether the contemporary approaches to meaning that Searle is critical of succeed in avoiding substantive accounts of the mind, but rather implicitly presuppose certain accounts. These presupposed accounts may well prove useless when one turns to asking how words are related to beings in the world. Frege, for one, posited his abstract realm of "thoughts" to guarantee the public non-subjective character of meaning, over against the privacy and fleeting character of ideas as he conceived of them. If language must be divorced from mind, so much the worse for those accounts of mind for someone like Searle or Aristotle.

One characteristic of the Thomistic tradition has been its willingness and sense of obligation to engage contemporary modes of thought. That is my intention here. I want to contribute toward moving my tradition forward by a critical engagement with one aspect of the *Linguistic Turn*. In doing so, what I argue will in some ways be implicitly and at times explicitly critical of aspects of my own tradition. So, for instance, I will make no use of the *verbum mentis*, one of the most venerable interpretations of St. Thomas, because I believe it is not part of his philosophical account of understanding and language.[16] Instead I will explore how each tradition approaches the crucial Aristotelian text that purports to enunciate how the capacity for words to be about the world involves the "mind" and what criticism involving this text might representatives of one tradition direct at the other.

In the first chapter, we will examine St. Thomas's commentary on the key passage from Aristotle. I will proceed far enough into the *Commentary on the De interpretatione* to bring to light the relevant issues necessary for understanding the passage and distinguish the way in which words are said to *signify* "passions of the soul," and how they are said to *signify* "things beyond the soul." In particular, we will examine the cognitive setting within which St. Thomas places the discussion. In the second chapter, we will consider a possible objection to the historical accuracy of my analysis. Norman Kretzmann claims that the interpretation of Aristotle's "semantic triangle" has been seriously misled by Boethius's Latin translation of the Greek, though he does not explicitly apply the objection to St. Thomas. I think this objection must be cleared away, but a note to the reader is in order here. Those readers who might not wish to examine the historical issues this closely can pass over the second chapter without too much difficulty for the rest of the work.

In the third chapter, I will try to set the historical stage for the recent criticism of the Aristotelian tradition. I will describe some aspects of early modern treatments of language and the contents of the mind, focusing upon Locke, Berkeley, and Hume in particular. That chapter ends with a discussion of the criticism that is leveled at these figures by Wittgenstein among others. In the fourth chapter, we will consider Jerry Fodor's recent rather self-conscious attempts to revive the Empiricist tradition by means of his account of mental representation and the *language of thought*, and in the fifth chapter, we will look at Putnam's characterization of the "Aristotelian" tradition, as well as his criticism of it. From the third, fourth, and fifth chapters will emerge a number of substantive philosophical presuppositions—three in particular which I will call the "Third Thing Thesis," the "Introspectibility Thesis," and the "Internalist Thesis"—presuppositions that determine the character of the mental objects that the critics find so troubling.

In chapters six, seven, and eight, we will turn to St. Thomas's account in light of these three theses. In the ninth and final chapter I make suggestions about how the Thomistic analysis that emerges may help us to move forward in our understanding of language, and what St. Thomas might mean when he writes that man speaks "for the sake of a more perfect existence."[17]

Chapter 1

Aristotle's Semantic Triangle in St. Thomas

Here the word, there the meaning.
The money, and the cow that you can buy with it.
—*Wittgenstein*

The Semantic Triangle in Boethius's Text

The *De interpretatione* begins without an explicit statement about the subject matter of the treatise. In his translation Hippocrates Apostle suggests that it is somewhat misleading to translate the title as *On Interpretation*, because in English this is too broad a summary of what is in fact treated in the work; it is concerned with propositions or enunciations, and their attributes.[1] Aristotle begins the treatise by setting out the order in which he will proceed. First he will discuss the two major divisions of an enunciation, noun and verb, and then proceed to discuss denial, affirmation, statement, and sentence.[2] The treatise then begins with what is commonly called Aristotle's "semantic triangle."

> Thus, those which are in articulated sound are signs (*notae*) of those passions which are in the soul: and those which are written are of those things which are in articulated sound. And just as letters are not the same for all, so neither are articulated sounds the same: but

the passions of the soul are the same for all, of which first (*primorum*) passions these [articulated sounds] are signs (*notae*), and the things of which these are likenesses are also the same. But these things were spoken of in those which were said of the soul, for that is another work.

There are a number of points to recognize in this passage. First, the "semantic triangle" takes its name from the three vertices—articulated sounds, passions of the soul, and things. Second, there is the clear ordering of written language to spoken as sign to signified. Aristotle does not argue for this order, but simply asserts it. It might be an expression of common sense for him, since we learn to speak before we learn to write,[3] and we put our words to writing when our audience has no access to our spoken words. Third, the relation between "passions of the soul" and "things" is said to be by likeness (*similitudines*). This is not said of the second relation between spoken words and passions of the soul, which provides the first indication that the two relations are not of the same kind. Fourth, there is the claim that, despite the difference of words among men, the passions of the soul that these words signify are the same for all, and the things of which passions of the soul are likenesses are the same.

Finally, there is the reference at the end to "those [things] which were said of the soul, for that is another work." This reference is commonly taken to point to the *De anima*. Translators Apostle and Edghill simply cite the *De anima*.[4] Ackrill is more specific in associating it with *De anima* III.3–8, but he notes that it might be the addition of an editor. St. Thomas also attributes the citation to the *De anima*.

> [Aristotle] excuses himself from a more attentive consideration of these things, because what nature the passions of the soul may be, and in what way they may be likenesses of things, was spoken of in the book *De anima*; for this work does not pertain to logic, but to the philosophy of nature.[5]

In the *Preface* to his *Commentary*, St. Thomas sets out many of the more general characteristics pertinent to enunciations considered as truths, relating them to the twofold operation of the intellect discussed by Aristotle in *De anima* III:

> As the Philosopher says in *De anima* III, there are two operations of intellect: indeed one which is called the understanding of indivisibles, through which namely the intellect apprehends the essence of each

thing in itself; moreover the other is the operation of the intellect composing and dividing.⁶

St. Thomas's emphasis on the unity of logic grounded in the prior unity of the acts of reason is underscored when in the *Preface* to his commentary on the *Posterior Analytics* he repeats almost exactly the same analysis.⁷ The *De anima* text of Aristotle reads as follows:

> [T]herefore understanding of indivisible things belongs to these things in which there is no falsehood, but then there is truth and falsehood in those where there is some composition of things understood, as of things which are one. . . . And so separate things are composed, as for example *asymmetry* and *diameter* or *symmetry* and *diameter*. But if [the combination is] of things done or of things to be done, time is co-understood and composed: for falsehood is always in a combination, if for example either white was composed with not white or not white with white; so also it happens for all that are called a division; so therefore it is not only true or false that Cleon is white, but also that he was or will be. But it is the intellect making each one one thing.
>
> But every saying something of something, namely, an affirmation, is either true or false; however not every act of understanding is true, but what is of the thing according to its quiddity (*quod aliquid erat esse*) is true, not however as something of something, rather as to see is always true of a proper object, however whether a man is white or not, is not always true.⁸

There is good reason to conclude that this is the text that St. Thomas has in mind since in the *De interpretatione* immediately after Aristotle makes reference to the *De anima* at 16a8–9, he mentions articulated sounds in which truth and falsity are found, in contrast to articulated sounds that are not true or false.

> However, just as sometimes there is some act of intellect in the soul without something true or false, but other times it is necessary for one or the other of these to be in it, so it is also in articulated sound. For truth and falsity belong to both composition and division. Therefore nouns themselves and verbs are like acts of intellect without composition and division. As for example 'man' or 'white', when nothing is added, for neither of these is thus far true or false. A sign of this

is this: 'goatstag' signifies something, but not yet what may be true or false, if 'to be' or 'not to be' is not added simply or according to time.⁹

St. Thomas's commentary on this text from the *De interpretatione*, at L. I, lc.3, 23–35, cites *De anima* III, and closely parallels his commentary on the *De anima* text at L. III, lc.11, 746–63. In both commentaries, St. Thomas explains that the simple or incomplex intelligible character is neither true nor false. However, the intellect by grasping the incomplex intelligible character can be *called* true, insofar as it is "adequate" to (*De anima* commentary), or "conformed" to (*De interpretatione* commentary), or "measured" by (*De interpretatione* commentary) some real being. In fact, the intellect in this act, like sight perceiving color, is never deceived. However, according to both commentaries, truth and falsity properly occur only in a composed or complex intelligible character when the complexity results from the intellect's act of judging that "something is" or that "something is not."

Even without St. Thomas's *Commentaries* the similarities in Aristotle's texts are striking. In the *De interpretatione*, Aristotle relates articulated sound in which there is neither truth nor falsity, that is, nouns and verbs in isolation, to acts of intellect in which there is neither truth nor falsity—they "are like acts of intellect without composition and division." By contrast, articulated sounds in which there is truth and falsity are like acts of intellect in which there is composition and division. The text in the *De anima* to which St. Thomas draws our attention has many of the same elements related in the same ways—truth and falsity generally related to compositions and divisions. The one clear difference is that Aristotle speaks in the *De anima* text of a simple act of understanding that is always true, namely, understanding the essence of a thing. In the *De interpretatione*, by contrast, Aristotle seems to deny that 'true' applies to such simple acts, much less that they are always true. But as I just noted, St. Thomas implicitly provides an explanation of this difference by explaining how 'true' is said in different ways in the various passages. There is also the parallel contrast in the *De anima* between enunciations of things that are one and enunciations in which time is "co-understood." In the *De interpretatione* he writes, "either simply or with reference to time." The example of 'white' and 'man' occurs in both passages.

J. L. Ackrill notices none of these parallels between the texts, despite the fact that he cites III.4–8 when he considers the reference to the *De anima* in the previous passage of the *De interpretatione*. The text of the *De anima* that I am now considering falls right in the middle of that reference, at

III.6.430a26–b5, b26–32. Instead, Ackrill objects to the "likenesses" of the initial passage, writing that "the suggestion that thoughts are likenesses of things is not acceptable even for simple thoughts like the thought of a cat. It is even less acceptable for thoughts that would be expressed in sentences."[10] He made clear earlier that the likeness that he has in mind here is akin to the likeness of a *picture of a cat* to a *cat*, so his difficulties with the text appear to be driven by concerns about mental representation.

Even if there had been no explicit reference in the *De interpretatione* to the *De anima*, the similarities of the two texts from Aristotle more than justify St. Thomas's use of the *De anima* to introduce his comments on the *De interpretatione*. In his commentary on the *De anima* text, he relates this first act of understanding indivisibles or simple intelligibles to understanding the *quod quid est* or the *quod quid erat esse* of a thing, but he does not use the term *essentia*. He provides understanding *man* or *ox*, and "other such simple things" as examples. This is also true in the *Commentary on the Posterior Analytics*, where he uses *quid est res*.[11] In the *Preface* to the *De interpretatione* commentary, however, he explicitly identifies this understanding of indivisibles as that "through which (*per quam*) . . . the intellect apprehends the *essence* of every single thing in itself."[12] Further, he identifies the first act with a *means* of understanding the essence of things in themselves; the act is that "through which" such understanding takes place. These acts of understanding indivisibles enter into the combinations or divisions of the second operation as its elements, "because there can only be composition and division of simple apprehensions." For this reason, the first operation is ordered to the second.[13]

This identification of the first act with a *means* of understanding will prove very important for grasping St. Thomas's analysis. The cognitive character of the remarks sets the background for an explanation of the subject matter of Aristotle's treatise. According to St. Thomas the subject matter of the *De interpretatione* is, citing Boethius, "significative vocal sound."

General or universal vocal utterances signify without mediation the intellect's concepts, the terminus of the first act or operation, and *via* the latter's mediation signify things (*res*).[14] The text from Aristotle, however, does not speak of "mediation" and "without mediation." St. Thomas provides an explanation on Aristotle's behalf of the necessity for the intellect's mediation of the signification of *things by words:*

> But here speech concerns articulate sounds signifying from human institution; and so it is necessary that here *passions of the soul* be understood as *conceptions of intellect*, which names, and verbs, and sentences

signify, according to the view of Aristotle: for it cannot be that they signify things themselves without mediation, which is clear from the mode of signifying: for this name *man* signifies human nature in abstraction from singulars, so it cannot be that it signifies a singular man without mediation. Hence the Platonists held that it signified the separate *idea* itself of man; but because in the view of Aristotle, this, according to its abstraction, does not really subsist, but is in the intellect alone, it was necessary for Aristotle to say that articulated sounds signify the conceptions of the intellect without mediation, and things by their mediation.[15]

One reason for the identification of *passions of the soul* with *conceptions of the intellect* in the context of "articulate sounds signifying from human institution" is that St. Thomas wants to rule out the possibility that "passions of the soul" might be taken as "affections of the sensitive appetite," since certain groans may more or less naturally, not by institution, signify pain. The most important part of the explanation, however, proceeds by associating generality with the *mode of signifying*, while associating singularity with the *res* (thing) signified. The explanation denies that a general articulated sound signifies a general *res*, where the context makes clear that *res* is taken *extra animam*, by contrast to what are "conceptions of the intellect." Instead, a general articulated sound signifies a nature "in abstraction from singulars," a nature existing only in the intellect. Further, the explanation asserts that general articulated sounds do indeed signify singular *res*, but only through the mediation of the intellect's conceptions. Thus, from the start, the *termini* of the intellect's first act or operation are identified with these general conceptions, and provide the basis for generality in speech. But they do not play the part of *res* (things) signified. They determine the *modus significandi* (manner of signifying).

'Res signified' is used loosely and informally here; it should not be taken as the more formal notion of the *res significata*, or simply the *significata* of a general articulated sound. The more formal notion of the *significata* of a general vocal utterance does not take as its signification a singular thing as such, for example, Aristotle, but rather the *form* or *nature* of a singular thing, for example, *human nature* in Aristotle, or Plato, or Xanthippe. St. Thomas writes:

[T]here are two things to be considered in any name whatsoever: namely that from which the name is imposed, which is called the quality of the name; and that to which the name is imposed, which is called the substance of the name; and the name, properly speaking, is said to signify the form or quality from which the name is imposed; on the other hand it is said to stand in place of [to supposit for] that for which it is imposed.[16]

Using the more formal notion of the *significata* of a general vocal utterance, one would then specify its *suppositio* in a particular utterance. For example, in "the man who was a student of Plato, and teacher of Alexander, and who founded the Lyceum wrote the *De interpretatione*," the *significata* of 'man' is the human nature of Aristotle, while its *suppositio* is the singular being Aristotle. In "the man who taught Aristotle, failed in his mission to Syracuse, and founded the Academy was not present at the death of Socrates," 'man' signifies the human nature of Plato and supposits for Plato.

However, if we look closely at the passage from the *De interpretatione*, we see that *signification*, in its application to *res extra animam*, is not being used in the technical sense associated with *significata*, but more informally. St. Thomas writes that "man signifies human nature in abstraction from singulars," which abstraction is then associated with the conception of the intellect. Nowhere in the passage does he write that 'man' signifies *human nature* in singular things through the mediation of the intellect's conception. Instead, he writes that the general articulated sound 'man' does not "signif[y] a singular man without mediation." However, having established that it signifies "without mediation" a conception of the intellect, he returns in the end to tell us that it *does* signify things "by the mediation of" the intellectual conception. Though he does not use "singular" to characterize "things" in this last line, it is fairly clear that he has in mind the singular men that he has just denied are signified without mediation, not *human nature* in those men. St. Thomas does make the requisite distinction later in the *Commentary* between signifying the *nature* in *res extra animam* and signifying the singular *res extra animam*, when he asks how a general word is to be "taken" in a particular enunciation. It is clear there that the *res extra animam* signified by the *nomen* (name) 'homo' is human nature, which may be that of Socrates, or Xanthippe, or Plato, or Aristotle, or Helen, all singular beings. However, it is important to keep in mind that here, early in the *Commentary*, his use is informal.

Before proceeding on, note two points about what is taking place in St. Thomas's *Commentary*. First, St. Thomas exhibits no scruple about restricting "passions of the soul" to the intellect's concepts. The fact that he restricts them in this manner, and that not just anything, like vision, hearing, and so on, counts as a "passion" in this text, is consistent with his view that the *De interpretatione* is ordered toward the discussion of scientific demonstration in the *Analytics*. This stress on the unity of the logical essays as a means of interpreting the passage is in marked contrast to the worries of Ackrill and Apostle, who suspect that the passage is wholly out of place.

Second, St. Thomas does not presuppose that words signify the intellect's concepts. Aristotle mentions intellect along the way, and so it becomes a problem for St. Thomas to address, and to explain its place in the text. He explains it by concentrating on the *modus significandi* of general words. However, there is no doubt whatsoever that general words signify things or *res extra animam*. If anything is presupposed in the analysis, that is. Further, it is presupposed that general words do not signify general *things*, which is why St. Thomas explicitly mentions Aristotle's criticism of Plato; he does not justify or even rehash it. The reader is supposed to take it as given. General words signify *things* that are individuals. There is no attempt at all to justify or argue that words signify things beyond the soul. Aristotle's mention of intellect is justified, not by appealing to the *things* signified by universal words, but to the *manner of signifying* (*modus significandi*) of general words. This point suggests that, as St. Thomas understands and interprets the text, were it not for the *manner* in which general words signify, intellect would not play a part in the analysis of signification. For St. Thomas, what needs justification in Aristotle is not *that* general words signify individual *res extra animam*, but *how* they do.

A Certain Ambiguity

The use of the same word, 'signification' and its cognates for both relations, *word* to *res extra animam* (things beyond the soul) and *word* to *passiones animae* (passions of the soul), is ambiguous and fits with the informal use that St. Thomas is employing here. However, it is important to recognize and emphasize that the *res* signified by the *word* is a *res extra animam*. By contrast, as St. Thomas's discussion indicates, the relation of a *word* to a *passio animae* is not the relation of a *word* to a *res signified*. It is the relation of a *word* to a *means* by which the *word* is related to a *res signified*. Still, despite this ambiguity, I do not wish to abandon the word 'signification' in order to clear it up. I will use the following conventions:

a) 'signification$_1$' indicates the relation of a *word* to a *passio animae*;
b) 'signification$_2$' indicates the relation of a *word* to a *res extra animam*.[17]

c) 'Similitude' indicates the relation of a *passio animae* to a *res extra animam*.

The point of making these terminological conventions is to avoid some of the ambiguity in the discussion to follow and to emphasize later why signification$_2$ is not the same kind of relation as either signification$_1$ or Similitude, as well as why it should not be reduced to a conjunction of the latter two. It also serves to indicate that the two uses of 'signification'—'signification$_1$' and 'signification$_2$'—are not wholly unrelated, but rather analogous.

It is important to avoid an imaginative picture of these relations as roads or conduits, as if by traveling along signification$_1$, through the intellect as a weigh station, and then along Similitude, the word gets to the *res extra animam*, the whole journey being equivalent to signification$_2$. In that picture, signification$_2$ would be a dyadic relation effectively reduced to the two dyadic relations, signification$_1$ and Similitude, and these latter two relations would be taken to be the same or roughly analogous.

On the contrary, signification$_2$ is akin to the sort of irreducibly triadic relation Walker Percy pursues in *The Message in the Bottle*.[18] There are relations that appear to be triadic, but that can be reduced to a system of dyadic relations, all of which are of the same basic kind. Take for example a simple system of two men pulling a rope. We might call the overall relation $F(r,m_1,m_2)$, which indicates the overall force exerted upon rope r, by man$_1$ and man$_2$. At first sight, it appears to involve a triadic relation. However, it is a fundamental presupposition of Mechanics that $F(r, m_1, m_2)$ can be reduced to a system of dyadic relations, namely $F(r, m_1)$ conjoined to $F(r,m_2)$. In this reduction, the latter two relations are of the very same kind, namely, relations of bodies exerting forces upon one another. The original apparently triadic relation just is the conjunction of the latter two, which can be thought of simply as $F(r, m_1 \& m_2)$, in which $m_1 \& m_2$ is treated, for the purposes of Mechanics, as a single body.

Now consider the discussion of signification$_2$. Let $s_2(w,r,p)$ represent the overall relation of signification between a word w, a *res extra animam* r, and involving a *passio animae* p. It might be tempting to think that this relation can be reduced to a system of two dyadic relations $s_1(w,p)$, the relation between the word and the *passio animae*, and $S(p,r)$, the relation between the *passio animae* and the *res extra animam*. On this reduction, one would imagine that upon hearing the word w, that is, upon its coming to mind,[19] in virtue of $s_1(w,p)$ one thinks of or has come to mind a certain *passio animae*. But when the *passio animae* comes to mind, in virtue of $S(p,r)$

one then comes to have the *res extra animam* represented to one, or comes to think of it. In this reduction, there seems to be a fundamental dyadic relation, call it sig(__,__), where once one relatum comes to mind the other follows. The fundamental function of this relation is *to have one thing brought to mind by having another thing brought to mind*. On this account, the signification$_1$ between *word* and *passio animae*, and the Similitude between *passio animae* and *res* are just instances of the same fundamental relation sig(x,y):

$s_1(w,p) = sig(w,p)$;
$S(p,r) = sig(p,r)$.

Further, the relation of signification$_2$ between *word* and *res* involving a *passio animae* is reduced to a dyadic relation sig(w,r) of the same fundamental kind as sig(w,p) and sig(p,r):

$s_2(w,r,p) = sig(w,r) = sig(w,p)$ & $sig(p,r)$

In this way of imagining the issue, the fact that Aristotle and St. Thomas first talk about sig(w,p) seems incidental, perhaps merely indicating a temporal series, or something else.

In the reduction above, I deliberately chose sig(w,r) rather than sig(w, p & r) in parallel to the original example from Mechanics for a number of reasons. Since on this reduction the relation between *word* and *res* turns out to be dyadic, the choice clearly suggests that the relata are w and r; a *word* signifies a *res*. But symbolically it also suggests that we can do without the p, that is, the *passio animae*. For the partisans of the Linguistic Turn it suggests this question—what role do *passiones animae* play in the relation between *words* and *res*? If a *word* can come to mind, and a *res* can come to mind, in order to understand our speech and know what we are talking about, what need have we of an entity *in anima* that also comes to mind occluding itself between the *word* and the *res*? Why can't the word's coming to mind take us directly, without the mediation of a *passio animae*, to the *res* signified, perhaps in a way solely dependent upon social conventions or practices?

But no such reduction is taking place in St. Thomas's discussion of signification. Suppose I say that Similitude is some dyadic relation S(p,r) between a *passio animae*, p, and a *res extra animam*, r. Suppose I say that signification$_1$ is some other dyadic relation $s_1(w,p)$ between a word, w, and a *passio animae*, p. Then signification$_2$ is a triadic relation $s_2(w,r,p)$ such that:

if $s_2(w,r,p)$, then $s_1(w,p)$ and $S(p,r)$,

where by $s_2(w,r,p)$ I mean w signifies$_2$ r by means of p. In this structure $S(p,r)$ and $s_1(w,p)$ differ in kind from each other—the relation of *word* to *passio animae* is not the same kind of relation as between *passio animae* and *res extra animam*. Consider the following text from St. Thomas's *Commentary*:

> [A]rticulated sounds [are signs] of passions [of the soul], such that no character of similitude is considered, but only the character of institution.... But in the passions of the soul it is necessary that the character of similitude to the represented things (*exprimendas res*) be considered, since they naturally designate them, not from institution.[20]

It makes clear that the relation of signification$_1$ is not the same as the relation of Similitude. Further, $s_2(w,r,p)$ differs in kind from both $S(p,r)$ and $s_1(w,p)$—the relation of *word* to *res extra animam* is not of the same kind as that of either *res extra animam* to *passio animae*, or of word to *passio animae*. Further, as we will see, the *passio animae* is not an intermediate object brought to mind or thrown up before the mind between *word* and *res*, but rather the manner by which the use of the *word* brings to mind the *res*.

With these clarifications in mind, a recurrent theme appears: general articulated sounds do not function as words signifying$_2$ a special category of beings, namely, universal *res extra animam*; they function as words signifying$_2$ singular *res extra animam*, in virtue of having a universal signification$_1$, that is, in virtue of being related by signification$_1$ to a universal act of understanding *res extra animam*. Keeping in mind again the informal use of '*res* signified' in this early passage of the *Commentary*, 'homo' signifies$_2$ *res extra animam* no less than singular words like 'Socrates', 'Xanthippe', or 'Plato' signify$_2$ *res extra animam*. The general articulated sound differs from the singular when we ask *how* or by what *means* such signification$_2$ takes place, that is, when we ask what signification$_1$ it has. Universality is associated with the *how*, not with the *what extra animam* is signified$_2$.[21] But this *how*, a *passio animae*, is at the same time and fundamentally a means of knowing *res extra animam*. Here St. Thomas does not tell us *how*, by contrast, a singular term signifies$_2$ a singular *res extra animam*.

So it is not trivial for St. Thomas to emphasize Aristotle's reference back to the *De anima* in the discussion of "significative vocal sound." How it is that *res extra animam* are talked about depends upon *how* it is that they are known; it depends upon *Similitude*, the relation of *passio animae* to *res*. By setting the context in this way, Aristotle immediately associates the analysis of words and statements with a logically prior discussion of the *means*[22] by which human beings understand *res* existing *extra animam*. We name as we know. This goal of subordinating an account of how language succeeds in being about *res extra verba* (things beyond the words) to an account of *how* we know *res extra animam* (things beyond the soul) has a long history, and remains for some present-day philosophers like John Searle and Jerry Fodor a necessity.

Some Distinctions of Terminology and *How* We Know

At this point it will be helpful to distinguish two *aspects* of the first operation of intellect. Let me introduce the term 'intelligible character' to designate the determinate form delimiting or structuring the first operation in act—I am attempting to pick out one of the senses of the multifaceted Latin word *ratio*. For any first operation of intellect, there will be an intelligible character functioning as the form or determinate structure of the act, and distinguishing the act formally from any other first operation differing with respect to that intelligible character. Considered simply as informing an act of intellect in this fashion, questions of truth and falsity do not strictly speaking apply in the first operation, though this is said with qualifications.

Intelligible character hearkens back to the discussion of *natures absolutely considered* (*absoluta consideratio ipsius*) in St. Thomas's early work *De ente et essentia*,[23] but the relevant theses remain with St. Thomas throughout his career and can be found in both the *Summa* and the *Commentary on the De anima*. A nature can be considered *absolutely*, in which case one considers or takes into account only those characteristics that pertain to it as such. In the case of man, *being rational* and *being animal* are considered, while *being here* or *being there*, *at this time* or *at that time*, *one* or *predicable of many*, *in re* or *in anima* are not considered; in the case of red, *being a color* is considered, but not *being the color of an apple*, and so on.[24] Or a nature can be considered with respect to two *ways* or *modes* of existing, one in singular things (*in singularibus*) as a principle of being, and another *in anima* as a principle of knowing. Characteristics pertain to the nature in virtue of these modes of existence, but not as such—*being here* or *being there*, *at this time* or *at that time*, *one* or *predicable of many*, *in re* or *in anima*.

It is important to be clear on these theses. St. Thomas is not asserting that there is some one *thing*, the absolute nature, existing now in reality, now in the intellect. He is asserting that given any nature one may ask two sets of questions. First, does the nature exist *in singularibus* or does it exist *in anima*? If the answer is that it is a nature existing *in singularibus*, then *that* existing nature can never exist *in anima* and be *that* nature. Vice versa for a nature existing *in anima*. However, when one considers two existing natures, one may find that they do not differ "considered absolutely." This set of questions determines the context for the second set of questions. Having determined between the modes of existence, one can then go on to consider what pertains to a nature in virtue of *the mode of existence* that it has, and what pertains to it *as such*. "Absolute consideration" is, as the phrase suggests, a manner of considering an existing nature; it is not a mode of *being* for a nature, a mode that would differ from *being in re* and differ from *being in anima*. Joseph Owens explains:

> The consideration of essence just in itself is not expressed as something that involves a proper *being* of the essence, but is a consideration according to the proper *ratio* of that essence. It is an absolute consideration of the essence, rather than a proper being of the essence. The text [of the *De ente*] seems to be framed carefully in a way that avoids any implication of *being* in the essence absolutely considered.[25]

By an *absolute consideration* one takes into account some characteristics of existing things that pertain to their natures *as such*, without having to take into account all of the characteristics those existing beings have. For example, these existing natures, the human nature of Socrates, the human nature of Xanthippe, and human nature existing *in anima*, do not differ when "considered absolutely." But there is no *thing* (*res*) that is at once the *absolutely considered nature*, the nature of Socrates, the nature of Xanthippe, and the nature of man *in anima*.

The only *existence* intelligible characteristics have is in singular things or *in anima*. Existing in a singular thing, intelligible characteristics constitute the *formal* principles of the being in which they exist, and so determine or specify it to be the being it is. For example, *rational* and *animal* in a man, which pertain to the nature *considered absolutely* and not according to its mode of existence, determine the man *to be* and *act* as a man is and acts. *In anima* the intelligible characteristics of a nature constitute the *formal* principles of an act of knowing, and so determine or specify that act to be the knowledge that it is.[26] *In anima*, *rational* and *animal* determine the soul to know what it does, namely, what is *rational* and what is *animal*, that is, men.

A nature taken as the form of the intellect's act of understanding is a principle of that act and is called the *intelligible species* insofar as it is considered *in anima*, not *absolutely* or *in re*.²⁷ 'Species' is not used here in its logical sense, where it is used for classification and is contrasted with 'genus', but rather in the sense of *form*. Let me emphasize this: *intelligible species* as I will use it in this work refers to a form taken as in-forming an act of intellect.

The *De Ente* and Predication

St. Thomas's debt in the *De ente et essentia* to Avicenna's *Metaphysics* is well known. In his eighth quodlibetal question, while organizing the distinctions slightly differently, he gives credit explicitly to Avicenna for having made them.²⁸ However, one respect in which his discussion differs from the Latin translation of Avicenna concerns this very discussion of the *being* attributed to the nature considered absolutely. Avicenna, like St. Thomas after him, does not attribute *being in re* or *being in anima* to natures considered *absolutely*, or *as such*, or *per se*. Using the example of the definition of horse, Avicenna says that "from itself, it is neither many nor one, nor existing in these sensibles nor in the soul."²⁹ However, taking another example, considered apart from existing in sensibles or in the soul, he does attribute to a nature *as such* a *certain being (esse)*

> for from this being (*esse*) it is neither a genus nor species nor an individual nor one nor many. But from this being it is animal as such and man as such.³⁰

Thus, according to Avicenna, there is a sense of *being* that is appropriate to the nature considered as such or absolutely, apart from the modes of existence it has either *in re* or *in anima*.

Though St. Thomas's discussion closely parallels Avicenna's, even using many of the same examples, he differs from Avicenna in the care he takes not to attribute any sense of *being* whatsoever to the nature *as such* or *absolutely considered*. Joseph Owens considers the place of *unity* in the discussion, in light of the *being* that Avicenna attributes to natures as such. In contrast to its proper *being*, Avicenna held that there was no *unity* to be attributed to it *as such*. But as the medieval discussion of *transcendentals* developed, *unity* came to be understood more and more as coextensive with *being*.³¹ So, by St. Thomas's time, it is not an option to deny *unity* to what has *being*. If one attributes a certain *being* to the nature as such, one must also attribute a certain *unity* to it. Where Avicenna attributed a proper *being* to the nature *as such* and denied a proper *unity* to it, St. Thomas denies both a proper *being* and a proper *unity* to it.

What implications does St. Thomas's position have for the predication of the nature? Taking a nature *per se* or *as such* is a mode of considering it, not a mode of its *being*. If the nature *as such* does not have a *being* proper to it, nothing prevents it from having the very same *being* as the individual in which it is found. Human nature is found in Aristotle and has his *being*. Human nature is also found in Plato and has his *being*. There is no danger, however, that Aristotle will turn out to have the same *being* as Plato since there is no claim that both have the *being* of the nature, or that, consequently, they and the nature are all *one being*. There is an asymmetry involved that blocks the problem. One does not say, "The individual has the *being* of the nature found in it," since the nature *of itself* has no *being* that the individual could have. Instead one says, "The nature has the *being* of the individual in which it is found." So there is no reason to expect that the nature will have the same *being* in diverse individuals, since diverse individuals have diverse *being*. The problem does not arise, since we cannot say that the nature found in an individual has the same *being* and *unity* as it has when found in another individual. Indeed St. Thomas makes just this point in the body of the *De ente* text, when he denies that unity belongs to the nature absolutely considered. But this is just a result of the absence of any being that might belong to the nature *absolutely considered*. There is no human nature possessing its own *being as such,* which also takes on the being of either Socrates or Plato or both and which is *one* being between them.

In St. Thomas, it is this lack of any *being* attributed to the nature *considered absolutely* that allows the nature predicated of an individual to be *one* with the individual. The *being* of human nature in Socrates is not the *being* of human nature in Plato, though Socrates and Plato are both human beings. In addition, the *being* of human nature in Socrates is not the *being* of human nature as it exists *in anima*—it is not the case, as they say, that "Socrates est species" (Socrates is a species).[32] However, the *being* of human nature in Socrates just is the *being* of Socrates; there is no impediment to a complete *unity* of the nature *in re* with the supposit of which it is the nature. According to St. Thomas, Socrates and his nature are simply *one*. Of course, this discussion has followed the texts in being restricted to substantial natures. Not everything predicated of a subject is a substantial nature, for example, 'white' said of Socrates. Not everything predicated of a subject has the same *being* and *unity* as the subject.[33] What St. Thomas's discussion opens up is the possibility that in some instances they do.[34]

This background is valuable for understanding St. Thomas's account of what is predicated in an enunciation, and its relation to the subject of that predication. *Extra animam*, the *essence* or *nature* is a principle of *being*. It is *what it is for a thing to be* (*quod quid est esse*) and *to act* from its *being*. The discussion from the *De ente et essentia*, the *Summa*, and the *Commentary on the De anima* tells us that some natures and principles of distinct things are found that do not differ when *considered absolutely*, that is, when considered apart from the actuality of their *being* (*actus essendi*) in distinct things. It also tells us that, considered apart from its actuality as an accidental operation of intellect, the nature of an act of understanding does not differ from the nature of the things understood by means of it.

But what does this lack of difference between what is understood *extra animam* and our understanding of it *in anima* have to do with what is predicated in an enunciation? St. Thomas writes:

> Therefore it is clear that the nature of man *absolutely considered* abstracts from being altogether, in such a way that it does not prescind from any of them. And it is this nature *so considered* that is predicated of all individuals.[35]

Primarily, it is the nature *absolutely considered* that is predicated of a real thing or many real things; so, only the elements of the nature *as such* are predicated along with it (*convenire*). These elements in any particular predication may only be "confusedly" and "potentially" understood in what is actually present in the predicate; they may wait upon an analysis of the nature *as such* in order to become actually understood in the predicate. In any case, "something in the head" is not predicated of a singular thing or many singular things. When we say that "Socrates, as well as Plato, is a man," we are not saying that either one *is* something *in anima*. The nature *in anima* is not predicated of the individual. It has a different mode of being from the nature *in re*, and so cannot be said *to be* it; the *being* of the one is not the *being* of the other.[36] So, *Socrates is an intelligible species* is not simply a false enunciation; it could not have been true.[37] Nonetheless, it is because the first act can grasp the nature *absolutely considered* and express it that the nature *as such*, not the nature *in anima*, can be predicated of Socrates and Plato and Xanthippe. The nature *in anima* is not *what* we predicate, but *how* we predicate.

For predication is something that is completed through the action of the intellect composing and dividing, nevertheless having as a foundation in reality the unity itself of those things, of which one is said of the other.[38]

Consequently, because it is the *means* for achieving predication in the *primary* sense, the nature *in anima* can be said to be predicated in a *secondary* or *analogous* sense.[39] Nonetheless, in the proper and primary sense it is the nature *absolutely considered* that is predicated.[40]

This contrast between the *proper* sense of signification and the *analogous* sense is why the use of general or common words in enunciations is subordinated to understanding in St. Thomas's work. A general word signifies$_2$ some being or beings *extra animam* because it signifies$_1$ or has a signification$_1$; it is subordinated to the *means* by which understanding of natures *as such* is achieved, and, consequently, the *means* by which natures *as such* are predicated. A word does not signify$_1$ a *quod* (a *what*) but an *a quo* (a *by which*). In signifying$_1$ the word does not "bring to mind" in the speaker *what* he is talking about, but the *means* for talking about it. In signifying$_1$, the word does not "bring to mind" in the listener *what* he is hearing about, but the *means* for understanding *what* he is hearing about. However, in signifying$_2$, the word does bring to mind a *quod*, and that *quod* is *what* is talked about. The "Aristotelian semantic triangle" cannot be reduced, because one angle in it is formed, not by a *quod*, but by an *a quo*, and its sides (signification$_2$, signification$_1$, and Similitude) are all relations of different kinds. Thus, the *De ente* provides an early and key text for understanding St. Thomas's appropriation of Aristotle to this effect.

Intelligible Characters and Concepts

It is the nature *absolutely considered* that I am calling the *intelligible character*. Let me introduce 'concept' as a technical term to designate the first act so informed or intelligibly delimited by the *intelligible character*,[41] that is, 'concept' names the first act of understanding.[42] Because an *intelligible character* informs or delimits a *concept*, it is a *principle* of a *concept*; but an intelligible character is not a *concept*, and 'intelligible character' is not a synonym for 'concept'.[43] It can be said that the *intelligible character* is the *intelligible content* of the concept, recalling again that it is this *intelligible content* that is *primarily* predicated, not the *concept*.

Because *concepts* are in-formed or delimited by *intelligible characters*, distinct first acts of intellect are distinguished from one another by the distinct *intelligible characters* that delimit them; in this sense it can be said that a first act of intellect *expresses* the *intelligible character* that in-forms or delimits it. One can then say that the *concept* is *how* a *res* is understood by one possessing intellect, and *how* a vocal sound signifies₂ the *res* that it does by one possessing and exercising the capacity to express understanding in vocal sound.⁴⁴ *How* a *res* is understood by one possessing intellect will differ according to the different intelligible characters that may in-form and be expressed in the first act of intellect. Consequently, *how* a vocal term signifies₂ a *res extra animam* will also differ according to the different intelligible characters that may in-form the first act of intellect. In virtue of the concept *green* achieved in simple apprehension, a green thing (*res*) may be understood and the vocal term 'green' will signify₂ the green *res* that is understood, while in virtue of the concept *tree*, a tree may be understood and the vocal term 'tree' will signify₂ the tree that is understood. Further, the *res extra animam* talked about by means of the words 'green' and 'tree' may be the very same thing, though differing in being *green* and being a *tree,* and being understood by means of the appropriately diverse concepts—"that green thing fell and is blocking the road," and "that tree fell and is blocking the road."⁴⁵

With these distinctions in mind, it is now possible to understand better *signification*₁, the relation between *word* and *concept. Signification*₁ indicates not *what* is being talked about, but *how* it is being talked about. It is now possible to appreciate more fully that *how* it is being talked about depends upon *how* the *res extra animam* is understood, that is, depends upon Similitude—signification₁ is subordinated to Similitude. "Any name would be aimlessly predicated of something unless through that name we might understand something concerning it."⁴⁶ Since a word is associated with a *means* of understanding a *res extra animam*, which *means* at the level of intellect is universal, a word has a *manner* or *way* of being about that *res extra animam*—it has a universal *modus significandi*.

Signification Again: Conventional Signification

St. Thomas follows Aristotle in asserting that signification₁, the relation between vocal sound and concept, conventionally differs among diverse linguistic communities. He makes an analogous use of the union of matter and form to describe the union of vocal utterance and concept. This analogous use of the form-matter distinction is more appropriately taken as the form of an artifact to its matter, than as a natural form to its matter.[47] Applying this distinction, a spoken word is the union of vocal sound and the concept it signifies$_1$. Recall that earlier when I introduced the conventions 'signification$_1$', 'signification$_2$', and 'Similitude', I said that it was important not to think of signification$_2$ as reducible simply to the conjunction of signification$_1$ and Similitude, with the latter two taken to be relations of the same basic kind. Here, discussing the instrumental function of articulated sounds, St. Thomas gives an indication of why the relations are different.

> [A]rticulated sounds [are signs] of passions [of the soul], such that no character of similitude is considered, but only the character of institution, as is the case in many other signs, as for example a trumpet is a sign of war. But in the passions of the soul it is necessary that the character of similitude to the represented things (*exprimendas res*) be considered, since they naturally designate them, not from institution.[48]

Earlier in the same passage he had noted that the passions of the soul are called similitudes because they constitute the knowledge or cognition of things, and "a thing is cognized by the soul only through some similitude of it existing in sense or in intellect."[49] As he analyzes it, the relation of Similitude that characterizes passions of the soul is a requirement of their role, namely, as *acts* of cognition of *res extra animam*. On the other hand, Similitude does not play a part in the relation of signification$_1$ because articulated sounds are not acts of cognition cognizing passions of the soul.

Whatever the institutionalized relationship of articulated sounds to passions of the soul is, it is not one of Similitude or likeness. But this also suggests that signification$_2$, the relation between vocal utterances and *res extra animam*, is not simply a relation of Similitude or likeness. Articulated sounds are no more acts of cognition of *res extra animam* than they are acts of cognition of *passiones animae*. Similitude or likeness just does not play a part in signification$_1$, and at best in signification$_2$ it does not characterize the relation between *res extra animam* and articulated sounds, but the necessary means for instituting such a relation.

The fact that the relationship between vocal word and concept is conventional is not meant to suggest that the speaker can choose any vocal sound whatsoever, and by some sort of "meaning-conferring act" make it meaningful by associating it with a concept. The view that is being expressed in St. Thomas's *Commentary* parallels other texts in which it is clear that the use of a term, at least in the beginning stages of inquiry, is to be taken from the ordinary use of the linguistic community.[50] A speaker does not simply choose an arbitrary sound in order to communicate something; he chooses a word because it is appropriate for what he intends to communicate. But this appropriateness does not proceed from the nature of the sound. It proceeds from and is bound within the limits set by the signification$_1$ that those sounds have in the context of their use. So, though used by an individual, words are not the tools of the individual, but of the community.

This Aristotelian account opens the door for recognizing, against the background of ordinary use, specialized and perhaps overlapping contexts determining both the signification$_1$ and the signification$_2$ of a word. One element of the new theory of reference developed in different ways by Kripke and Putnam[51] was the claim that the use of a general "natural kind" term by the linguistic community involves an implicit reference to a scientifically describable species. So, consider the use of 'fish' by the whaler when he says to his mates, "Well, do you think we'll catch the big fish today?" Describing his view, Kripke notes,

> Scientific discoveries of species essence do not constitute a 'change in meaning'; the possibility of such discoveries was part of the original enterprise. We need not even assume that the biologist's denial that whales are fish shows his 'concept of fishhood' to be different from that of the layman; he simply corrects the layman.

This is an innocuous claim if the whaler or "layman" is engaged in making a scientific identification. Thus the whaler means the same thing as the biologist does in his use of 'fish'; but given what he means by the term, he misidentifies the beings in the world that are referred to by that term.

But is the scientific identification and classification what the whaler is engaged in within the typical contexts in which he would say, "Let's pull in the big fish today"? If so, it is as if the whalers know what they mean by their words, yet do not know what in the world they are after; they stand and wait for the college professor to come to their assistance. The assumption here is that ordinary or "lay" use is an effort at scientific use, and most likely a failed one at that. But that assumption was built in to the example at the beginning when such general terms were described as "natural kind terms," and that both the scientist and the lay person were using, not the same general term, but the same "natural kind term."

On the contrary, St. Thomas is clear that the discussion in the *De interpretatione* is ordered toward scientific use as a development out of ordinary use, not as a corrective to it. What reason do we have to assume that the "lay" use of a general term is intended to be the use of a "natural kind term"? The whalers know perfectly well what in the world they are after, and their practice in that context is to signify$_2$ it with the vocal 'big fish'. Outside of the specialized context of scientific inquiry, there is no need to attribute to "lay" use the desire to signify$_2$ the determinate characteristics of natural kinds classified within the sciences. That is not to say that lay persons do not have true and false scientific beliefs, and express them. It is only to say that lay persons need not have true and false scientific beliefs whenever they use general terms that are also used as natural kind terms by scientists. The only time it would be appropriate for the biologist to correct the whaler is when the whaler has given up his whaling for biology.

Aristotle and St. Thomas think scientific inquiry begins from ordinary use. Having determined that there exists a determinate nature signified$_2$ by a word as commonly used, they proceed to determine the principles and properties of the nature. However, faced with the diversity of the ordinary use of a word, such as 'fish', it may turn out that there is no particular nature determined by all the ordinary uses. In that case, in order to proceed with the science, it may prove necessary for the biologist to restrict the use of the word to one of the natures present in the ordinary use. The biologist, interested as he is in classifying natural kinds, would indeed be in error if he called whales fish, since the linguistic practice of biology restricts the use of the word 'fish' to talk about a class of natural kinds of animal that does not include whales as a genus. But such a restriction need not be a correction. A biologist who sets out to correct the whaler does not know the difference between his own practice and the practice of whaling. It is easy to imagine a case in which a philosophically sophisticated whaler would respond to such a correction by saying, "Well, I suppose if I were a biologist I would talk the way you talk, but I'm not interested in classifying natural kinds; we whalers know what we are talking about."

Finally it is no part of the Aristotelian thesis that a speaker considers a *meaning* separate from a sound,[52] and then by a "meaning-conferring act"[53] decides explicitly or implicitly to render an arbitrary sound meaningful.

The language of *intention* and *choice* may appear misleading in this context if it is associated with signification$_1$. Instead, one might say that in choosing or intending to talk about some *res* in a context (signify$_2$), the speaker must implement appropriate means (signify$_1$). The means he implements are not chosen, but are set by the *usus loquentium* in that context for talking about those *res*. So, the actual occurrent signification$_1$ of an articulate sound does depend upon the speaker's intention to signify$_2$. John Haldane expresses the thesis in this way:

> Of course, linguistic meaning is related to public use and is sustained by interpersonal conventions, but what these animate are speakers' expressive and communicative intentions.[54]

On the one hand, if he wishes to talk about *such and such*, a speaker must utter certain words. On the other hand, without his intention to talk about *such and such*, he will not actually utter just those words necessary to communicate *such and such*, and implement their attendant signification$_1$. When I choose my words to talk about things, in some sense I have already chosen the concepts that provide their forms.

Natural Similitude

Despite the conventional relationship between *vocal word* and concept, the relationship between *concept* and *res*, namely, Similitude, is natural and the same for all human beings.[55] In part, this claim ought to be clear from the theses gathered from the *De ente et essentia*. However, since the signification$_1$ between the vocal term and the concept is itself conventional, and the concept is *how* the term signifies$_2$ *res extra animam*, the signification$_2$ between vocal term and *res* is conventional. Thus, if the conventional signification$_1$ of a vocal utterance changes and is subordinated to another concept, the utterance's signification$_2$ will change accordingly.

The thesis that concepts are natural Similitudes of *res* is not a doctrine of innate ideas, as if the concepts are naturally present in or emanate as likenesses of things from the nature of the intellect. The concepts signified$_1$ by vocal terms are *passiones animae*. It is natural for the intellect to be informed by the *intelligible characters* that structure and delimit *res* to the extent that it is "potentially all things."[56] However, it does not follow that the *actual* in-formation of the intellect proceeds solely *from* the natural principles of the intellect, simply because it is *in accord with* the natural principles of the intellect. So it does not follow that the Similitude between concepts and *res* proceeds solely *from* the natural principles of the intellect. By analogy, the in-forming of organic material in the generation of a dog, material that is potentially but not actually a dog, proceeds *in accord with* the natural principles present in the organic material, but that in-formation does not proceed solely *from* the natural principles of the organic material.

Commenting upon the text, St. Thomas indicates that the term 'passio' is used here to indicate receptivity from something other; these passions of the soul have the origin of their formal characteristics "*ab ipsis rebus*" ("from the things themselves"), which things, acting as agents upon the soul, cause them. *Passiones animae* are Similitudes of *res extra animam*, because the natural principles of the *res* in-form the intellect's act. The Similitude is a result of the natural principles of the intellect only insofar as the intellect is a capacity to receive and be in-formed by the principles of things *extra animam*.

This discussion leaves out the agent intellect's role of abstracting the intelligible species from sensible things, which we will take up later.

Sameness or Identity of Concepts and Words

According to Aristotle and St. Thomas, *passiones animae* are the same for all men despite the diversity of utterances by which they are conventionally signified$_1$. Further, the "things of which passions of the soul are likenesses are also the same."[57] What does this *sameness* or *identity* consist in; conversely what does *diversity* of concepts consist in? Here, obviously, I am not interested in the senses of *same, identical,* and *different* pertaining to the *mode of being* of concepts, that is, considered as acts of intellect, for example, *being a product of abstraction,* or *being the means of universal predication,* or *being present in Aristotle's intellect.* Instead, what differentiates one concept from another, and, conversely, what does sameness or identity of concepts consist in *absolutely considered*?

It is clear that concepts are not differentiated from one another by their conventional relation to words in linguistic practices; on the contrary, vocal utterances *qua* words are differentiated by the concepts they signify$_1$.[58] Instead, concepts are the *same* when they have, or express the same nature *absolutely considered*—the condition of their identity is their *intelligible character.* In the sense of interest here, concepts existing in different intellects or employed at different times are the *same* because their *intelligible character* is the same. They differ when they have or express diverse *intelligible characters*. But these *intelligible characters* are *what* they are in virtue of being principles of being for *res extra animam*.[59]

The recognition of formally identical concepts, in distinct intellects or in the same intellect at different moments of exercise, is no more controversial than recognizing identical substantial forms in Plato and Aristotle, even though, as substantial forms of numerically non-identical beings, they are numerically non-identical substantial forms. But since it is the *intelligible character* that delimits and differentiates a concept, and the *intelligible character* is a principle of being in the *res* of which the concept is a Similitude, it follows that for formally identical concepts in distinct intellects there is formal identity in the relations of Similitude between the numerically distinct concepts and the *res*.[60] It should also be clear that because they derive their formal identity from the concepts they signify$_1$, distinct token uses of a word are formally identical when they signify$_1$ formally identical concepts.

It is now possible to be more accurate about the formal identity of words. It is less than precise to say that the type identity of words is determined by the concepts they signify$_1$. The *type identity* of words is determined by the *intelligible character* of the concepts they signify$_1$. But those intelligible characters of concepts are, absolutely considered, the same as the *intelligible characters* of the *res extra animam* of which the concepts are Similitudes. Consequently, the *type identity* of words is determined by the intelligible characters of the *res extra animam* that they signify$_2$. Recall that for St. Thomas a concept is not *what* is signified in the proper sense of the term signification$_2$. What is signified in the proper sense are *res extra animam*. It is now clear that the *type identity* of words is determined ultimately by those *res*. According to St. Thomas, intellect was brought into the discussion to provide an account of *how* general words signify$_2$ *res*. That leads one to speak of signification less properly and analogously, namely signification$_1$. Signification$_1$ indicates the relation of a word to the *means by which* it signifies$_2$; not the *what* that it signifies$_2$, but rather the *how*.

It is not necessary for St. Thomas to explicitly write that he is making use of analogous terms here. The fact that the uses are instances of analogy in St. Thomas is *prima facie* evident in the use of the two forms of the Latin verb *significare* in 'res significata' and 'modus significandi', the technical phrases he employs to distinguish between the signification of *res extra animam* and *passiones animae*, and which form the basis for my conventions 'signification$_2$' and 'signification$_1$'. Nuances of Latin usage evident to a medieval author and reader may well pass unnoticed by the early twenty-first–century reader of Latin. It is not the simple grammatical difference between the perfect passive participle 'significata' and the gerund 'significandi' that indicates the analogous use, but what the grammatical distinction is used for, the first to modify the *thing*, the second to indicate the *manner*.

The *De ente et essentia* makes it clear that the *intelligible character*, that is, the nature absolutely considered is of itself neither singular nor universal. In *res extra animam* it is singular; *in anima* it takes on the intention of universality. Thus, the semantic triangle takes shape against the background of the *De ente et essentia*. The three vertices are clear enough: (1) the word, (2) the *what* of signification, and (3) the *how* of signification. The tie that binds them together and makes of them a triangle is the nature *absolutely considered*, the intelligible character, or in Latin the *ratio*.

Conclusion

Because formally identical concepts can exist in distinct intellects, numerically distinct occurrent uses of words can be formally identical when they signify$_1$ those concepts. Thus, the *intelligible character* that in-forms the first operation of intellect provides the basis for a description of how different token occurrences can be occurrences of the same type word, though those token occurrences be spatially and temporally distant, and uttered by different speakers. This analysis forms the basis for recognizing the possibility of the social practice of *scientia*, a practice that is subordinated to understanding, but also involves communication of that understanding. It also provides an appropriate setting for considering in subsequent chapters an issue that presents a direct challenge and stumbling block to Aristotle's account of *scientia*—the problem of *mental representationalism*.

Chapter 2

THREE RIVAL VERSIONS OF ARISTOTLE

"Well, but do you not see Cratylus, that he who follows names in the search after things, and analyses their meaning, is in great danger of being deceived?"

—*Socrates*

A New Interpretation and a Challenge to the Traditional Reading of the Text

In the previous chapter, I emphasized the cognitive background against which St. Thomas places Aristotle's text, in particular, both the similarity of the initial passages of the *De interpretatione* to parallel passages in *De anima* III.6 and Aristotle's explicit reference to the *De anima*. This was part of the larger task of explicating the relations that commentators have traditionally found in the text between words and mental impressions, mental impressions and things, and an implied relation of words to things. In "Aristotle on Spoken Sound Significant by Convention" Norman Kretzmann argues that this traditional approach is fundamentally misguided. He believes that nothing explicitly or implicitly in Aristotle's text suggests a relationship "relating spoken sounds or written marks to actual things:"[1]

> [T]his text makes better sense and fits its context better if it is interpreted as playing a more modest role. If it contains no claim at all, explicit or implicit, about a relationship of spoken sounds to actual things, then it is not even a sketch of a general theory of meaning.[2]

This opposition to a semantic element in the text marks a shift in his understanding. In an earlier piece on the history of semantics for the *Encyclopedia of Philosophy*, he was quite clear that "[Aristotle's] semantics of words . . . is to be found mainly in *De interpretatione*, Chapters 1–3." Outlining Aristotle's semantics in those initial chapters, he wrote, "[I]t seems that, according to this account, words signify things in virtue of serving as symbols of mental modifications resembling those things."[3] He abandons this aspect of his earlier view in the later article: "When we are told that spoken sounds are symbols and signs of mental impressions and that mental impressions are likenesses of actual things we are given no license to infer anything at all about a relationship between spoken sounds and actual things."[4] In addition, he begins the later article by telling us that "a few sentences near the beginning of *De interpretatione* (16a3–8) constitute the most influential text in the history of semantics,"[5] because of its fate in the hands of Aristotle's traditional interpreters, not because of what he now claims Aristotle actually intended to show.

Kretzmann believes Aristotle only intends to show that there is a priority exhibited between words considered as *signs* that are naturally related to mental impressions, and words considered as *symbols* that are conventionally related to those mental impressions. Aristotle's mention of the *natural* likeness of mental impressions to things is nothing more than a support, by contrast, to the *conventional* relation between *symbols* and *mental impressions*.[6] The relation of priority embodied in this contrast is, Kretzmann concludes, "one of the strengths in the Aristotelian account of conventional signification," "this combination of what seem to be complementary opposite types of signification."[7] Kretzmann believes that his new interpretation, without, and in contrast to, the traditional semantic triangle that he had earlier attributed to Aristotle, properly displays the strength of Aristotle's contribution. "Such semantic theory as is in that text is there, I think, only to the extent to which it contributes to the support for conventionalism."

Kretzmann believes that Aristotle's statement "actual things are the same for all" is "innocuous in itself," but that it

> can work together with the reference to *De anima* to give the first half of Chapter 1 the look of a summary statement of the foundations of

knowledge and communication, and it is that look which has deceived so many.[8]

But even more than the way in which these elements of Aristotle's passage can be misread, Kretzmann thinks the traditional interpretation has been seriously misled by flaws in Boethius's Latin translation of the *De interpretatione*. Most interpreters have followed Boethius's translation and adopted a mistaken "primacy of any semantic relation of spoken sounds to mental impressions over any semantic relation they may bear to actual things," and in a footnote Kretzmann writes "as far as I know, I am the only exception."[9]

Kretzmann's alternative interpretation is based more directly upon the Greek text and presupposes a resolution to a disputed question among the editors of modern editions of the Greek text of the *De interpretatione*. Among Greek scholars there is a debate about whether Aristotle's text contains an adjective πρώτων ('of first things'), the genitive plural of πρώτον, or an adverb πρώτως ('primarily' or 'in the first place').[10] In what follows, we will see how Kretzmann's interpretation requires that this editorial debate be decided in favor of the adverb.

To press his case about the traditional misreading of the text, Kretzmann makes much of the presence of the adverb πρώτως, and the distinction between 'symbols' (σύμβολα) and 'signs' (σημεῖα) in the Greek text. Kretzmann prefers Ackrill's Oxford translation, writing that it is "the only one in English that shows an understanding of the text."[11] Here is Ackrill's translation:

> now spoken sounds are symbols of affections in the soul, and written marks symbols of spoken sounds. And just as written marks are not the same for all men, neither are spoken sounds. But what these are in the first place signs of—affections of the soul—are the same for all; and what these affections are likenesses of—actual things—are also the same. These matters have been discussed in the work on the soul and do not belong to the present subject.[12]

Note that Ackrill's translation does not differ that much from Boethius's Latin text that I provided in the previous chapter, except for the adverbial phrase "in the first place," and the appearance of "symbol" in the discussion of written and spoken words. The most recent edition of Boethius's Latin text, by contrast, has the phrase "of which ... first [things]" (quorum ... primorum) with the adjective "first." In addition, Boethius had translated both σύμβολα and σημεῖα as *notae*, which fateful translation

Kretzmann charges with "hiding [Aristotle's] difference [between a sign and a symbol] from the view of Western philosophers for seven centuries or more."[13] This time period places the Boethian misreading of the text well into the thirteenth century and beyond. Kretzmann notes that even recent translations directly from the Greek, other than Ackrill's, committed the same error, often making use of phrases like 'symbols or signs', and more simply 'signs', where the Greek has 'symbols'.[14]

The difference is apparent in the first and the third sentences of the passage, respectively. In my translation of Boethius, I have 'signs' throughout, corresponding to Boethius's *notae*. Ackrill has, on the other hand, 'symbols' in the first sentence corresponding to σύμβολα, and 'signs' in the third sentence corresponding to σημεῖα. Jean Oesterle notes the difference in the Greek text between σύμβολα and σημεῖα, but downplays it when she writes, "sign" and "symbol" are later meanings which have become technical and therefore less directly convey what Aristotle intends here.[15] Kretzmann disagrees, "I am going to proceed on the hypothesis that this terminological difference reflects a real difference Aristotle recognized."[16] This is a hypothesis that survives from his earlier history of semantics article, where he wrote in much stronger terms:

> No doubt the traditional misreading of those passages during and after the Middle Ages is largely the result of the fact that in his otherwise faithful rendering Boethius obliterated the Aristotelian distinction between symbols and symptoms, translating both σύμβολα and σημεῖα as *notae*.[17]

The Greek adverb πρώτως forms the *presupposition* for Kretzmann's entire discussion of the difference between 'signs' and 'symbols'. It is presupposed because Kretzmann believes that he and the tradition share the adverb, but that they differ on how to interpret it. For him, placing an adjective like πρώτων in the text makes less sense than the adverb, and he believes that the Greek manuscript tradition is overwhelmingly in favor of the adverb.[18] Further, he believes that the tradition has overwhelmingly based its misinterpretation on an adverb; it has understood the adverb to mean that words are primarily or directly related to concepts, and only secondarily or indirectly related to things. This presupposition is also the basis for reading Kretzmann's criticism of the tradition as an initial pass at the problems of language and mental representationalism. The claims and objections in contemporary philosophy directed at the Aristotelian "semantic triangle" are more plausible to the extent that the history of the textual interpretation of Aristotle bears them out. If Kretzmann is correct about the adverb and its traditional interpretation, the plausibility of the charge is clear that this is what "commentators on this text have regularly done, usually remarking (as if it were obvious) that Aristotle maintains that words stand directly for thoughts and indirectly for things."[19]

Kretzmann's Thesis

There are then two major elements of Kretzmann's position that I wish to examine, first what I will call the thesis, and second what I have called the presupposition. The thesis is divided into two parts. The first part holds that 'sign' and 'symbol' are not synonymous in the text. It was Boethius's translation of the Greek words for 'symbols' (σύμβολα) and 'signs' (σημεῖα) by the single Latin word *notae* that is responsible for "hiding [Aristotle's] difference [between a sign and a symbol] from the view of Western philosophers for seven centuries or more." John Magee in *Boethius on Signification and Mind* provides a very detailed discussion of the philological arguments against this first part of Kretzmann's thesis,[20] but we are concerned with a second part, namely, how the presumed difference between the two bears upon the correct reading of Aristotle's text.

According to Kretzmann, there is a technical distinction between 'symbols' and 'signs':

> for x to be a symbol of y is for x to be a notation for y, to be a rule-governed embodiment of y in a medium different from that in which y occurs.[21]

In the case at hand, the verbal medium is presumably different from the mental. A sign, on the other hand, is a natural indication, like a medical symptom.[22] The part of Aristotle's text that says spoken sounds are signs of mental impressions would then have the sense of "spoken sounds are (in the first place) effects indicative of their concurrent causes, mental impressions."[23] The adverbial phrase "in the first place" comes from the presupposed adverb, and indicates a first *function* of spoken sounds; spoken sounds indicate as *symptoms* their natural and concurrent causes, *mental impressions*. The *symbolic* function of a spoken sound is then implicitly *in the second place* another function, logically distinct from the first, and in a sense constructed upon it. The mere sign, *primarily* a "natural indication," is *secondarily* invested with its symbolic function.

Kretzmann notes that his symbol relation is symmetric. If spoken words are notations for mental impressions, and rule-governed embodiments of mental impressions in a medium different from that in which the mental impressions occur, then mental impressions are notations for spoken words, and rule-governed embodiments of spoken words in a medium different from that in which the spoken words occur. Despite the fact that Aristotle does not write that mental impressions are symbols of spoken words, Kretzmann does not believe that this is an untoward implication of his interpretation; it is rather what Aristotle is philosophically committed to by what he does write. Thus, Kretzmann distinguishes between "encoding symbols," and "decoding symbols." Spoken sounds are "encoding symbols of mental impressions," while "mental impressions are decoding symbols of spoken sounds."[24] Aristotle is discussing the former, "encoding symbols of mental impressions." The language user encodes his mental impressions into spoken sounds, while the listener decodes those very sounds into mental impressions. Another way of putting it is that the system of mental impressions gets *translated* into the system of spoken sounds, while the system of spoken sounds gets *translated* into the system of mental impressions. The symbolic function of language encodes mental impressions in order that they may be transmitted to another; this function is contrasted with its merely natural significative or symptomatic function. Kretzmann does not use "translation." It will become clear later in this work why I draw the parallel between encoding and decoding and translation.

The role played by Kretzmann's interpretation of the presupposed adverb, and what it makes possible for his thesis about signs and symbols, is clear. By contrast with the traditional interpretation, his interpretation of the adverb πρώτως indicates nothing about an indirect or secondary semantic relation of words to things. It indicates an order of priority between two relations by which a vocal utterance may be related to a mental impression, *first* or *primarily* the sign or natural symptom relation, *secondarily* the symbol relation. The presupposed sense of *primacy* in both the traditional interpretation and his own is a point of intersection between them through which they can be compared. It is due to the adverbial phrase "in the first place," Ackrill's translation of πρώτως. Kretzmann seems to agree with the tradition that this "in the first place" calls for an implicit "in the second place." The difference between the parties concerns to what this implicit "in the second place" refers.

Kretzmann argues that the inability of the tradition to even consider his interpretation of πρώτως is rooted in its failure to recognize a distinction between σύμβολα and σημεῖα. He had already argued that the Latin tradition's failure to distinguish between these two terms has its origin in Boethius's translation of both by *notae*. Boethius's failure to distinguish the terms "precludes recognition of the [primacy of the sign-relation over the symbol relation]," and "most have adopted [the primacy of any semantic relation of spoken sounds to mental impressions over any semantic relation they may bear to actual things] with no sense of having rejected a competing interpretation." This point seems correct. If commentators failed to see a distinction between 'sign' and 'symbol', they could hardly be expected to rank them in terms of primary and secondary. By contrast, recognizing the non-synonymy of σύμβολα and σημεῖα allows Kretzmann to provide what he takes to be the correct interpretation of πρώτως. All Aristotle is interested in showing is that words as symbols are conventionally related to mental impressions, as opposed to the sign relation which is natural, and the natural likeness of mental impressions to actual things.[25]

Problems for Aristotle Raised by Kretzmann's Thesis

If Kretzmann's interpretation is correct, it appears to raise difficulties for Aristotle closely related to and no less troubling than the problems raised by the traditional interpretation.

No one now contests the presence of σύμβολα and σημεῖα in the Greek text, and the Latin *notae* in Boethius's translation. So consider Kretzmann's description of the importance of the difference between the terms σύμβολα and σημεῖα:

> [E]lsewhere in Aristotle and in others before him and after him, the words 'σημεῖον' and 'σύμβολον' differ in being associated broadly with natural and with artificial indications, respectively. A medical symptom may be considered the paradigm of a σημεῖον, and an identity token (especially one of two irregular broken halves of a potsherd or a seal on a document) may be considered the paradigm of a σύμβολον. This natural/artificial division is the philological basis of my hypothesis.[26]

Kretzmann explains just what he means by a symbol when he describes it as a rule-governed encoding or decoding in a distinct medium by reference to a parrot's vocalization of sounds that would normally be symbols, if spoken by a human being.[27] And he provides a characterization of the natural sign relation he has in mind when he explains why written words are symbols but not signs of spoken words. For Kretzmann's Aristotle, in order for a spoken sound to be a symbol it must "in the first place" (πρώτως) be a sign, an effect that indicates the mental impression that is its cause. Earlier he had characterized a symptom as involving "a regular natural association of occurrence,"[28] and the examples he thinks are indicative are those of medical symptoms concurrently indicating underlying pathologies. Then on top of this *primary* regular natural association of occurrence a *secondary* symbolic function can be constructed, one that "encodes" the mental impression in a distinct medium, the spoken sound, thereby rendering the sign a word.

Because Aristotle uses 'symbol' but not 'sign' in characterizing the relation of *written* word to *spoken* word, he must be claiming that a written word can be a symbol of spoken words without *first* being a sign of spoken words. "They [written words] are not symptoms of spoken sounds because they are regularly produced in the absence of spoken sounds."[29] *Regular association* then is the dominant element distinguishing the natural causal *sign* relation between sounds and mental impressions from the conventional encoding *symbol* relation. And the point about written words being produced in the absence of spoken suggests that what Kretzmann has in mind is that mental impressions are necessary conditions for the production of verbal sounds; verbal sounds are symptoms of mental impressions because they are not regularly produced in the absence of mental impressions. Whether the mental impressions are sufficient conditions for the production of verbal sounds is another matter. Presumably Kretzmann does not intend to suggest that according to Aristotle a verbal sound is produced whenever there is an occurrence of a mental impression in the soul.

Kretzmann believes that a parrot does not use spoken words or symbols because there is no *primary* sign relation involving the parrot upon which to build the *secondary* symbol relation. A "parrot may produce spoken sounds of which impressions in your mind may be the (decoding) symbols," that is, you *could* take them as words. But that would be a mistake because the parrot, making many of the same sounds that human beings use as words, does not conventionally "encode" mental impressions into those sounds. Those sounds are not *primarily* natural signs or symptoms of mental impressions in the mind of the parrot. The sounds coming from the parrot are not real words, because they "are not symptoms of the occurrence of [the same kind of] mental impressions [as occur in human minds]."

This contrast between the parrot and the human being immediately raises the issue of the character and role of the mental impressions involved, since whatever mental impressions, if any, may be associated with a parrot's vocalizations, they are not *of the same kind* as occur in human minds. Thus, even if Aristotle is not giving a semantic sketch of how words are meaningful, the passage in concert with Kretzmann's example of the parrot immediately suggests the question of the role of the mental impressions. If a necessary condition for a sound being a word is that it be an encoding of *human* mental impressions, without doubt this has implications for semantics. The character of the "natural likeness" to things that passions of the soul bear is essential to understanding their symbolic function. Perhaps the special encoding relation to human mental impressions does not *constitute* the meaning of words, but Kretzmann's example of the parrot is designed to show that lacking it, a sound is not a word, and thus, *a fortiori* it has no meaning, no semantic content.

Aristotle does suggest that the vocalizations of animals are naturally related as signs to something.

> I say 'by convention' because no name is a name naturally but only when it has become a symbol. Even inarticulate noises (of beasts, for instance) do indeed reveal something, yet none of them is a name.[30]

Notice that this text does not suggest that before becoming a symbol a vocal sound is a sign of passions of the soul. Still, Kretzmann believes this text supports his thesis since, "the notion that a spoken sound *becomes* a symbol is well suited to the view that it is *primarily* a symptom."[31] If the claim about the noises of beasts is applied to Kretzmann's example of the parrot, it gives one reason to believe that the parrot's vocalizations do "reveal something," though not as conventional symbols. The contrast between 'conventional' and 'natural' suggests that the vocalizations reveal something as natural signs. Whatever may be revealed by these natural signs in parrots, Kretzmann's Aristotle must hold that they are not the same mental impressions as exist in human minds, and that are naturally signified by human vocalizations that sound the same as the parrots. Otherwise, one might have good reason for taking a parrot to be speaking human language; the parrot would not be parroting but speaking. Kretzmann says that since the parrot's vocalizations "are not symptoms of the occurrence of such mental impressions [as occur in your mind as decoding symbols of those spoken sounds] they are not produced by the parrot as (encoding) symbols."[32] But then what do they reveal as natural signs?

Kretzmann might respond that in the case of a parrot's "parroting" human speech this text does not apply. The vocalizations do not reveal anything as natural signs, since they are *articulate* sounds, while the text from Aristotle above refers to the "inarticulate noises" of beasts revealing something. The *inarticulate* noises that parrots make *in the wild* may reveal something as natural signs, but not the articulate sounds they are trained to make by human trainers.

This response doesn't really help Kretzmann's Aristotle. In the text from Aristotle that Kretzmann appeals to in order to make the point that "the notion that a spoken sound *becomes* a symbol is well suited to the view that it is *primarily* a symptom," the natural revelation of something by beasts or man in the "primary sign relation" is associated with a *lack of articulation* in the sounds produced. If the text is supposed to support Kretzmann's point about the secondary relation of conventional symbolization, it also implies that in humans the primary sign relation is as "inarticulate" as it is in beasts. Articulation pertains to the secondary conventional relation, not the primary natural one. So Kretzmann cannot then appeal to articulation as a way of distinguishing *the natural sign relation* in truly human vocalizations from the natural sign relation in beastly vocalizations. Articulation is a product of symbolization; it distinguishes the natural sign relation from the conventional symbol relation.

So Kretzmann cannot reply to the problem raised above by appealing to the production of articulate sounds by parrots; on Kretzmann's interpretation of Aristotle, if the parrots are capable of articulation it is only because they are capable of the conventional encoding that he attributes to human speakers. But Kretzmann has denied the consequent. So, even when they sound like the words we speak, the sounds produced by parrots are not articulate sounds, inasmuch as they are not conventional symbols. Just as Kretzmann claimed it is a mistake to take the parrot's sounds to be words, encoding symbols, it is equally a mistake to take them to be articulate. They are as inarticulate as the sounds they may produce in the wild. And this judgment rests upon what is going on "in the minds" of humans versus what is (or is not) going on "in the minds" of parrots. Once again, the character of the natural likeness of the "passions of the soul" in human beings determines the semantic character of the symbols we produce. That relation is not simply another example of a natural relation, as Kretzmann claimed, merely mentioned to give support to the contrast with the conventional relation. It is inseparable from the character of that conventional relation—semantic.

Further, in general what need has Aristotle of the underlying *natural* relation between sounds and mental impressions? Why not simply recognize the *conventional* relation in humans, and the lack of it in parrots? Here one has difficulty understanding why Kretzmann's Aristotle feels any need to introduce the natural sign or symptom relation at all. Aristotle could have made the point that vocal words are conventionally significant without first asserting that they are "natural" signs of mental impressions. We saw that by the "natural symptom" relation, Kretzmann means nothing more than that vocal utterances do not occur in the absence of mental impressions; mental impressions are necessary conditions for verbal utterances. But how does that fact serve to distinguish the natural sign relation from the conventional symbol relation? Even if there were no underlying "natural" relation, presumably it would still be true that a conventional symbol never occurs in the absence of a mental impression; a mental impression is just as much a necessary condition for a purely conventional symbol as it is for a natural symptom. So that necessary condition cannot serve to distinguish the natural symptom relation from the conventional symbol relation. Had Kretzmann suggested that a mental impression was a sufficient condition, that whenever a mental impression occurs a verbal utterance occurs, the situation might be different. Such a natural relation might well be distinguishable from a conventional one. But he shows no sign of intending that further condition, and it would be a highly controversial, not to mention implausible claim. So what explanatory role does the underlying *natural* relation that Kretzmann attributes to Aristotle play, and what justifies that role?

Kretzmann contrasted written words as symbols of spoken words with spoken words as symbols of mental impressions by noting that written words regularly occur in the absence of spoken words, while presumably spoken words never or rarely occur in the absence of mental impressions. But for Aristotle is that enough to mark a distinction between the conventional and the natural, a relation of regular concurrence, where the signified is a necessary but not sufficient condition of the sign? Consider for a moment whether written words are "in the first place" natural symptoms of mental impressions. It may well be the case that a written word is observed in the absence of a spoken word and the absence of a mental impression because the writer of that word is now dead, for example. But Kretzmann quickly dismisses the possibility that written words are natural signs of mental impressions because "they persist past the time of their production as spoken sounds do not."[33] But that seems too great a restriction on the notion of a natural sign. Why should it be a necessary condition for a natural sign that it not persist beyond the duration of its production? It is very unlikely that Aristotle would agree to that necessary condition. Tree rings persist long after the duration of their production and function as natural signs of the climactic conditions that obtained in their production. To cite an example that Aristotle himself gives in another context, urine functions as a natural sign of health or disease that persists after the duration of its production.[34] If the person's health has changed, the urine remains a natural sign of the state of health at the time the urine was produced.

Surely there is a greater parity that Aristotle would or should have recognized between producing spoken words and producing written words than whether they persist beyond the duration of their production: the *writing* of words does not occur in the absence of mental impressions. So the *writing* of words is just as much a natural sign of mental impressions as is the speaking of them. Kretzmann's rejection of them as natural signs of mental impressions is based upon an arbitrary restriction of the notion of a natural sign that Aristotle would not accept. And of course written words conventionally encode mental impressions as well. So, even if one grants that written words are not natural signs of spoken words, if Kretzmann is correct about spoken words being "in the first place" natural signs and "in the second place" conventional symbols of mental impressions, one would have expected Aristotle to have pointed out that this is equally true of written words. Instead Aristotle restricts himself to saying that written words are symbols of spoken words. These considerations cast serious doubt upon Kretzmann's attribution to Aristotle of two distinct relations between vocal utterances and mental impressions, "in the first place" the natural symptom relation, and "in the second place" the conventional symbol relation. We have little reason for thinking that the sign relation is anything other than the conventional symbol relation, with 'sign' functioning in the text as simply a synonym for 'symbol'.

Now suppose we turn to the issue of the identity and character of the mental impressions mentioned in Aristotle's text. The difficulties associated with them stand out just as much in Kretzmann's interpretation as they did when the sketch was understood to be of a semantic theory. Once Kretzmann grants that mental impressions are "decoding symbols" of vocal utterances, the question of *access*, cognitive or otherwise, to the mental symbols in our own minds or the minds of others, is particularly important. We have already seen that the character of the natural relation between passions of the soul and things is crucial for understanding the symbolic function of vocal utterances. But even as Kretzmann downplays the import of the role of similitude in the character of the mental impressions, they retain the character of internal objects upon which the mind acts to produce meaningful words. Encoding them reveals one's mind to another; decoding them reveals another's mind to oneself, unless the other happens to be a parrot.

Three Rival Versions of Aristotle   53

Indeed, Kretzmann's use of the parrot is reminiscent of Locke's. Locke had also used the vocalizations of parrots as an example of a contrast with real speech. According to Locke, parrots do not speak because they do not have in their minds the same mental representations as exist in human minds. For this reason, among others, I do not think Kretzmann's criticism of the Latin tradition can be read in abstraction from the larger philosophical tradition of mental representationalism.

This picture remains in place even if, as Kretzmann maintains, the *similitude* of mental impressions to things does not play an essential role in the sketch. A standard account of what is meant by 'symbol' in contemporary philosophy is "a sign which is constituted a sign merely or mainly by the fact that it is used and understood as such."[35] Kretzmann's interpretation of Aristotle would add to this that in the case of vocal utterances the sign must be a natural symptom of what it signifies. If mental impressions are "decoding symbols" it seems that they must be "used and understood as such," "rule governed," where the rules are understood to be conventional. Presumably "law governed" would not do, because it would suggest that the encoding and decoding are natural relations governed by natural laws, in which case it would be difficult again to distinguish them from the underlying natural sign relation Kretzmann attributes to Aristotle. But the notion of a "rule" seems to suggest a rule follower who applies the rule in such a way that it is distinct from merely being "covered" by a natural law.

Thus, Kretzmann's talk of "encoding" and "decoding" certainly suggests that the mental symbols are "used and understood as such," and not inaccessible, cognitively or otherwise, to the intellect that so uses and so understands them. Kretzmann's analysis suggests that for Aristotle they are mental objects upon which the mind's activities bear; indeed, they are symbolic objects upon which the intellect acts. A symbol is a *symbol for* someone who uses it as such. So Kretzmann's interpretation commits Aristotle to the position that the mental impressions he mentions function as symbols for the intellect that has them. In addition, Kretzmann's interpretation of Aristotle implied that a sound takes on "articulation" only to the extent that it is the cryptologist's *encoding* of her mental impressions, her mental symbols. Encoding and decoding presuppose a language upon which to operate. They do not constitute a language from what is not already a language. A cryptologist encodes *from* a language, and decodes *into* a language. So the spoken language is only a language because it encodes one, presumably the language that the mind speaks to itself.

It is this picture of the mind, its objects and how it operates upon them, in concert with the vexed issue of how the mental objects are in turn related to the world that gives rise to the contemporary criticism of Aristotelianism. If Kretzmann's later interpretation is correct, Aristotle's text seems to fare no better than on Kretzmann's own earlier interpretation, which reflected much more closely the traditional interpretation he would now criticize. Indeed, it fares worse. Kretzmann's interpretation attributes to Aristotle's text an additional untenable distinction between sounds as signs of mental impressions and the very same sounds as symbols of those mental impressions, a distinction that does not advance our understanding of "in the first place" over the "traditional" interpretation. It is to the texts of Aristotle and the "traditional" interpretation that we now turn, to consider the presupposition of Kretzmann's argument.

The Text of Aristotle and the Commentary Tradition

The Greek Text of Aristotle

According to Kretzmann the traditional interpretation has attributed to Aristotle the view that spoken words *directly* signify mental impressions, and only *indirectly* signify things, a claim based on the presence of an adverb in Aristotle's text. Kretzmann provides an alternate interpretation of that presupposed adverb. The presence of this adverb is no small matter for Kretzmann, for his interpretation cannot be sustained without it.

There is a dispute among editors about the presence of the adverb πρώτως or the adjective πρώτων in the Greek text. There are two main modern editions of the Greek text to consider. The first edition is the text produced in the nineteenth century by Immanuel Bekker to be found in the *Aristotelis Opera*. The second edition of the text was produced over a hundred years later by L. Minio-Paluello for Clarendon Press in 1949.[36] The Bekker text has the adverb πρώτως, while Minio-Paluello's text has the genitive plural adjective πρώτων.[37] There are a number of different translations of Aristotle's text into English. Hippocrates Apostle,[38] apparently operating from the Bekker text, translates the adverb as "primarily," as does Harold Cooke in the Loeb text and translation.[39] John Ackrill in general uses the Minio-Paluello text.[40] However he translates the disputed text with the adverbial phrase "in the first place." Kretzmann points out that Ackrill seems to have followed the Bekker text, without an explicit indication that he has adopted the variant at this point. Noting this controversy between the Bekker and Minio-Paluello texts, Kretzmann opts for Bekker against the more recent Minio-Paluello. He explains:

Since the manuscript testimony is overwhelmingly in favor of the adverbial form here, the only reason for adopting the adjectival form to be found in Minio-Paluello's edition is that the adverb makes no sense. Since it seems to me to make good sense, and better sense than the adjective, I follow Bekker's edition (and Ackrill's translation).[41]

John Magee, for one, has disputed at length Kretzmann's claims about the overwhelming evidence of the manuscript tradition and Minio-Paluello's text.[42] Still, Edghill in the Oxford translation of 1928 follows Bekker, and translates the whole phrase "the mental experiences, which these directly symbolize, are the same for all." Note the use of 'directly' here, in place of what others have generally translated 'primarily'. This use supports Kretzmann's claim about the tradition and states explicitly what 'primarily' or 'in the first place' only suggest.

The influence of this tradition of translation can have a powerful and important influence upon our understanding of St. Thomas. Consider the Blackfriars' translation of Question 13, article 1 in the *Prima Pars* of the *Summa Theologiae* which begins the discussion of how we speak about God. Citing the passage from the *De interpretatione*, the translation reads:

> Aristotle says that words are signs for thoughts and thoughts are likenesses of things, so words refer to things indirectly through thoughts.[43]

That St. Thomas would begin his response to the question of whether God can be named by us by citing the passage from the *De interpretatione* provides some sense of the importance of this text to him. In addition it is by extrapolating from St. Thomas's discussion in Question 13 that most readers are exposed to his "philosophy of language." Few pursue the reference back to a close reading of his commentary on Aristotle's *De interpretatione*. The first phrase ending in "of things" is the translator's paraphrase of the quotation from the *De interpretatione* that appears in the Latin text of the *Summa*. The second phrase ending in "thoughts" is the translation of St. Thomas's explanation of the Aristotelian quote. The translator translates "mediante conceptione intellectus" as "indirectly through thoughts." It is no great leap to conclude from this translation that St. Thomas is committed to the thesis that words are directly related to thoughts, perhaps that they even directly refer to thoughts. Indeed this is precisely what Bernard Lonergan claims is St. Thomas's position in the commentary on the *De interpretatione*. Paraphrasing St. Thomas he writes, "directly, we are talking about objects of thought, *inner words,* and only indirectly, only in so far as our *inner words* have an objective reference, are we talking of real things."[44] The Blackfriars' translation of the *Summa* and Lonergan's paraphrase of St. Thomas accord perfectly well with the sense of Aristotle's text as it is found in Bekker's edition and Edghill's translation.

The association of 'primary' with 'direct', and conversely, though often implicitly, 'secondary' with 'indirect', is particularly important. For the most part, Kretzmann relies upon the use of 'primary' versus 'secondary'. But he also uses 'direct' and 'indirect' when he writes that "commentators on this text have regularly [interpreted the text in this way], usually remarking (as if it were obvious) that Aristotle maintains that words stand directly for thoughts and indirectly for things." According to Kretzmann, this commentary tradition goes well into and beyond the thirteenth century. At least as found in the Blackfriars' translation, St. Thomas should be included among these commentators that Kretzmann has in mind.

I have found only one translation that follows Minio-Paluello's genitive plural adjective at this point. It is a translation by John Crossett that Apostle includes as an alternate in an appendix to his own adverbial translation. Crossett's translation is "the primary [affections] which take place in the soul . . . of which these symbols are signs, are the same for all." Clearly he takes the adjective to modify the mental impressions or affections, not a function that words have "in the first place."

While Kretzmann might be able to sustain the distinction between 'signs' and 'symbols' solely on a grammatical distinction between σύμβολα and σημεῖα, if there is an adjective in the text he cannot sustain his claim that the sign relation is primary and the symbol relation secondary. He has the adverb 'primarily' or the adverbial phrase 'in the first place' modifying the verb 'to be' in "what these are in the first place signs of—affections of the soul—are the same for all." Vocal utterances are primarily signs of mental impressions. Through the verb 'to be' the sense and force of the adverb transfers to the subject, the vocal utterances or words referred to by the demonstrative pronoun 'these' (ταῦτα).[45] The emphasis is on what these vocal utterances are or do in the first place. *In the first place* they *are* signs of mental impressions; *in the first place* they *do* signify mental impressions. Kretzmann also believes that 'in the first place' or 'primarily' requires an implicit 'in the second place' or 'secondarily'. That implicit 'in the second place' is supplied by the second function of vocal utterances; *in the second place* vocal utterances *are* encoding symbols of mental impressions; *in the second place* they *do* symbolize mental impressions.

But suppose that there is a genitive plural adjective πρώτων in the text as the most recent critical edition maintains; what does the adjective modify? It simply cannot modify the vocal utterances through the demonstrative pronoun (ταῦτα), nor the signs (σημεῖα), since they are in the nominative case. The only thing for the adjective to modify is the mental impressions or affections in the soul, which in the Greek are referred to by the genitive plural παθημάτων and which Ackrill translates as "of affections" in "spoken sounds are symbols of affections in the soul." Modifying Ackrill's translation of the passage, as if he had translated the adjective, it would then read "but these are signs *of the first [affections]*, which affections are the same for all, and what these are likenesses of—actual things—are also the same." The sense and force of primacy is now directed to the affections of the soul, which are called 'first', whether or not one wants to say that they are signified or symbolized. Instead of emphasizing what vocal utterances are or do, the emphasis is upon the status of the affections of the soul, what they are.

Now, if one is going to apply 'first' to affections of the soul as related to vocal utterances, what is one going to apply an implicit 'second' to? Not the vocal utterances, since they are no longer modified by or the object of a note of primacy, and our attention is now drawn to the things related to the vocal utterances. And this 'second' cannot apply to the relation of symbolization that Kretzmann sees in the text. Since the relation of signification is no longer modified by or the object of a note of primacy, it no longer makes sense to say with Kretzmann that the relation of symbolization is secondary. One needs an appropriate contrast to the affections of the soul that are the same for all, and that are called for some reason "first things." What are these second things? Whatever one comes up with to fill that role, it is clear that Kretzmann cannot sustain his case if there is a genitive plural adjective where he believes there is an adverb. He must hold that the critical edition of Minio-Paluello is in error in order for his interpretation to have any plausibility at all.

A much greater difficulty for the presupposition of Kretzmann's analysis is what to make of his claim about the tradition of *misreading* the text, which he claims is based upon an adverb in the Greek. He argues that the tradition misreads the adverb because it is misled by Boethius's translation. He argues that the failure of the translators and commentators influenced by the Latin tradition to recognize a distinction between σύμβολα and σημεῖα produces the failure to even consider the interpretation of πρώτως that he develops. But he has already argued that, in the case of the medieval Latin tradition of commentary, this latter failure has its origin in Boethius's translation when he "obliterates" the distinction with the Latin term *notae*. But this raises the issue of what the Latin text looks like, since the medievals were not misled by the Greek text. They might still be wrong about the interpretation of Aristotle as a whole, but if there was no adverb in the Latin, they could not have been misled in their interpretation of it by the "obliteration" of the distinction between signs and symbols. So it matters a great deal whether the commentators had a Latin text with an adjective or an adverb in it.

The Latin Text of Aristotle

The dispute about an adjective or the adverb in the Greek text of Aristotle finds a parallel in the modern editions of Boethius's translation, with the dispute now between the Latin genitive plural adjective *primorum* and the adverb *primo*. The 1888 Leonine edition of Boethius's text has both the adjective *primorum* and the adverb *primo*, while the 1989 Leonine has only the adjective. The 1989 editors are following Minio-Paluello's critical edition of Boethius's Latin translation in the *Aristotelis Latinus*, which has "quorum autem hae primorum notae."[46] The genitive plural *primorum* goes naturally with the *quorum*, not the nominative plural *notae*; this corresponds well with Minio-Paluello's Greek version of Aristotle's text. Gerald Verbeke provides a critical edition of both William of Moerbeke's translation of the Greek text of Aristotle, and Moerbeke's translation of the Greek text of Ammonius's *Commentary*. Kretzmann, while noting that Moerbeke produced the translation of Ammonius's *Commentary* for St. Thomas, does not take note of Moerbeke's translation of Aristotle. These two translations by Moerbeke, however, present an interesting twist on the story. In the translation of the *De interpretatione*, William has the adverb *primum*,[47] in "quorum tamen haec signa primum, eaedem omnibus passiones animae,"[48] where Boethius has the adjective *primorum*. Thus William's text of Aristotle corresponds well with the Bekker Greek text, and it is the only Latin text of Aristotle with an adverb alone. Verbeke explicitly contrasts William's translation with Boethius's at this point, writing that his choice of the adverb suggests "that he had another reading of the text in front of him, in order to choose between 'πρώτων' and 'πρώτως'."[49]

In the body of his *Commentary*, however, Ammonius quotes the passage in Aristotle. But translating Ammonius here Moerbeke has the adjective in "dicit enim 'quorum tamen haec signa primorum.'" The embedded quote is Ammonius's quotation of Aristotle. The genitive plural adjective stands out in Moerbeke's Ammonius where Moerbeke's Aristotle had the adverb *primum*. Later, explaining the quotation, Moerbeke's Ammonius repeats the adjective *primorum* twice, as modifying *quorum*. I will show later that when Ammonius comments on the text, he holds that the adjective associated with the *quorum* refers to the soul's concepts, or as it is in the genitive plural in the text *conceptionum*. So it seems that the Greek text of Ammonius that Moerbeke is working with has the adjective πρώτων, which is also supported by the sense of Ammonius's comments, where the Greek text of the *De interpretatione* that Moerbeke is working with has the adverb πρώτως. To the credit of the accuracy of his translation of both, William preserves the difference.

In his critical notes on Ammonius Verbeke indicates that Busse's 1895 critical edition of the Greek Ammonius[50] has the adverb πρώτως, where William has the adjective *primorum*.[51] However, he also points out that one of the codices that Busse used, but chose not to follow at this point, has πρώτων. It is not likely that Busse would take codices of William's Latin texts seriously as providing evidence for the adjective in the Greek, for the only one that seems to have been available to him also seems to have been in poor shape, and Busse's judgment of it was that "it was made by a man poorly informed, in an uneven and not very accurate language."[52] In general, editors prior to Minio-Paluello appear not to have relied very much upon ancient Latin translations of Aristotle for evidence about what the Greek text contained. By contrast, in 1949 Minio-Paluello was exhibiting a very different attitude, maintaining as Verbeke puts it "that certain ancient versions are particularly useful to discover the original sense of the Greek text."[53] Minio-Paluello used various Latin texts of the *De interpretatione*, including William's, to construct his critical edition of the Greek. Nonetheless, when it came to deciding between the adverb or the adjective he did not follow William. Where William chose the Latin adverb *primum* for the text of the *De interpretatione*, Minio-Paluello chooses the Greek genitive plural adjective πρώτων. So Busse does not follow William in constructing the Greek text of Ammonius, while at this point Minio-Paluello does not follow William in constructing the Greek text of Aristotle. John Magee supports Minio-Paluello and provides a compelling argument that the genitive plural adjective is the correct reading. As he points out, "all of the Boethian manuscripts render the genitive plural," and Minio-Paluello's Greek text is "based upon the indirect but secure evidence of the oldest translations of the treatise,"[54] texts that antedate by far any of the extant Greek manuscripts that are available.

This paleographical discussion should give one pause, even if it does not finally settle the issue. We are, all of us, heavily dependent upon the editorial decisions of experts in the texts we use. Though one might still question it, the most recent critical edition of Aristotle's Greek contains an adjective not an adverb. But there is little or no question among editors that Boethius's translation contained an adjective. Kretzmann believes that for the Greek text the adverb makes much better sense than the adjective.[55] But this focuses upon the Greek text of Aristotle, which I maintain is immaterial to Kretzmann's charge against the traditional Latin interpretation that renders the genitive plural adjective. The trouble is that in charging the Boethian Latin tradition with misinterpreting Aristotle's text, Kretzmann never addresses the fact that Boethius's text contains an adjective where Kretzmann needs it to contain an adverb to make the charge. Even if Boethius had not obliterated the distinction between symbols and signs that Kretzmann posits, the adjective would remain. The only reason for Kretzmann to dispute the consensus about Boethius's Latin text would be to sustain his charge about the Latin tradition's misinterpretation of the adverb. If Kretzmann's interpretation is not possible given an adjective, the traditional interpretation *prima facie* does not appear to be a misinterpretation; indeed, at this point it appears to be the only game in town.

The Traditional Interpretation: Ammonius's and Boethius's Commentaries

The Latin text of Ammonius as provided by Moerbeke is a very literal translation of the Greek. It is in the Latin form that the work was known by St. Thomas, so I will consider Ammonius in the Latin. I mentioned the opposition between Moerbeke's translation of Aristotle and his translation of Ammonius. Here is how they read. Moerbeke's translation of Aristotle is "but of which things these are *in the first place signs,* are passions of the soul the same for all,"[56] while his translation of Ammonius is, "for he says '*of which first things* these are signs.'"[57] I indicated earlier that for Ammonius the "first things" refers to the intellect's concepts or mental impressions. Commenting on the text, he contrasts these "first things" with actual things, presumably extra-mental things:

> 'these' meaning those which are in articulated sound, that is names and verbs, of which first things these are signs (but he means of concepts, and through them things are signified, not however immediately, but through the mediation of concepts, yet the concepts are not signified through other media, but primarily and immediately), of which first things therefore the signs of which are in articulated sound.[58]

Notice the association of 'first things' with 'primarily' and 'immediately'. The adverb 'primarily' is particularly interesting, since it introduces Kretzmann's adverb into the discussion, even though it is not in the text of Aristotle as Ammonius quotes it. Further, both the adjective and the adverb are associated with 'immediately'. Concepts are "primarily and immediately" signified by vocal utterances. But notice what is contrasted with the concepts "primarily and immediately" signified—things, presumably things other than concepts. These things other than concepts are not signified "primarily and immediately," but rather through the mediation of the "first things" signified—concepts. Presumably there is an implicit reference to the *second things* signified, things *extra animam*. Were he to make explicit what is implicit, one would imagine Ammonius's text adding "of which *second things* these are signs," and "of which *second things* therefore the signs of which are in articulated sound."

This contrast with *things beyond the soul* amplifies the sense of 'first things'. Concepts are *first things* immediately signified, by contrast to *second things* beyond the soul signified by the mediation of the first things signified. The structure of contrasting terms appears to exhibit just the structure Kretzmann criticizes. Indeed, when Ammonius began his discussion of the passage earlier, he summarized but did not quote Aristotle, and he used an adverb in the summary.

> Through this [passage] Aristotle teaches us, what the things are that are signified principally (*principaliter*) and immediately (*immediate*) by these [nouns and verbs], both that they are concepts, and through these media, however, things (*res*), and that it is not necessary that anything else be put in addition to this as a medium between the concept and the thing.[59]

It is only later that the *adjective* makes its appearance, when Ammonius is commenting phrase by phrase while explicitly quoting Aristotle. However, the important thing about this importation of an adverb into the paraphrase is that it is bound to what is being paraphrased, a genitive plural adjective in the text that Ammonius is looking at, an adjective that modifies the passions of the soul, or concepts as Ammonius has it.

Boethius himself wrote two commentaries on Aristotle's text, and much the same thing happens in them as happens in Ammonius's commentary. Of course his translation of Aristotle has the adjective *primorum*. John Magee believes that in rendering Aristotle's text in this way, Boethius was resisting the weight of "the commentary [tradition which] (at least by Ammonius' time) demanded an adverb."[60] Magee himself argues for the genitive plural reading of the Greek Aristotle. In fact he writes of the manuscripts with the adverb in them, that

> these variants most probably came into being in connection with what was presumed by the commentators to be the meaning of the text rather than with the mistakes made by the scribes who copied it.[61]

Magee believes that most of the ancient texts had a genitive plural in them, but that commentators, for philosophical reasons, had difficulty making sense of the adjective, and so began to substitute an adverb in their texts and commentaries. Magee is correct when he writes that "it speaks well of Boethius' reliability as a translator that he retains *primorum* in his translation," despite the weight of this tradition of philosophical interpretation.

Boethius's virtues as a translator of Aristotle are amplified by contrast with his commentaries, for in the latter, like Ammonius, he joins the commentary tradition that Magee cites. In particular, he uses the adverb 'principaliter' explicitly contrasted with 'secundo loco'. Like Ammonius, Boethius gives an interpretation of the *adjective* that imports an adverbial sense into it. In his second commentary, he writes:

> for those which are in articulated sound signify things and concepts, indeed principally concepts, and the things which these concepts comprehend by a secondary signification through the mediation of the concepts,[62]

while in his first commentary he writes:

> for an articulated sound signifies both the concept of a thing (*intellectum rei*) and the thing itself (*ipsam rem*). As when I say 'stone' it designates both the concept of the stone and the stone itself, that is the substance [of the stone] itself, but first the concept, though, to be sure, it signifies the thing in the second place (*secundo loco*). Therefore, not all things which articulated sound signifies are passions of the soul, but only those which are first [signified]; for [articulated sound] signifies a concept first, and in the second place a thing. But concepts are passions of the soul, of which first things [passions of the soul] the signs are articulated sounds, and the passions of the soul are the same for all men.[63]

In these passages Boethius appears to make the same sort of moves as Ammonius did in interpreting the adjective *primorum* of the text. Both of them believe that in the passage Aristotle is describing how words signify things, against Kretzmann's claim that Aristotle is doing nothing of the kind. With Ammonius, Boethius believes that 'first things' refers to the passions of the soul. With Ammonius, he contrasts those *first things,* passions of the soul, with *second things,* things beyond the soul.[64] And again with Ammonius, he introduces an adverb *principaliter* to interpret Aristotle, an adverb that he does not see in his own translation of Aristotle's text.

Even if Kretzmann's own interpretation is not a viable option, Boethius strengthens the charge against the traditional interpretation, that it makes the relation of words to things secondary to the primary relation of words to thoughts, by supplying the "secondary" or "in the second place" modifier which to this point has only been implicit in Ammonius. Words are "in the second place" or "secondarily," indirectly, and not immediately, but by the mediation of passions of the soul, related to these things. Finally, as with Ammonius, the sense of 'first things' applied to concepts is amplified by considering its correlative and contrasting term—things beyond the soul *secondarily* signified. Kretzmann's argument with the traditional Latin interpretation would have been much stronger had he assigned the responsibility for the error to Boethius's commentary, rather than his translation. Yet it would also have been an argument much harder to make, since Ammonius does not rely upon Boethius's translation or his commentary to give much the same reading of the adjective. And of course Ammonius's Greek has the grammatical distinction between 'signs' and 'symbols'. Yet Ammonius writing in his native Greek makes nothing of the distinction that Kretzmann does between σύμβολα and σημεῖα. If anything, it would seem that Boethius's commentary may well have simply been following the standard Greek line of commentary when it assigned primacy to the relation of words to thoughts over words to things.

The interpretation of *primorum* that Ammonius and Boethius give is also important for understanding another parallel between their commentaries. Both of them make use of the structure of *immediate* and *mediated* signification. This discussion of *mediation* is not in the text of Aristotle at 16a3–8, but both Ammonius and Boethius associate it with the primary-secondary structure. The primary signification is associated with the immediate signification, while the secondary signification is associated with the mediated signification. In this context, 'immediate' appears to be a synonym for 'primary' and 'by the mediation of' for 'secondary'. Placing the terms in this sort of proximity, it is no great leap to associate 'primary' and 'immediate' with 'direct', and 'secondary' and 'mediated' with 'indirect'.

I thought it appropriate earlier to mention Locke when considering Kretzmann's interpretation and example of the parrot. I think it equally appropriate here. Consider Locke's statement that:

> *Words in their primary and immediate Signification, stand for nothing, but the* Ideas *in the Mind of him that uses them,* however imperfectly soever, or carelessly those *Ideas* are collected from the Things, which they are supposed to represent.[65]

In light of Ammonius's and Boethius's commentaries, Locke's phrase "*primary and immediate Signification*" jumps off of the page. To be sure, he makes no mention here of whether words have any signification other than the *primary and immediate* one he is describing. Nonetheless, well over a millennium after Ammonius and Boethius wrote their commentaries the presence of the phrase "primary and immediate" is striking. Locke was by no means commenting upon the text of the *De interpretatione* when he wrote this.[66] Still it certainly suggests the sort of continuity of a tradition that contemporary critics find so troubling, and that Kretzmann faults for misinterpreting the text. That tradition may have a longer and more complicated history than even he realizes. I think it also adds weight to my reading of Kretzmann's analysis as taking place within the recent controversy over *mental representationalism*.

Though I used Ammonius and Boethius originally to provide evidence of the genitive plural adjective in the text of Aristotle, paradoxically their own *commentaries* on the disputed text both provide evidence that supports the *substance* of Kretzmann's charge, if not the *appearance*. They manage to interpret the adjective in a way that embodies the structure that Kretzmann criticizes for holding a "primacy of any semantic relation of spoken sounds to mental impressions over any semantic relation they may bear to actual things." No indication is given that the semantic relation between words and mental impressions differs in kind from the semantic relation between words and things. All we are led to believe is that the one is primary and immediate, and the other secondary and not immediate; the one is direct, the other indirect. This is not so difficult to understand. Adjectives and adverbs can certainly play synonymous roles in sentences. In fact, in Ackrill's translation 'first' is an adjective, yet it functions in the context of an adverbial phrase in "what these are in the first place signs of." Consider the English sentences "the first person to cross the finish line was Roger," and "the finish line was crossed first by Roger." In the former, 'first' functions as an adjective modifying 'person', while in the latter it functions as an adverb modifying 'crossed'. Yet 'first' plays roughly the same role in the two sentences, and no great semantic difference hangs upon the grammatical difference between the adjective and the adverb in the switch from the active to the passive construction. Much the same thing seems to happen to Aristotle's text in the commentaries of Ammonius and Boethius.

Consequently, one could object to the emphasis that I have placed on the textual fact of the adjective in the Latin tradition, and charge that I have been making a distinction without a difference. Before addressing that charge, it is important not to lose sight, however, of one difference that remains. While Kretzmann's interpretation cannot stand on the basis of an adjective, but requires the adverb, it is clear from this examination of Ammonius and Boethius that this traditional interpretation can stand on either an adverb or an adjective in the text. They certainly could have given the same commentaries had there been an adverb in the texts of Aristotle. It is all the more interesting, then, that Boethius and Ammonius were interpreting an adjective. They do not need an adverb in the text to provide their accounts of what the text means. Indeed this gives weight to Magee's suggestion that ancient editors may well have begun to substitute an adverb to fit what was taken to be the standard interpretation. Thus, the traditional interpretation criticized by Kretzmann is neutral with respect to settling the editorial or textual dispute.

Is it a distinction without a difference when one turns to others in the Latin tradition of commentary? To be sure, Kretzmann could still argue that to interpret the adjective in this way is a bad interpretation of Aristotle, and yet have no alternative interpretation from which to make that argument, or put in its place. Still, if it is the only interpretation in play, it is *prima facie* the best. I will now show that St. Thomas has an alternative interpretation of the adjective to put in its place. In that difference, he provides at least one significant instance of a departure from the "continuous tradition" that stretches at least from Ammonius and Boethius through Locke and beyond, and so is not subject to the charge directed against that tradition.

St. Thomas's Commentary

St. Thomas provides two accounts of the place of *primorum* in the text, at L. I, lc.2, 19 and 20.[67] In the first instance at passage 19 he identifies the "quorum ... primorum" with the passions of the soul. He writes, "*of which first things,* that is of which first passions, *these,* namely articulated sounds, are *notes,* that is *signs."* Consistent with this identification, in the Latin of his subsequent paraphrase he changes the neuter gender of the adjectival phrase in the quotation from *quorum primorum* to the feminine *quarum primarum* corresponding to *passionum.* In the first account of *primorum,* he tells us that 'first' subordinates words to the passions of the soul. Passions of the soul are related to words as first to second, "because words are only put forth to express the interior passions of the soul." Several times in the prior lessons he has stated that the purpose of words is to express the passions of the soul, for example, in passage 12 where he says that they are instituted as means satisfying the social character of men, as well as passage 16. This social necessity of words is a point of agreement between Ammonius and St. Thomas.[68] What Ammonius does not do, however, is interpret *primorum* in light of this social necessity of language. There is no need here to provide an implicit 'second'; St. Thomas writes it explicitly—passions of the soul are related to words as first to second.

What stands out in this passage is that one of the vertices of the semantic triangle in Ammonius's and Boethius's commentaries is not present in St. Thomas's, the *res extra animam.* For St. Thomas the primary-secondary axis maps onto the axis of passions of the soul and words, not the axes involving words, passions of the soul, and things. Indeed, it is interesting to note that in this particular passage St. Thomas uses neither *significatio* nor *significare,* though he does use *notae* and *signa.* Rather, when he explains why words are secondary in comparison to *passiones animae,* he uses *exprimendum.*

He had already explained four lessons earlier in passage 15 the place of *passiones animae* in signification. In passage 16 he moved on to reply to an objection of Andronicus that the work is not Aristotle's because of the identification of *passiones animae* with concepts. In passage 17 he turned to the explanation of the signification of writing. In passage 18 he explained the non-natural, that is, conventional signification associated with written and spoken words. Here in passage 19 there is no suggestion that he is picking up again the discussion of signification in passage 15, or adding anything to the explanation of signification that took place there. To suppose that he is would be to take passages 16–18 as an interruption in the train of the commentary. Nothing in the text suggests this; everything suggests that the discussion of the signification of words in isolation is over. Now he is contrasting words with *passiones animae;* words are conventional and differ among communities of men, while *passiones animae* are the same for all men. Thus passage 19 is a natural continuation of 18. His

Three Rival Versions of Aristotle 67

explanation of *primorum* in this passage has nothing to do with the explanation of signification; it does not amplify it; it does not specify anything new. *A fortiori*, in this first analysis of *primorum* there is no overshadowing of the signification of *res extra animam* by the signification of *passiones animae*, as secondary to primary, the overshadowing criticized by Kretzmann, the overshadowing that does take place in Ammonius and Boethius.

Immediately following in passage 20 St. Thomas considers an objection to the thesis in passage 19 that passions of the soul are the same among all men. This objection gives him the opportunity to provide a second account of *primorum* as he finds a second role for it to play in Aristotle's text. The objection is that the *passiones animae* cannot be the same for all men as Aristotle claims, because different men have different opinions. Though St. Thomas does not specify the implicit assumptions, the context of the discussion makes it clear that the objection is strongest if the different opinions are taken to be contrary opinions about the same thing. Boethius had responded to this objection that Aristotle is speaking of the intellect's concepts, and that the intellect is never deceived. As St. Thomas understands him, Boethius denies the assumption of the objection as applied to the intellect generally. Summarizing Boethius he writes, "if someone differs from the truth, he does not understand it [at all]."[69] Note that the passage in Boethius that St. Thomas is referring to here is not the same passage in which Boethius discusses *primorum*. Boethius's discussion of *primorum* is explicitly about signification, while the passage in Boethius that St. Thomas has in mind now is about the claim that concepts are the same for all.[70]

According to St. Thomas Boethius's reply does not address the difficulty, since it does not address the difference between the first act of the intellect in which there is never falsehood, and the second act of intellect, the act of combining or dividing, in which falsity as well as truth can be found. St. Thomas writes:

> But, because there can be something false in the intellect, according as it composes and divides, but not according as it cognizes the *quod quid est*, i.e., the essence of the thing, as is said in *De anima* III; this [Aristotle's statement] ought to be referred to the simple conceptions of the intellect which simple articulated sounds [*voces incomplexae*] signify, which are the same among all men.[71]

In 16a9–16, the passage in the *De interpretatione* immediately following the one now under consideration (16a3–8), Aristotle writes:

However, just as sometimes there is some act of intellect in the soul without something true or false, but other times it is necessary for one or the other of these to be in it, so it is also in articulated sound. For truth and falsity belong to both composition and division. Therefore nouns themselves and verbs are like acts of intellect without composition and division. As for example 'man' or 'white', when nothing is added, for neither of these is thus far true or false.[72]

As I pointed out in the previous chapter, there are numerous and explicit parallels with the commentary on the *De anima*, *De anima* III.6.430a26–431a4 in particular, as noted by St. Thomas in his commentary. There Aristotle contrasts what is simple or undivided in thought and in which there is no falsity, with "combination of thoughts as forming a unity," "where there is both falsity and truth."[73] He does not, however, call what is simple and undivided "first thoughts." Just two chapters later, however, at *De anima* III.8.432a10 he asserts that "imagination is different from assertion and denial; for truth and falsity involve a combination of thoughts." He proceeds to ask "but what distinguishes the first thoughts (πρῶτα νοήματα)[74] from images,"[75] since he believes they are in fact distinct. What is of interest here is the reference to "first thoughts." These first thoughts are clearly the thoughts that enter as elements into the combination of thoughts he has just mentioned. But this appears to be the distinction between simple and complex discussed two chapters earlier in *De anima* III.6, with 'first' substituting for 'simple'.

Taking his cue from Aristotle's mention of the *De anima*, St. Thomas stresses again the cognitive background of the whole discussion. Particularly important is the contrast between the simple concepts of the first act apprehending the essence of things in which no falsity can arise, and the second act involving combinations and divisions in which truth and falsity first appear. St. Thomas believes that the objection does have merit, *pace* Boethius, as directed to the second act of intellect in which St. Thomas believes opinions are formed. However, he does not simply reject Boethius's reply. He believes the objection is not to the point of this particular text in Aristotle, because Aristotle's assertion, that the *passiones animae* are the same among all men, refers to the first act of intellect, not the second. Boethius is right if we understand his reply as directed to the first act of intellect; he is wrong if we understand it as directed to the intellect as such, since the intellect has a second act that can be deceived.

St. Thomas's analysis of the text expands and completes the comments he had made earlier in the *Preface*. There he had arranged the logical works according to the order of the acts of intellect. The *Categories* precedes the *De interpretatione* as the first act of intellect precedes the second act, because simple terms signify simple acts of intellect, while statements signify the complex acts of intellect expressed in the second act. Here in the body of the commentary, he writes that *primorum* is used to indicate the simple concepts, namely first acts of intellect by contrast to the second acts:

> this ought to be referred to the simple concepts of the intellect which incomplex articulated sounds signify, which are the same for all: because if one truly understands what a man is, whatever else he apprehends other than man, he does not understand as man. But such are the simple concepts of the intellect, which articulated sounds *in the first place* signify. . . . And so he expressly says, "*of which first things these are signs,*" in order that it may be referred to the *first concepts* signified *in the first place* by articulated sounds.

The whole context, involving Boethius's attempted solution, makes it clear that *first* acts of intellect are to be understood against the background of the more general division of acts of intellect into *first* and *second*, that is, *simple* versus *combined* or *divided*. Notice in these texts the identification of the intellect's "simple concepts" with its first act.

Thus, in these texts St. Thomas makes use of the distinction between first act and second act in order to interpret the text involving *primorum*. In that light, consider the plausibility of the modification I made to Ackrill's translation above, "but these are signs *of the first [affections]*, which affections are the same for all, and what these are likenesses of—actual things—are also the same." The *first things* [affections in the soul] signified by words are the first simple acts of intellect. But here, just as I have already shown in passage 19, the contrast is not with *res extra animam* as secondarily signified. Rather the contrast is with the second complex acts of intellect, combining and dividing, making true and false judgments.

At this point it is fair to say that everyone seems to think the note of *first* in Aristotle's text cries out for an implicit *second*, with some like Ammonius leaving it implicit, and others like Boethius and Kretzmann making it explicit. What are the possible candidates to which to apply this implicit *second*? Boethius and Ammonius, of course, apply it to *things beyond the soul*. Kretzmann applies it to a conventional relation between vocal utterances and mental impressions that is built upon a first relation of natural signification. St. Thomas clearly provides a third possibility.

Notice the importance here of the shift from 'first things' to 'first affections of the soul' or 'first passions of the soul' in St. Thomas. The first phrase leaves enough ambiguity for the second things to be *res extra animam*, as in Ammonius and Boethius. The second phrase is more restrictive. Not just any *thing* will be a possible candidate for the *second things*, but rather some other passion of the soul or mental affection, other than the "first passions." For St. Thomas, it will be "passions of the soul" that are not the same for all by contrast with those that are the same for all, the second acts of intellect by contrast to the first.

As Magee points out, both Ammonius and Boethius are aware of the textual parallels between the initial passages of the *De interpretatione* on true and false enunciations and true and false thoughts, as well as *De anima* III with regard to the first and second acts of intellect and truth and falsity in intellect.[76] However, neither of them interpreted the *primorum* in light of those parallels. Magee's otherwise fine analysis is marred here by reading St. Thomas through a number of secondary authors.[77] Relying upon them, Magee attributes to St. Thomas precisely the traditional interpretation that one sees in Ammonius and Boethius. Had he looked more closely at St. Thomas's text itself, he would not have made that mistake. In this second discussion of *primorum*, it is again clear that for St. Thomas it has nothing to do with explicating or amplifying the discussion in passage 15 of how words signify *things beyond the soul*. In particular, when he explains signification in passage 15, St. Thomas inherits and makes use of the structure of 'without mediation–by the mediation of' that he would have seen in Boethius and Ammonius. But that discussion is over. He makes no further use of it here in passage 20. The discussion of 'without mediation–by the mediation of' is completely separate from both discussions of *primorum* in passages 19 and 20. I will discuss later what the separation of the two discussions implies for understanding St. Thomas's use of 'without mediation–by the mediation of', as opposed to that of Ammonius and Boethius.

It is interesting to note that in St. Thomas's analysis of *primorum* the adverb *primo* makes an appearance twice, when he writes "but such are the simple concepts of the intellect, which articulated sounds *in the first place* (*primo*) signify," and "in order that it may be referred to the first concepts signified *in the first place* (*primo*) by articulated sounds." But *primo* was not in the text upon which he was commenting, at least in the judgment of Minio-Paluello and the 1989 Leonine editors. Boethius and Ammonius also introduced the adverb into their discussions. This use of an adverb in St. Thomas's commentary might well be an inheritance from his reading of them, or from Moerbeke's translation of Aristotle. If St. Thomas has them in mind when he introduces it, it is clear that he introduces it only to proceed to differ from them in his use of it. Because of the different context of interpretation, its appearance has an entirely different sense now than it had in Boethius and Ammonius. Words, that is, "simple vocal utterances" (*voces incomplexae*), *in the first place* signify the first act of intellect, contrasted with *in the second place* the second act of intellect, the act of combining and dividing. The force of the adverb presupposes the already interpreted adjective, not vice versa. St. Thomas seems to be saying to his predecessors, albeit implicitly, "Yes, you can say that words signify affections of the soul in the first place, but that doesn't require the addition that they signify *res extra animam* in the second place."[78] Rather it requires a distinction between the passions of the soul that words signify in the first place and the passions of the soul that they signify in the second place.

Then what are the words that will *in the second place* signify the second act of intellect? They are the *enunciations*, that is, statements consisting of at least a noun and a verb to be discussed in the very next passage at 16a9–16, and the proper subject matter of the *De interpretatione*. This contrast with the *second act of intellect* at the end of lesson 2 then sets up a natural transition to lesson 3, in which St. Thomas will discuss this next passage of Aristotle that addresses enunciations in which truth and falsity are found. There St. Thomas addresses again the discussion of *De anima* III.6, when he writes in passage 24:

> It ought to be considered that, just as was said in the beginning, the operation of intellect is twofold, as is treated in III De anima, in one of which neither the true nor the false is found, but in the other of which it is found. And it is for this reason that he says that *sometimes in the soul* there is *understanding without the true or the false*, but *sometimes* from necessity it has *one or the other* of these.[79]

St. Thomas does not make the claim explicit here that there is a natural transition between the two passages 16a3–8 and 16a9–16, set up by the discussion of *De anima* III. But he doesn't need to; he has already made that point explicit twice before in the commentary. This passage cannot be read in isolation from the discussion of the order of logic in the *Preface*, and the original division of the text enunciated in lesson 1. In both passages, the transition from simple words to enunciations, based upon a transition from the first act of the intellect to the second, is quite clear.

One clear advantage of St. Thomas's interpretation over the others is the continuity that it sees between the discussion of the *Categories* and the *De interpretatione*, a continuity not at all remarked upon by the other interpreters.[80]

St. Thomas's Interpretation of the Semantic Triangle in Aristotle

While St. Thomas departs significantly from the tradition that he inherits, he also displays an important continuity. He does believe with Ammonius and Boethius that Aristotle is sketching how words are related to things.

In his discussion of 16a3–4 and his explanation of the signification of *res extra animam* in passage 15, St. Thomas employs the 'without mediation–by the mediation of' pair that he would not have read in Aristotle's text, but appears to inherit from Ammonius and Boethius.[81] But that discussion is independent of the discussion in passages 19 and 20 of *primorum*, his discussion of 16a6. In passage 15 he takes it for granted that words signify *res extra animam*, which is why according to him it is merely implicit in Aristotle's discussion. Earlier in passage 12 dividing the order of the text, he writes that Aristotle is interested in placing articulated sounds in a certain order of signification (*ordinem significationis*), namely, written words, articulated sounds, and passions of the soul, "from one of which a fourth is understood," namely, things. "For a passion is from the impression of some agent; and so passions of the soul have their origin from things themselves."[82] In passage 15 he simply takes for granted what he has already written in passage 12. Nowhere in the *Commentary* does St. Thomas display a felt need to justify the place of *res extra animam* in the order of signification. He takes it for granted as "understood" that *res extra animam* have a place in the order of signification. Words, including general words, signify things beyond the soul.

Instead, St. Thomas is at pains to give an argument justifying Aristotle's claim that words signify passions of the soul; for St. Thomas that is what should not be taken for granted, and why he thinks Aristotle explicitly mentions it. To justify it, he makes an appeal to the *modus significandi* of general words. It is here that he introduces as integral parts of his interpretation the paired terms 'without mediation–by the mediation of'. However, the independence of this discussion in passage 15 from the later discussion of *primorum* in passages 19 and 20 casts a different light on the sense of the terms than the sense they have in Ammonius and Boethius. For Ammonius and Boethius 16a6, involving *primorum*, is a discussion

Three Rival Versions of Aristotle 73

of how words, concepts, and things are related, and they use the pairs 'primary-secondary' and 'without mediation–by the mediation of' as if providing synonymous expressions. The sense in that context is easily associated with the direct-indirect pair of Edghill, and Kretzmann's own use of 'direct' and 'indirect'. But for St. Thomas 16a6 is not where Aristotle discusses the relations of words, concepts, and things. That discussion has already taken place at 16a3–4, and is now over.

Because of the independence of the two discussions in St. Thomas, there is no *prima facie* reason to make the same associations that are made in Ammonius and Boethius between 'primary' and 'without mediation', and 'secondary' and 'by the mediation of'. Indeed, the opposite is suggested by the fact that St. Thomas probably saw these associations in both Ammonius's and Boethius's interpretations, and yet separated them. On the other hand, when he discusses signification at 16a3–4 and how words, *res extra animam*, and passions of the soul or concepts are related he sees the need for the structure of *without mediation* and *by the mediation of*. The passage in St. Thomas's commentary reads:

> But here speech concerns articulate sounds signifying from human institution; and so it is necessary that here *passions of the soul* be understood as *conceptions of intellect*, which names, and verbs, and sentences signify, according to the view of Aristotle: for it cannot be that they signify things themselves *without mediation*, which is clear from the mode of signifying: for this name *man* signifies human nature in abstraction from singulars, so it cannot be that it signifies a singular man *without mediation*. Hence the Platonists held that it signified the separate *idea* itself of man; but because in the view of Aristotle, this, according to its abstraction, does not really subsist, but is in the intellect alone, it was necessary for Aristotle to say that articulated sounds signify the conceptions of the intellect *without mediation*, and things by their *mediation*.[83]

If St. Thomas feels the need to justify anything, it is the cognitive vertex of the triangle even though it is explicitly mentioned by Aristotle—the claim that passions of the soul are signified by words. St. Thomas indicates that general terms do indeed signify *res extra animam*. But because there are no general *res extra animam*, that is, Platonic forms, only particular *res extra animam*, the generality of the term has to be accounted for in some other way than the thing signified (*res significata*). It is accounted for by the *modus significandi* of general terms which is identified with the intellect's first act *by which* (*a quo*) the intellect understands particular things, an act which is general in its application. We have already seen that St. Thomas identifies these general acts with concepts. Thus words stand to concepts, not as to *things* signified, but as to *means* by which *things* are signified.

What stands out here by its absence is just the 'primary-secondary' pair as it figures so prominently in Ammonius's and Boethius's interpretations. The phrases 'without mediation' and 'by the mediation of' stand on their own and are not explicated by 'primary' and 'secondary'. On the other hand, in a fashion clearly opposed to Ammonius and Boethius, St. Thomas comments on *primorum* in his discussion of 16a6. Where they took the discussion at that point to involve words, concepts, and *res extra animam*, he takes the discussion at 16a6 to involve words, first acts of the intellect, and second acts. Thus, careful attention to these differences in St. Thomas's actual commentary, against the backdrop of Ammonius's and Boethius's commentaries, makes it difficult not to conclude that he simply does not want to associate his discussion of the signification of concepts (without mediation) and the signification of things (with mediation), with the traditional discussion of the primary or direct signification of concepts and a secondary or indirect signification of things. In other words, it is clear that St. Thomas does not want to associate the need for the mediation of concepts with a supposed primary relation of words to concepts, over a secondary relation of words to things.

Someone might object that even if this is all true, isn't it still drawing a textual distinction without a philosophical difference? The objector might say, "I grant you that St. Thomas drops the language of 'primary' and 'secondary' signification, and separates the discussion of 'without mediation' and 'by the mediation of' from the discussion of *primorum*. But doesn't the very language of 'without mediation' and 'by the mediation of' suggest 'primary' and 'secondary'? Perhaps St. Thomas just eliminated a redundancy in the text he was dealing with. The sense is still there even if the words aren't."

No. In and of itself the language of 'without mediation' and 'by the mediation of' does not necessarily suggest the senses of 'direct' and 'indirect', or 'primary' and 'secondary'. Etymologically 'immediate' is just the negation of 'mediate', as 'immortal' is no more than the negation of 'mortal', and 'immaterial' of 'material', 'impotent' of 'potent', and so on.[84] 'Immediate' simply means 'without mediation'. Nothing mediates the relation of *word* to *concept*, while the *concept* mediates the relation of *word* to *res extra animam*. But why should one think that is equivalent to saying that the relation of *word* to *concept* is *primary* or *direct*, and the relation of *word* to *res extra animam* is only *secondary* and *indirect*?

In order to see why 'immediate' and 'mediate' are not necessarily synonymous with 'primary' and 'secondary', or 'direct' and 'indirect', consider the sort of mundane analogy from artistry that Aristotle and St. Thomas often employ. A carpenter in making a bench may use a hammer and a saw. Nothing mediates the relation of the carpenter's hand to his instruments—it is *immediate*. Further, the instruments do mediate his relation to the work to be produced, that is, characteristics of his proper effect differ according to the tools he uses. The shape of the boards, for example, is brought about by his use of the saw, not the hammer, while the way in which the boards are joined is brought about by his use of the hammer, not the saw. If he is using a hammer he can strike pegs into joints, but he cannot cut wood. If he is using a saw he can cut wood, but he cannot strike pegs into joints. If he tried to do the latter things one would say he is abusing his instruments, and that his work will suffer for it. So the character of his instruments mediate the artisan's relation to his work.

Does it follow, however, that because his relation to the work to be produced is *mediated* in this way by the instruments he uses, and that his relation to his instruments is *not mediated* by anything, or is *immediate*, that qua artisan he is *primarily* and *directly* related to the instruments that *mediate* his action, and only *secondarily* and *indirectly* related to the work to be done? On the contrary, the proper effects of the instruments are subordinated to the proper effect of the agent—the work to be done that proceeds from his art. Indeed the material and instruments used are determined by the work to be produced.[85] The artisan is characterized or defined in terms of the work he produces, not in terms of the instruments that *mediate* his doing it. A sculptor, blacksmith, and carpenter all may use hammers. But the hammers they use differ according to the different materials they work with, and the materials they work with are in turn determined by the work to be done. It is the work to be done that defines them *qua* artisans of certain kinds. Their relation to the works they produce is not *indirect* or *secondary*, even though it is *mediated* by the tools and materials they use. A blacksmith is not indirectly or secondarily related to the making of horseshoes, fire irons, stirrups, and so on, simply because he uses a forge, hammer, and anvil to make them. Further, the relation of artisans to their tools is not *direct* or *primary* by contrast with an *indirect* or *secondary* relation to the work to be done, simply because their use of them is *not mediated* by anything else. 'Immediate' may in some contexts mean 'direct' or 'primary', but as this example shows, certainly not in all contexts.

In particular, when the context is the use of something *for* something, 'mediated' or 'by means of' simply and typically means 'using', or 'the way' in which something is brought about. It does not mean 'indirectly' or 'secondarily'. On the other hand, and by contrast, as a negation 'immediately' or 'not mediated' or 'without mediation' simply means 'nothing is used'. And this is the context of St. Thomas's discussion. I showed in the previous chapter the need to distinguish two analogous senses of 'signification' in the text, namely, the two-term relation between words and concepts indicated by 'signification$_1$', and the three-term relation between words, concepts, and things indicated by 'signification$_2$'. *Res extra animam* are signified$_2$ *by means of* concepts. Concepts are *used* in the signification$_2$ of *res extra animam,* and so concepts *mediate* that relation. On the other hand, if concepts are going to be *used* in this fashion, that is, if they are going to *mediate* the relation of signification$_2$, they must themselves enter into a relation with words. That relation is signification$_1$.

So, all St. Thomas intends by writing that signification$_1$ is *immediate* is that nothing in addition to words and concepts needs to be used to establish signification$_1$. This lack of mediation is contrasted with signification$_2$ which, at least in the case of general words, must be *mediated* by something in addition to *words* and *res extra animam,* mediated namely by concepts. Nothing in that context implies that it is the sort of context in which 'immediate' means 'direct' or 'primary', and 'mediated' means 'indirect' or 'secondary'. Indeed, the fact that in his commentary St. Thomas excludes the Ammonian and Boethian uses of these terms from the discussion of 'immediate' and 'by the mediation of' confirms this conclusion. In St. Thomas, there is no textual reason to conclude that the relation of words to *res extra animam* is not direct, even though that relation is established *in a certain way* by our understanding of *res extra animam*—the *modus significandi* of general words. On the other hand, there is no suggestion here that the vocal utterance bears two relations to the concept, a natural symptom relation, and a conventional symbolic encoding relation, as Kretzmann believes. St. Thomas's is a third distinct interpretation of the text.

Conclusion

In the *Summa* discussion of the divine names, immediately after the passage from the *Peri hermeneias* quoted and explained in I.13.1, St. Thomas writes "we name as we know." It is the burden of the later chapters of this work to show that for St. Thomas what we directly know and thus directly name are not concepts, but *res extra animam*.[86] Careful attention to St. Thomas's actual commentary does not simply shed light on the textual question of how St. Thomas interprets a disputed text in Aristotle. It goes beyond this, and provides a foretaste of the complexity of the more philosophically substantive charges directed at the Aristotelian tradition that I will examine in subsequent chapters. Locke's reference to the "primary and immediate" signification of words indicates that the tradition Kretzmann wishes to criticize was powerful enough to have lasted even into modern philosophy. Thus, this chapter provides at least an initial historical understanding of the plausibility of the charge of mental representationalism directed at the Aristotelian tradition in its account of language. When contemporary philosophers make this charge, it cannot simply be dismissed as completely unfounded. There is evidence for it in Ammonius and Boethius, even if, as I have just argued, there is no evidence for it in St. Thomas's commentary on Aristotle. Certainly contemporary scholars performing historical-textual analyses suggest the plausibility of the charge, even if they miss the boat on St. Thomas. It is not surprising then that contemporary philosophers who may depend upon these historical-textual analyses, rather than provide them themselves, make a similar charge in their more substantive philosophical claims. In short, the substantive philosophical charge made by inheritors of the *Linguistic Turn* against "traditional philosophy" is not without merit. Simply showing that St. Thomas is not subject to the charge at the level of his textual interpretation of Aristotle does not bring an end to the matter, since his substantive philosophical discussion might still fall prey to it. It gives one hope, but it does not absolve one from the responsibility of responding to the more philosophically substantive charges.

Chapter 3

LANGUAGE AND MENTAL REPRESENTATIONALISM
Historical Considerations

> [T]hose who inquire without first going over the *difficulties* are like those who are ignorant of where they must go; besides, such persons do not even know whether they have found or not what they are seeking, for the end is not clear to them.
> —Metaphysics

Among some contemporary philosophers there is a vision of the opening passage in Aristotle's *De interpretatione* as a standard that planted a seed which grew relatively continuously in Western philosophy, flowered within British Empiricism, and continues to influence the philosophy of language to this day. Consider Michael Dummett's remarks:

> A continuous tradition, from Aristotle to Locke and beyond, had assigned to individual words the power of expressing 'ideas', and to combinations of words that of expressing complex 'ideas'; and this style of talk had blurred, or at least failed to account for, the crucial distinction between those combinations of words which constitute a sentence and those which form mere phrases which could be part of a sentence.

Perhaps the most important of all the contributions made by *Grundlagen* to general philosophy is the attack on the imagist or associationist theory of meaning. This is another of those ideas which, once fully digested, appear completely obvious: yet Frege was the first to make a clean break with the tradition which had flourished among the British empiricists and had its roots as far back as Aristotle. The attack that was launched by Frege on the theory that the meaning of a word or expression consists in its capacity to call up in the mind of the hearer an associated mental image was rounded off by Wittgenstein in the early part of the *Investigations,* and it is scarcely necessary to rehearse the arguments in detail, the imagist theory now being dead without a hope of revival.[1]

Ackrill, in his translation and commentary on the *De Interpretatione,* makes oblique reference to this sort of criticism:

> Aristotle probably calls ['affections of the soul'] likenesses of things because he is thinking of images and it is natural to think of the (visual) image of a cat as a picture or likeness of a cat. But the inadequacy of this as an account or explanation of thought is notorious. Again, what is it for a spoken sound to be a 'symbol' of something in the mind?[2]

Ackrill goes on to minimize the place of this thesis in the work, writing that "Aristotle does not often appeal to psychological experiences or facts to explain or support what he says about names, verbs, statements, etc." His use of "psychological" to characterize what is going on here appears to be indicative of the fears of many philosophers that Aristotle's semantic triangle reduces the study of logic to a branch of philosophical psychology or philosophy of mind, associating it in part with theses which have long since been shown to be untenable.

Kretzmann, on the other hand, seems to want to protect Aristotle from these kinds of criticism by distancing Aristotle's text from its tradition of interpretation. He believes that the "traditional misreading of [these] passages" in the West is a product of Boethius's unfortunate translation of the Greek words for 'symbols' (σύμβολα) and 'signs' (σημεῖα) by the single Latin word 'notae', thereby "obliterating the Aristotelian distinction between symbols and symptoms." He believes that Boethius's translation, in conjunction with the adverb 'primarily' (πρώτως) has led, or misled, the tradition into attributing the view to Aristotle that our words are primarily or directly about mental impressions and only secondarily or indirectly about things.

Though he does not attribute responsibility for the problem to Aristotle, Robert Sokolowski states the difficulty succinctly, "[I]n a Lockean spirit we have allowed words to range only over the domain of our ideas, and we have tacitly taken ideas to be some sort of internal things. But philosophically this is terribly naive."[3] Our contemporary critics often use "mental representationalism" as a blanket term to refer to the philosophy of mind to which the account of language is yoked. John Haldane provides a useful paraphrase of *mental representationalism*:

> [i]t is the view that the immediate objects of cognitive acts or states are internal entities: *species, ideas, images, sentential formulae* and such like, which may or may not stand in some further referential relation to objects and features in the world; and that it is the former, *inner*, relational attitudes which constitute the essential 'object-directed', or intentional character of cognitive states.[4]

Alexander Broadie points out that as early as the eighteenth century Thomas Reid criticized Hume specifically, and with him the whole tradition back to Plato, for holding theses such as Haldane describes.[5]

A Historical Setting for the Problems of Language and Mental Representationalism

The intention of the survey that follows is to provide a minimal but adequate background for understanding what might be called "the received view" of *representationalism* in modern philosophy, the view that animates in many ways the Linguistic Turn generally, and Hilary Putnam's discussion particularly.

What does the tradition that the critics reflect upon look like? Consider the following comments of Bertrand Russell who, though not discussing the signification of terms, writes:

> The view seems to be that there is some mental existent which may be called the "idea" of something outside the mind of the person who has the idea.... [I]n this view ideas become a veil between us and outside things—we never really, in knowledge, attain to the things we are supposed to be knowing about, but only to the ideas of those things. The relation of mind, idea, and object ... is utterly obscure, and ... nothing discoverable by inspection warrants the intrusion of the idea between the mind and the object.[6]

Russell speaks of ideas, a term I have not introduced, but the gist of the quotation stands as is for concepts.

The first thing to note here is the equation of "object" and "outside things." There are difficulties raised by this equation of "object" and "outside things," at least as it bears upon understanding St. Thomas, in particular the latter's use of the Latin word 'obiectum'. Russell's sense is akin to what Quine has in mind:

> [W]e are prone to talk and think of objects. Physical objects are the obvious illustration when the illustrative mood is on us, but there are also all the abstract objects, or so there purport to be: the states and qualities, numbers, attributes, classes. We persist in breaking reality down somehow into a multiplicity of identifiable and discriminable objects, to be referred to by singular and general terms. We talk so inveterately of objects that to say we do so seems almost to say nothing at all; for how else is there to talk?[7]

This sense of 'object' stresses the independence of things—what exists "in itself," and it gives force to the common dichotomy between what is "objective" and what is "subjective." It is roughly synonymous with 'thing'. In St. Thomas, however, *obiectum* is used precisely to characterize a *res* (thing) in its relationship to a power, and for my purposes here a cognitive power. It is not synonymous with *res* simply, but connotes '*res* with respect to some formal aspect in virtue of which it is subject to a cognitive power'. But here in Russell 'object' is used as a synonym for "outside things" as used in the quote, where "outside thing" is contrasted with "mental existent." Sokolowski and Haldane also make reference to these "mental existents."

Presumably Russell's worries should not be confined simply to so-called internal mental existents interposed between the knower and the object known; they seem to have more to do with the thesis of *mediated* knowledge of things, and what that *mediation* consists in. His difficulty as stated ought to be aimed at any third thing interposed between the mind and the thing known, whether it is "internal," or "external," or "ideal," or an "objective" but "not actual" thing, or a text, and so on. For example, if one held that all our knowledge of non-textual things is mediated by publicly accessible texts, written or verbal, wouldn't Russell's worry still stand?

In fact, at least according to Michael Dummett, Russell's overall concern was the question of mediated and representative knowledge, rather than whether or not the mediating object was "internal."[8] In that respect, his concerns are not new; it was precisely upon this point, among others, that Aristotle criticized the Platonists' theory of *Ideas,* a criticism often discussed and reechoed by St. Thomas. However, as stated, there are two key elements to Russell's difficulty: first, the existence of an *internal* mental entity which is directly known and which mediates the knowledge of things *outside* of or *external* to the mind, and second, the subsequent skepticism about knowledge of the *external* world.

Locke

The first element receives further illumination in the light of themes stemming from, among others, Descartes, Locke, Berkeley, and Hume. Having cast doubt upon the deliverances of his senses, Descartes frames the discussion particularly clearly. He holds that "through the intellect alone I perceive such ideas about which I am able to make a judgment."[9] The "ideas in me are as images, which can easily fall away from the perfection of the things from which they are taken."[10] The intellect gazing upon these internal representations, a judgment can be made "that the ideas which are within me (*in me*) are likenesses or conformities of certain things beyond me (*extra me*)."[11] For Locke, it is immediately evident to the consciousness of every man that "that which his Mind is employ'd about whilst thinking [are] the *Ideas* that are there."[12] These ideas are the immediate objects of thought and reason. Indeed there are no other immediate objects of the mind for Locke than these ideas, and "our knowledge is only conversant about them."[13] It is by the "intervention" of these ideas immediately known that the mind can have knowledge of things other than ideas. "*Our Knowledge* . . . is *real,* only so far as there is a conformity between our *Ideas* and the reality of Things."[14]

Locke recognizes that there is a skeptical problem latent in this analysis of knowledge. If knowledge is real to the extent that the ideas which we immediately know conform to things which we do not immediately know, can we know that conformity? As he puts it, "[H]ow shall the Mind, when it perceives nothing but its own *ideas,* know that they agree with Things themselves?"[15] In the very asking of the question, Locke makes clear that the mind only perceives its own ideas. He solves the skeptical problem not by an *ad hoc* appeal to, or comparison with the things themselves, as if he could get outside the circle of ideas and compare the thing to the idea of the thing. Rather he appeals to the *phenomenological* properties of the perceived ideas. That is, by reflecting upon *simple* ideas that appear before the mind, we recognize, as he had earlier shown,[16] that these ideas are not subject to the will or productive power of the mind, either in the fact of their original occurrence in the mind or in the phenomenological character of their appearance, that is, being a sweet taste as opposed to bitter, red vision as opposed to yellow, and so on.

Thus, on the one hand, the mind can be confident that the simple ideas are produced by something other than the mind and its ideas, by "things operating on the Mind in a natural way."[17] Such simple ideas "represent to us Things under those appearances which they are fitted to produce in us." On the other hand, complex ideas other than those of substances, ideas such as are used in mathematical and moral reasoning, are "*Archetypes* of the Mind's own making";[18] consequently, there is no need for them to correspond to anything beyond the mind. But in the case of complex ideas of *substances,* both because they are supposed to represent things beyond the mind, that is, something beyond the mind is the "Archetype" for the ideas of substances, and also because their combination is subject to the will and productive power of the mind, we can only be sure and have a "confidence" about a small number of them that they do in fact represent objects or things beyond the mind. But we cannot have certainty about the vast majority of our complex ideas of substances that they constitute real knowledge. The small measure of certainty about the reality of complex ideas of substances is given us when "all our complex *Ideas* of them [are] such, and such only, as are made up of such simple ones, as have been discovered to coexist in Nature."[19] So in the case of complex ideas of substances it seems Locke's certainty is produced by two phenomenological factors, the characteristics of the simple ideas, plus the aspect of those simple ideas being perceived at least once to "coexist" together.

In any case, Locke very often compares the internal ideas to internal pictures before the mind,[20] which pictures are for the most part inadequate to represent things to the mind. He believes they are inadequate for two reasons:

> *Ideas* have in the Mind a double reference: 1. Sometimes they are referred to a supposed real Essence of each Species of Things. 2. Sometimes they are only design'd to be Pictures and Representations in the Mind, of Things that do exist, by *Ideas* of those qualities that are discoverable in them.[21]

Such ideas are inadequate representations or pictures in the first instance, because we can have no knowledge of the "inner constitution" of things. This denial, of course, forms the basis for Locke's discussion of real and nominal essences.[22] In the second instance, they are inadequate because we cannot hope to have more than a passing acquaintance with all the different characteristics that give rise to the perceptions we have of things. Thus any idea of an object will consist of just a few simple ideas of the array of powers had by that object, and leave much, indeed most of it un-represented or un-pictured.[23]

What is crucial in his account is that the object of one's knowledge, what one is thinking about or what one knows, is primarily the private stock of ideas internally present to one's own consciousness.

> Whereas, in truth, the matter rightly considered, the immediate object of all our reasoning and knowledge, is nothing but particulars. Every man's reasoning and knowledge is only about the ideas existing in his own mind . . . and our knowledge and reason about other things, is only as they correspond with those of our particular ideas. So that the perception of the agreement or disagreement of our particular ideas, is the whole and utmost of all our knowledge.[24]

Notice that Locke stresses that it is the "perception of the agreement or disagreement of our particular ideas" with themselves that "is the whole and utmost of all our knowledge," *not* the perception of the agreement or disagreement of our particular ideas with extra-mental things. Again, we do not judge that our ideas correspond with "other things," since we have no outside access to those other things. Rather, outside things happen to correspond to our ideas. By directly knowing our ideas we happen to know the outside things indirectly. He compares the mind to a "closet wholly shut from light, with only some little openings left, to let in external visible resemblances, or ideas of things without: [would the pictures coming into such a dark room but stay there.]"[25] The ideas, at least in their simplest forms and excluding those that arise from the mind's reflection upon its own activity, have their origin from things external to the mind. However, the things external to the mind are not present to the understanding.[26] For the received view of *representationalism,* what is important in the account is that an external thing is represented internally by an appearance that functions like a picture of it. Thought attends directly to this appearance.

The use of "picture" here is one area where the "received view" of Locke would have to be examined very closely. One has to keep in mind his account of the origin of ideas, much of which is familiar and standard.[27] All ideas have their origin in either sensation or reflection upon the mind's activities. These ideas exist in the mind, while qualities exist in bodies beyond the mind. The qualities are divided into *primary* and *secondary.*

The *primary* qualities are "utterly inseparable from the Body," which "produce simple *Ideas* in us, *viz.* Solidity, Extension, Figure, Motion, or Rest, and Number." On the other hand, secondary qualities are "nothing in the Objects themselves, but Powers to produce Sensations in us by their *primary Qualities, i.e.* by the Bulk, Figure, Texture, and Motion of their insensible parts, as Colours, Sounds, Tastes, *etc.*"

All of this is well known and familiar, but what one has to keep in mind is that Locke's use of the "picture" metaphor has to be understood in light of what he says about primary and secondary qualities. It is only the simple ideas of primary qualities that can strictly and non-metaphorically be called "resemblances" of the bodies that produce them.

> [T]he *Ideas of primary Qualities* of Bodies, *are Resemblances* of them, and their patterns do really exist in the Bodies themselves; but the *Ideas, produced* in us *by* these *secondary Qualities,* have no resemblance of them at all. There is nothing like our *Ideas,* existing in the Bodies themselves. They are in the Bodies, we denominate from them, only a Power to produce those Sensations in us: And what is Sweet, Blue, or Warm in *Idea,* is but the certain Bulk, Figure, and Motion of the insensible Parts in the Bodies themselves, which we call so.²⁸

Complex ideas are said to be representative because they combine many simples that are representative in this way. Complex ideas of substances, to whatever extent they incorporate simple ideas of secondary qualities, will fall away from the strict sense of representation that Locke has in mind here. So generally he does not believe that ideas are pictures in the sense in which we would say that a snapshot of a tree is compared visually to a tree, the snapshot "looking" like a tree.

For Locke, the signification of words is then restricted to the only things we directly know and can directly talk about, the ideas that fill up our minds. Because the contents of one's mind are wholly internal and private, for Locke words are "voluntary signs" having as their "primary and immediate signification" ideas, serving to communicate them to others. Locke's use of "immediate" and "primary" is strongly reminiscent of Boethius's translation and commentary. The whole passage is:

> *Words in their primary and immediate Signification, stand for nothing, but the* Ideas *in the Mind of him that uses them,* however imperfectly soever, or carelessly those *Ideas* are collected from the Things, which they are supposed to represent.²⁹

For Locke everything that exists is a particular existence. A proper or "peculiar name" signifies an idea which is "peculiar" or particular in its representation of a particular existence. But if everything that exists is a particular existence, what is the signification of general names? Notice that the problem is of *what* general words signify, not *how* they signify, and arises because all things are particular existences.

> *General and Universal,* belong not to the real existence of Things; but *are the Inventions and Creatures of the Understanding,* made by it for its own use, *and concern only Signs,* whether *Words,* or *Ideas.* Words are general, as has been said, when used, for Signs of general *Ideas;* and so are applicable indifferently to many particular Things; And *Ideas* are general, when they are set up, as the Representatives of many particular Things; but universality belongs not to things themselves, which are all of them particular in their Existence, even those Words, and *Ideas,* which in their signification, are general.[30]

General words primarily signify ideas which are general or universal in their representation of particular existences. So the problem of the signification of general words is solved. Notice that *how* they signify "many particular things" is solved by telling us *what* they primarily and directly signify—a particular idea that is "set up" as the representative of many particular things.

This passage from chapter 3 of Book III of the *Essay* is important for a number of reasons. There are clear similarities and dissimilarities with the analysis in St. Thomas. I don't want to suggest that Locke is commenting on Aristotle, as St. Thomas is. Reference to the *De interpretatione* occurs nowhere, though there are plenty of references to the "Scholastics."[31] But the accounts appear similar because of the place of universality in the analyses. Universality applies to the concept in St. Thomas's account, because the concept does not differ formally from its object, and thus may take within its scope a number of material particulars that do not differ with respect to that form. In Locke's account, universality applies to ideas, which:

> become general, by separating from them the circumstances of Time, and Place, and any other *Ideas,* that may determine them to this or that particular Existence.[32]

An idea that is a particular existence in the mind is general in its representative function because it represents many particular existences, by leaving out characteristics in virtue of which the particulars differ. Consequently, a word is general when it signifies such a general idea. Despite this appearance of similarity, it is important to keep in mind that Locke does not appeal to the *form that does not differ* between concept and object.

Second, it is also important to keep in mind how the accounts differ more subtly, but crucially. They differ in the way in which this discussion of general words plays a part in their larger discussions. St. Thomas is concerned to explain the place of intellect in Aristotle's text. Intellect is not presupposed by the account of signification. Aristotle mentions it, and so it becomes a problem for St. Thomas to address. General words do not signify$_2$ a universal thing, but signify$_1$ a concept whose intelligible content does not differ from a number of individual things. For St. Thomas, general words do not pose a *problem* to be solved against the background of an account of signification involving mind that has already been given.

Nothing in St. Thomas's text suggests that particular words like 'Aristotle' require intellect to be involved. Perhaps they do. But the text is silent on them.[33] Thus, intellect is not presupposed in St. Thomas's account of signification. Nothing in his *Commentary* on 16a3–8 justifies the place of intellect in the account of signification other than general words.

When one turns to Locke, on the other hand, one finds the presupposition quite different. He does not need to justify the place of the mind and its ideas in the account of signification, as St. Thomas had to justify the place of intellect and its concepts. It is presupposed. He lays that down long before he gets to the problem of general words. Words *primarily and immediately* signify ideas because the latter are the only things that the mind *primarily and immediately* knows.

> Words being voluntary Signs, they cannot be voluntary Signs imposed by him on things he knows not. That would be to make them Signs of nothing, Sounds without Signification.[34]

If extra-mental things cannot be known, they cannot be signified by words. But extra-mental things are only known to the extent that they correspond to what a mind directly and immediately knows, ideas existing within it. Thus, in general, a man can only signify extra-mental things, or ideas in the minds of others, if they correspond to ideas that he immediately and directly knows in his own mind. The task of explaining this, however, was already completed in his second chapter; in his third chapter, by contrast, he is simply explaining how the signification of general words differs from that of "peculiar" or particular words.

Thus, for Locke the signification of general words is a problem for an account that essentially involves mind, an account that has already been given by the time he gets to the problem of general words. General words differ from particular words, not, as in St. Thomas, because they involve the intellect in the *means* by which they signify extra-mental things, but rather in the character of the intra-mental *thing* they *primarily and immediately* signify—a general idea instead of a particular idea. The place of mind in Locke's analysis of general words is presupposed, not justified. For Locke, an idea, by primarily being an intra-mental *object* or *thing directly signified*, becomes secondarily a *means* for a word to signify indirectly extra-mental things which are represented by that idea. But it will soon become apparent that there are problems even with this thesis. In short, St. Thomas presupposes that words, particular or general, signify extra-mental things, and pursues how the intellect may be involved in that signification; by contrast, Locke presupposes that words, particular or general, signify intra-mental things, and pursues *whether* they signify extra-mental things.

Finally, there is a striking difference between the two that is easily missed. Locke writes that *universal* or *general* do not pertain to existence, but "*concern only Signs*, whether Words, or Ideas." In the process of doing so, he identifies 'sign' as applying indifferently to words or ideas. In his discussion St. Thomas does not call the concepts *signs*; they are similitudes. Neither, for that matter does Aristotle. For Aristotle and St. Thomas 'sign' properly speaking applies indifferently to written or spoken words. Indeed, St. Thomas explicitly warns in lesson 2, passage 19 against treating words and concepts indifferently as signs. Here I simply want to make the point that it is Locke who treats words and concepts indifferently as signs, not St. Thomas.

E.J. Ashworth has documented the extent to which many of the theses involved in Locke's discussion were the common coin of logical discussion at Oxford during his time there. In particular, she stresses Locke's indebtedness to scholastic debates of the sixteenth century for several theses.[35] In this way at least, Locke, though not explicitly commenting on Aristotle, appears to be carrying on the tradition embodied in Ammonius's and Boethius's commentaries, and that forms the basis for the plausibility of the charge of mental representationalism against the whole of that tradition. His mention of "primary and immediate" signification is perhaps the most striking evidence of Locke's debt to that tradition.

Some have tried to see in Locke's account of language intimations of the sort of private language account that Wittgenstein argued against. Indeed,

Norman Kretzmann believes that the early chapters of Book III, "present a classic formulation of what Wittgenstein was later to criticize as the notion of a 'private' language." Genevieve Brykman, on the other hand, argues that such an interpretation is misguided. Locke stresses the social nature of language. He is "thoroughly concerned to make it obvious *that we succeed in communicating*." She finds it "paradoxical to state both that language is the great instrument of knowledge and the main tie of society, and that we are, each of us, imprisoned in our private ideas."[36] She attributes to Locke the distinction between the formal and objective reality of ideas, found in Descartes's third meditation and Antoine Arnauld.[37] Further, she believes that his description of ideas as perceptions indicates the "formal sense" of ideas, while his description of them as "objects of knowledge" betrays the "objective sense," that is, the sense in which they are representative of things. The objective or representative sense guarantees the public nature of knowledge and language, and undercuts the "private language" charge.

This is clearly an instance in which the "received view" of Locke is undergoing critical examination. But, with regard to the question of mental representation, it is not clear that her interpretation will help Locke, since one way of casting the difficulty is whether ideas as he describes them can have this "objective" or representative role, that is, can ideas simultaneously have formal and objective reality, if the formal is what is known first as an object of perception? Recall that it is upon the phenomenological aspects of ideas in what Brykman identifies as the "formal sense" as perceptions that we are to base our certitude about their objective reference, what she identifies as their "objective sense."

Locke's thesis on the need for language suggests that were the contents of each mind accessible to other minds, there would be no need for language. He writes:

> Man, though he has great variety of thoughts, and such, from which others, as well as himself, might receive profit and delight; yet they are all within his own breast, invisible and hidden from others, nor can of themselves be made to appear;

and,

> because the scene of ideas that makes one man's thoughts, cannot be laid open to the immediate view of another. . . . therefore to communicate our thoughts to one another, as well as record them for our own use, signs of our ideas are also necessary.[38]

The suggestion is that the point of language is not to communicate directly about extra-mental things, but to communicate our internal ideas about extra-mental things, because we do not have access to the inner thoughts of others. Thus, it seems that according to Locke the need for signs, that is, words, is in some fashion due to a deficiency on the part of human knowledge. A speaker needs signs, in the hope that such signs will bring to the mind of the listener similar objects of knowledge as exist in the speaker's own mind, objects just as internal and private to the listener's understanding, but representing perchance the same inaccessible external object as the speaker's representations do. (It is difficult not to think of Kretzmann's metaphor of encryption and decryption here.)

For Aristotle and St. Thomas, the place of language does not appear to make up for a deficiency in human knowing, but as the union of two perfections in human nature—the ability to actively assimilate external and public things to oneself, that is, "to become all things," in conjunction with the social and political character of human nature, namely, the capacity for pursuing natural perfection through common life. Thus, St. Thomas, in the context of commenting on the *De anima* writes, "nature . . . uses air that has been breathed in for the formation of words, which is for the sake of a more perfect existence."[39] The importance of this text is that a study of the soul (*de anima*) intimates the social and political existence of human beings, even as it does not in its mode of abstraction study their social and political existence, the proper subject matter of the *Politica*. Study of the soul itself points to its own incomplete understanding of the full reality of human being. Language is based upon the perfection of human nature, not its imperfection.

In Locke's use of "voluntary signs" and "immediate signification," there is, of course, some terminological similarity with the discussion in Aristotle, Boethius, Ammonius, and St. Thomas, a similarity that perhaps further facilitates the assimilation of Locke to the "long tradition stemming from Aristotle," even though he does not present himself as commenting upon Aristotle. Consequently, one might expect Locke to use "mediate" or "by the mediation of" to characterize the signification of external things by words. But it presently appears how substantively different his account is from St. Thomas's in particular, since for Locke words properly speaking signify *only* the ideas *internal* to the mind, in particular only the ideas in the speaker's mind. "[T]his is certain, their signification, in his use of them, is limited to his *Ideas,* and they can be signs of nothing else."[40] He does speak of men giving a "secret reference" to words beyond the ideas of the mind, namely to the ideas present in the minds of others, and to things as they are *external* to the mind, so some sort of signification of extra-mental things occurs. He seems to think that it is illicit, "it is a perverting the use of words, and brings unavoidable obscurity and confusion into their signification, whenever we make them stand for any thing, but those ideas we have in our minds."[41] Words do not, or ought not to signify *external* things, but the *internal* representations of *external* things. In an interesting passage on the names of simple ideas and substances, he writes that "the *Name of simple* Ideas *and Substances,* with the abstract *Ideas* in the Mind, which they immediately signify, *intimate* also *some real Existence,* from which was derived their original pattern."[42] Clearly Locke does not want to cut words off from things, which seems to be why he says that they "intimate" real existence, at least in the case of simple ideas of substances. The difficulty is with *signification,* and whether it applies to external things, and whether it is mediated by ideas. The sense of "immediate" and any use of "by the mediation of" seems to be determined by this "intimation of things," or by the "perversion" by which words are made to *signify* extra-mental things. If neither this "intimation" nor "perversion" took place, Locke's use of 'immediate' to characterize the relation of words to ideas would seem to be superfluous.

Berkeley

Berkeley and Hume submit the structure of Locke's analysis to sustained criticism, in particular his confidence in an extra-mental world, and how we know and talk about it. Berkeley argues in part against the skepticism attendant upon Descartes's and Locke's accounts of ideas, inasmuch as they maintained a distinction between ideas existing in and directly perceived by the mind, and a mind-independent material world. He agrees with Locke that the origin of all our ideas is either sensation or reflection upon the mind's own activities, but thinks that no material thing could be known to be the cause of these ideas.

> [W]hat reason can induce us to believe the existence of bodies without the mind, from what we perceive, since the very patrons of matter themselves do not pretend, there is any necessary connection betwixt them and our ideas. . . . It is granted on all hands (and what happens in dreams, frenzies, and the like, puts it beyond dispute) that it is possible we might be affected with all the ideas we have now, though no bodies existed without, resembling them.[43]

Language and Mental Representationalism 93

The reference to dreams makes it clear that he has Descartes for one in mind, when he points out that the "patrons of matter" do not think there is any necessary connection between ideas and their material causes. Locke would also seem to be implicated here, at least with respect to complex ideas of substances.

Locke tries to avoid the skepticism about an external world by an examination of the *phenomenological* properties of the ideas that he perceives, but this does not give him a necessary connection between ideas and material things. Even if it did, he does not believe that any particular complex idea of a substance has a necessary connection to any particular extra-mental thing sufficient to distinguish it from some other extra-mental thing. There is no necessary connection between the inner constitution or "real essence" of things and the complex ideas or nominal essences existing in our minds and representing things to us, and under which we sort things into classes. Because the real essence or inner constitution of things is quite unknown to us, a complex idea of a substance is at best a nominal essence.[44] So any number of things that differ in inner constitution, say the chemical element *Au* and *iron pyrites*, may produce the same complex idea or nominal essence in our minds, say *yellow metal*.

Berkeley, by contrast, solves the problem of skepticism by denying that there is any material world existing beyond the mind, and productive of our ideas of it.

> But philosophers having plainly seen that the immediate objects of perception do not exist without the mind, they in some degree corrected the mistake of the vulgar, but at the same time run into another which seems no less absurd, to wit, that there are certain objects really existing without the mind, or having a subsistence distinct from being perceived, of which our ideas are only images or resemblances, imprinted by those objects on the mind.[45]

Ideas come from sensation, but not because they are produced by some material being existing beyond the senses with qualities or powers capable of producing those sensations. Ideas cannot represent such a world by likeness since "an idea can be like nothing but an idea."[46] All that exist are spirits and ideas. If ideas are not produced by the mind in which they exist, they must be produced by another spirit, not a material thing.

> If [sensations] are looked on as notes or images, referred to *things* or *archetypes* existing without the mind, then are we involved all in

scepticism. . . . All this scepticism follows, from our supposing a difference between *things* and *ideas,* and that the former have a subsistence without the mind, or unperceived.⁴⁷

If there is no mind-independent material world, then there can be no skepticism about it. On the other hand if all that exists are minds, spirits, and ideas existing in them, and the ideas are directly and immediately perceived, then there need not be any skepticism about them.

Further, the mind has no power of framing the sort of abstract ideas of things that Locke described, in which "the mind being able to consider each quality singly, or abstracted from those other qualities with which it is united, does by that means form to itself abstract ideas."⁴⁸ This denial forces Berkeley into an account of the signification of general words that differs in an important respect from Locke's. A word is not general because it signifies a single idea in the mind, a particular existence, which idea Locke describes as general in its representation because it is an abstract image. On the contrary, Berkeley believes that words and ideas can be signs or represent other things, but those other things are only other ideas, which are just other particular existences, certainly not general ideas. A sign, whether it is a word or idea, becomes general not by signifying a single general idea, but by signifying indifferently a number of particular ideas. If one were to ask whether the word 'man' signifies a single idea of *man*, which is in turn a sign of all particular men by leaving out all particular qualities in virtue of which the particular men differ, Berkeley would say no. If one were to ask, is it then a sign of a single particular idea to the exclusion of all other particular ideas, he would again say no. It is a sign of this particular idea, or that particular idea, or this other particular idea, and so on, indifferently and excluding no "particular ideas of the same sort."⁴⁹ Berkeley's goal is not to deny that there are general signs, including general ideas, but rather that there are general *abstract* ideas, and that these are signified by general words. This is one point where Locke and Berkeley differ.

It is not surprising, however, to recognize similarities. Again, as in Locke, the mind is presupposed to the account of general words. Also, as in Locke, *sign* applies indifferently to words and ideas. But there is a difference. Berkeley appears to recognize no difference *qua* sign between ideas and words. His description of how a particular idea or word becomes general suggests that for Berkeley a sign is something that is *made* to represent, or to stand for something else. He uses the example of a geometric demonstration on a line, as that particular line becomes general, by being made a sign, so the name *line*, which taken absolutely is particular, by being a sign is made general. And as the former owes its generality, not to its being the sign of an abstract or general line but of all particular right lines that may possibly exist, so the latter must be thought to derive its generality from the same cause, namely the various particular lines which it indifferently denotes.[50]

For ideas, *being a sign* is just as much a matter of being *made* to represent as it is for words. Locke, on the other hand, had said words were voluntary signs, but he did not say that ideas *qua* signs are voluntary.

For Berkeley, *universality*, whether of a word or idea, is no attribute of the word or idea; it consists in the relation the word or idea bears to the many things it may indifferently be *made* to represent. It is also important to note that on Berkeley's account a universal word can bear this *relation to many*, without the interposition of a *universal* idea. For Locke, on the other hand, a word becomes general by signifying a general idea, which idea in turn represents things. However, on Berkeley's account, even if there is an idea representing indifferently a number of particular things, he makes no suggestion that a word must signify that idea as a necessary condition for it to signify indifferently those things. The quote suggests that both idea and word "derive [their] generality from the same cause, namely the various particular [things] which [they] indifferently denote." 'Sign' applies indifferently to both words and ideas, and there is no suggestion that one is subordinated to another; *qua* universal sign, they are on an equal footing.

For Berkeley there is no question that words signify ideas existing in the mind. However, there is no use of "immediate" and "primary" to characterize the signification of ideas by words, as there was in Locke. Obviously this is not because he wishes to differ from Locke on this point, and attribute the "immediate" and "primary" signification of words to something else. The use of "immediate" and "primary" in Locke serves to distinguish what is "not immediate" and what is secondary, the signification of extra-mental material reality. But in Berkeley there is nothing for words to signify other than ideas. If "immediate" and "primary" demand their correlatives "mediated" and "secondary," what is there that can function as secondarily signified, contrasted with ideas primarily signified? What point could there be in Berkeley's use of "immediate" and "primary?" None. So he eliminates it.

He does disagree with Locke's claim that the "chief end" of language is to communicate ideas. Words may also have the end of communicating *passions* like love, hate, dread, and so on, even when no idea accompanies the impression. But these latter, even if not ideas, are states of mind. More importantly, he does not say that the words that "communicate" these passions signify them. 'That beast' applied to Hitler may well communicate hatred, but it does not signify it, and it need not on that occasion of use bring to mind an image of a beast. Presumably words like 'love', 'hate', and 'dread' signify respectively *ideas* of *love, hate,* and *dread,* but they need not in their utterance *communicate* the passions of *love, hate,* and *dread.* "Signification" seems to be reserved for words taken as "marks" for ideas, not the passions.[51]

Hume

Hume in large measure follows Locke in his account of ideas as the *internal* objects of knowledge.[52] However, he differs from him in drawing out the implications of the theses even more rigorously, casting doubt, like Berkeley, upon whether these *internal* objects really are representations of *external* objects at all. He is clear that "no beings are ever present to the mind but perceptions."[53]

> The most vulgar philosophy informs us that no external object can make itself known to the mind immediately, and without the interposition of an image or perception. That table, which just now appears to me, is only a perception, and all its qualities are qualities of a perception.[54]

Our conviction that *external* objects resemble these *internal* representations arises because the *use* of internal representations is a necessary condition for describing external objects.[55] However this necessity does not arise from any characteristic of ideas themselves. Both Locke and Berkeley thought a phenomenological examination of some of one's ideas shows that they must be produced by something other than the mind, in Locke's case *matter,* and in Berkeley's another *mind* or *spirit.* However, they did not claim that such an examination provides determinate characteristics of those other things, just certainty of their existence. Hume, however, disagrees—"to form the idea of an object, and to form an idea simply is the same thing; the reference of the idea to an object being an extraneous denomination, of which in itself it bears no mark or character."[56] A phenomenological examination of one's sensations and ideas will turn up no "mark or character" that indicates or refers to an object distinct from the sensation or idea.

This is an important clarification for Hume to make; one cannot tell the difference between the formation of an idea *of* an object, and the formation of an idea simply. To use a phrase that will appear later in the chapter, ideas "just stand there." An idea has a certain "in itself" character. It is individuated "in itself" and can be described "in itself." Any relation to any other object is "an extraneous denomination of which *in itself* it bears no mark or character."

In a criticism clearly reminiscent of Berkeley's, Hume submits Locke's conviction that there are external objects represented by our resembling perceptions to a sustained critique. He argues that if all we can actually know are *internal* sense impressions and less vibrant more attenuated ideas, there is no intelligible sense that can be given to the claim that there are objects *external* to the mind that *cause* resembling and representative perceptions in minds.[57] Not only is the *causal* claim questionable, but the very notion of an *external* object resembling our *internal* representations is the "monstrous offspring" of a philosophical system. He carries through this critical analysis for the impressions and ideas of, among other things, *substance, cause, self, mind,* and *personal identity,* considered as ideas representative of, or related to, externally existing objects. To suppose a "double existence" of *external* objects resembling *internal* perceptions is a philosophical "palliative" thrown at our desire to possess objects existing independently, and enduring unperceived by episodic mental acts.

Reminiscent of Locke's dark room, though without the connotation of a luminous external world, Hume compares the mind to a stage.

> The mind is a kind of theatre, where several perceptions successively make their appearance; pass, repass, glide away, and mingle in an infinite variety of postures and situations. . . . The comparison of the theatre must not mislead us. They are the successive perceptions only, that constitute the mind; nor have we the most distant notion of the place where these scenes are represented, or of the materials of which it is composed.[58]

It seems a mere *façon de parler* for Hume to continue to speak, as he occasionally does, of the contents of the mind as *representations* of *external* objects, or as *referring* to *external* objects, for there is no good reason for thinking that they *represent* or *refer to* anything *external*.

Hume does not often speak of signs or signification. His preferred term is 'meaning'. Further, he does not have a general treatment of 'meaning' and words; most of what he says is as commentary on Locke and Berkeley. In general a word has a meaning when it is associated with a sensation or idea. He agrees with Berkeley against Locke that there are no general ideas in the Lockean sense. There are just particular ideas. In fact, with Berkeley's criticism of Locke clearly in mind he calls him a "great philosopher," and writes that "I look upon this [Berkeley's account of general ideas] to be one of the greatest and most valuable discoveries that has been made of late years in the republic of letters."[59]

Despite his admiration, Hume puts an interesting twist upon Berkeley's account of general or universal words. A word becomes general because we apply it "when we have found a resemblance among several objects, that often occur to us." This produces in us a "custom," or habit such that "the hearing of that name revives the idea of one of these objects, and makes the imagination conceive it with all its particular circumstances and proportions." Thus as in Berkeley, and against Locke, a word is general because it is indifferently applied to a number of objects, not because it is subordinated to a general Lockean idea. But in Berkeley a general word is also not subordinated to a general Berkelean idea. For Berkeley, the general idea and the general word are on an equal footing; they are not in need of one another in order to signify indifferently a number of particular things.

It is at this point that Hume subordinates the general idea to the general term! The custom attending the use of a general term causes the imagination to conceive the idea of some object signified by that word. This idea is completely particular and individual:

> 'Tis certain *that* we form the idea of individuals, whenever we use any general term; *that* we seldom or never can exhaust these [other] individuals; and *that* those, which remain, are only represented by means of that habit, by which we recall them, whenever any present occasion requires it.

The mind cannot survey all the ideas of objects that may be signified by a general term, but they are in some sense represented virtually in the *power* of the habit or custom associated with the term, that is, the power to bring them to mind on a particular occasion. But because a particular idea is "annex'd" to that general word, and its occurrence in the mind on the occasion of the utterance of the word is caused by the "custom" associated with that word, that particular idea can serve as a general representation of all the *other* particular things signified by that word.

> 'Tis after this manner we account for the foregoing paradox, *that some ideas are particular in their nature, but general in their representation.* A particular idea becomes general by being annex'd to a general term; that is, to a term, which from a customary conjunction has a relation to many other particular ideas, and readily recalls them in the imagination.[60]

It is necessary to be careful here. Hume is not asserting that words "immediately" or "primarily" signify things, and only "mediately" and "secondarily" signify ideas. As with Berkeley, those terms play no part in the discussion. That they do not is not surprising considering Hume's general skepticism about objects other than ideas. Words signify *ideas* of particular things.

One does see, however, a complete reversal of Locke on how words are related to general ideas of things. For all three philosophers, words signify ideas. That is the presupposed account. For all three the signification of general words becomes a problem for their respective accounts of ideas. It is in their solution to this problem that they differ. In Locke the generality of words is subordinate to the generality of ideas. In Berkeley, the generality of words and ideas are on a par. In this he serves as an intermediate point between Locke and Hume. In Hume the generality of ideas is subordinate to the generality of words.

Preliminary Conclusions

Russell's first thesis concerning the existence of an *internal* mental entity that is directly known, and that mediates the knowledge of things *outside* of or *external* to the mind, finds real correlates in Locke, as well as Berkeley and Hume if one takes away the external things. His second thesis about the inevitability of skepticism about the *external* world finds justification in all three, but especially in Berkeley and Hume. Berkeley raises the question about a necessary connection between mental representations and external objects. Hume then denies that there is any such connection, inasmuch as any such relation seems to be purely extraneous to what the representations are "in themselves." The turn to language in the last hundred years, and what it can be understood to have been a turning from, has to be seen in light of this "classical representationalism."

Historical Criticism of Mental Representationalism

Reid

As the rough contemporary of Locke, Berkeley, and Hume, Thomas Reid, critical of the theory of ideas, held that it is futile for those who accept Descartes's account of ideas to try to show that Berkeley's conclusions are absurd.[61] Reid sees Locke and Berkeley as merely carrying on the skeptical program initiated by Descartes, a skeptical program that cannot be resolved once it has begun. Indeed, for Reid, Hume simply finished what Locke and Berkeley had started.

> Mr. Locke had taught us, that all the immediate objects of human knowledge are ideas in the mind. Bishop Berkeley, proceeding upon this foundation, demonstrated, very easily, that there is no material world. But the Bishop, as became his order, was unwilling to give up the world of spirits. Mr. Hume shews no such partiality in favour of the world of spirits. He adopts the theory of ideas in its full extent; and, in consequence, shews that there is neither matter nor mind in the universe; nothing but impressions and ideas.[62]

For Reid, all of this is sufficient reason to abandon the theory of ideas. In a sense, he agrees with Hume on the conditional *if internal representations are the sole direct objects of knowledge, then we can have no knowledge of external objects;* on the basis of his "common sense" philosophy he denies the consequent where Hume had affirmed the antecedent. Thus, for him, internal representations are not the sole direct objects of knowledge.

The Turn to Language: Husserl and Frege

While most of what follows will center on Wittgenstein, the recent criticism of mental representationalism did not begin with him. There are certain foreshadowings of his criticism in others, especially Husserl and Frege. In the late nineteenth century both of them, concerned for the objectivity of truth, meaning, and the reference of language, reacted strongly against *classical representationalism* and the identification of the meaning of terms with internal mental representations. Besides the classical Empiricists, both of them had in mind their latter-day descendants, especially the *psychologism* of John Stuart Mill in his *Logic*.[63] Husserl writes:

> If I hear the name 'Bismark' it makes not the slightest difference to my understanding of the word's unified meaning, whether I imagine the great man in a felt hat or coat, or in a cuirassier's uniform, or whatever pictorial representation I may adopt. It is not even of importance whether *any* imagery serves to illustrate my consciousness of meaning, or to enliven it less directly.[64]

He devotes a large section of *Investigation* II to the analysis and critique of Locke, Berkeley, and Hume on these questions of the representational character of thought, criticizing them for making ideas interior objects of perception.

Noting that in Locke such interior presentations are usually intentional experiences, that is, "each idea is an idea of *something*, it presents something," Husserl criticizes him for confusing the "presentation with what is as such presented, appearance with what appears, the act or act-phenomenon, as a really immanent element in the stream of consciousness, with the object intended."[65] In this passage one can see a clear parallel with Russell's criticism. Something, an "act or act-phenomenon" is intruding itself in between the mind and the object. Further, with regard to universal terms, he writes that "naming by means of universal names does not . . . consist in picking out particular universal ideas from such complexes of ideas, and attaching them to words as their 'meanings.'" In a passage that foreshadows the problems brought out by Wittgenstein, he writes of Hume,

> the same individual idea fits into many circles of similarity, though in each *definite* thought-context it only represents ideas from *one* such circle. What circumstance picks out this circle of representation in a given context, what limits the representative function of the individual idea in this manner and so makes unity of sense possible?[66]

For Husserl the fundamental problem in the "historical treatment of abstraction," or general representation, is that "*the sign did duty for the thing signified.*" The sign stands in place of or substitutes for what is intended because the latter is not present. Very recently Robert Sokolowski echoes Husserl's criticism when he writes, "an idea is not an internal entity and in an important sense it is really not other to the thing of which it is the idea."[67]

When we turn to Frege, we find that he wishes "to separate sharply the psychological from the logical, the subjective from the objective."[68]

> What is a content of my consciousness, my idea, should be sharply distinguished from what is an object of my thought. Therefore the thesis that only what belongs to the content of my consciousness can be the object of my awareness, of my thought, is false.[69]

It was in part the subjective and private character of mental states that led Frege to posit his "third realm" of thoughts to account for the "objective" meaning of language. Colin McGinn questions whether Frege's appeal to *senses* can in fact overcome the empiricist problem of interpreting the inner subjective images that come as objects before the mind, the problem of interpretation to be discussed below in the treatment of Wittgenstein. "According to Frege, you succeed in referring to or thinking of something by virtue of standing in a relation of 'grasping' to a sense, which determines some entity as reference."[70] Frege simply substitutes one *intentional object* for another, an abstract non-mental *sense* for a mental *idea*, an *intentional object* that, like a mental idea, is supposed to determine all future use of the sign for which it is a sense. According to McGinn,

> a sense can be no more *intrinsically* related to use than any other sign-like item can. We are tempted to think otherwise because we have the picture of directions for use being somehow inscribed in (or on) the sense, but of course this just is the old interpretational conception in another guise.[71]

Recall that Russell's worry need not be confined solely to mental intermediaries, but any intermediaries that are wholly interposed as objects or "third things" between knower and known. On Frege's behalf Michael Dummett responds that this aspect of Frege's account need not be emphasized.[72] These objections of Husserl's and Frege's and the common concern with the skepticism perceived to be attendant upon "mental representations" form a historical point of intersection and fruitful area of inquiry between their philosophical heirs.

Wittgenstein

In the *Philosophical Investigations* Wittgenstein can be understood to have indirectly submitted classical representationalism to criticism for both its "internalism" and its "picturing."[73] In particular, Wittgenstein criticized the thesis that the referential function of language can be accounted for by *internal* representations conceived of as images of the objects referred to in conjunction with meaning conferring acts that correlate or assign these internal images to external sounds or written symbols. "[N]othing is more wrong-headed than calling meaning a mental activity! Unless, that is, one is setting out to produce confusion."[74]

The concern in the *Investigations* with the pictorial determination of reference finds its context in the *Tractatus's* earlier picture theory of sentence truth,[75] but also the Fregean thesis that the *sense* of a term is "the mode of presentation" of its referent, or as it has come to be put in many recent discussions "sense determines reference."[76] Suppose one identifies the sense or meaning of a term with an *internal* mental entity that is thought to determine reference by picturing the object referred to and presenting it to the mind. Then if sense is to determine reference, one wishes to know what it is about pictorial representation that is sufficient for constituting the meaning, and determining the reference of the term. How is it that one knows an object is referred to from an examination of its *internal* mental representation?

> Is there such a thing as a picture, or something like a picture, that forces a particular application on us. . . . What is essential is to see that the same thing can come before our minds when we hear the word and the application still be different. Has it the *same* meaning both times? I think we shall say not.[77]

Suppose a Renaissance painter has his wife and baby pose as models for a painting of the Madonna and child. The painter's intention is that it represent Mary and the child Jesus, but anyone who knows his wife and child, without knowing his intention, would take it to be representing the latter, not Mary and Jesus. Nothing about the painting as such determines unambiguously what it represents.

If it is replied that it naturally represents his wife and child by visual resemblance, while conventionally and symbolically representing Mary and Jesus, then one can still ask whether by an examination of the picture alone one could determine unambiguously that it is of this particular woman and child, rather than, say, her closely resembling sister and child. Further, as John Haldane points out, "the representational properties of images are either *particular*, in which case generality is left unexplained, or else they are *general* and singular reference remains unaccounted for."[78] Perhaps the picture is not intended to represent any particular mother and child, but all of them. How can one determine this?

It seems that the painter determines the object represented by the painting, in some sense already knowing what he wants to represent; he *intends* it that way. Or perhaps the observer determines what is represented in view of his own context and concerns, and irrespective of the intent of the painter—"only when one knows the story does one know the significance of the picture."[79] The various elements of the painting, for example, perspective, style of clothing, background landscape, and so on, have a use within the context that conveys what is represented. But considered apart from that context of use, they can be made to represent any number of things. Outside of the context of use and considered "in themselves," representations are in a sense faceless icons.

I want to proceed with caution here. Even granting Wittgenstein the point, it is not obvious that "natural" similarities play no part. It seems that contexts in which no natural similarities play a role in representation, or very few anyway, would have to be very specialized and extraordinary. What is the context of use like, and what are the interests of the painter or observer, in which a line drawing of a large cube with a smaller cube on top of it is a better representation of Madonna and Child than a Renaissance painter's wife and child? In any case, it is not clear that Wittgenstein was claiming anything so radical as that natural resemblance plays absolutely no part in representation.[80]

But when the picture is made into an *internal* mental image, the difficulties appear compounded. Who is the painter or observer establishing the referential properties of the internal presentation, determining it to an object? Not the mind acting on the image, for by supposition the mind does not have independent knowledge of the reference by which to fix the representative character of the image, but it seeks to determine the reference from the image. If the mind itself fixes the reference of the image, then we are faced with a disjunction—the mind performs this function either without the need of an additional image or with an additional image. If we opt for the first disjunct, it is unclear why we needed the original image, since we are granting that the mind can engage the world referentially without an image. If we opt for the second disjunct an infinite regress of homunculi conferring reference on internal images looms.[81]

Wittgenstein's criticism centered upon recognizing that representations likened to pictorial resemblances require a context of interpretation in order to represent determinate objects, which internal mental images must do if they are going to determine reference. But "any interpretation still hangs in the air along with what it interprets, and cannot give it any support. Interpretations by themselves do not determine meaning."[82] What is required is a custom and context of use to determine the meaning of linguistic characters, which meaning in turn determines the interpretation of the images that may or may not accompany speech. Here one might usefully recall Hume's account of how a particular idea comes to have a universal representation from its association with a universal term. In any case, according to Wittgenstein, what is clear is that "when I think in language, there aren't 'meanings' going through my mind in addition to the verbal expressions: the language is itself the vehicle of thought."[83]

There is some debate about how to interpret Wittgenstein's actual intention in the *Investigations* as it bears upon representations and the meaning of words. One interpretation suggests that Wittgenstein's intent is to completely externalize meaning in the social practices of one's community. Saul Kripke takes this position. John McDowell, quoting Kripke, summarizes the latter's understanding of the problem, "Kripke's reading of how the regress of interpretations threatens the very idea of understanding turns on this thesis: 'no matter what is in my mind at a given time, I am free in the future to interpret it in different ways.'"[84] Kripke bases his reflection upon, among other things, Wittgenstein's example of the interpretation of rule-following in mathematics, for example, the addition function '+', in particular 'add 2', as well as Nelson Goodman's famous 'grue' predicate. How do we know that two or more individuals will not provide divergent interpretations of the rule 'add 2' in the future at some heretofore unexpressed stage of the series; perhaps they are using two distinct functions, 'plus' and 'quus', that just happen to coincide for all the values that have so far been computed. There is no guarantee that the coincidence will not come to a precipitous end. Then we shall find out that when commanded to 'add 2', the individuals had only the appearance of grasping the same meaning in the command—in their respective heads they had different interpretations of the same command before their minds. This being true of two or more individuals, it is also true of the future use of a particular individual, who is, as Kripke says, "free in the future to interpret it in different ways." Applied to the tradition of *representation* this requires taking *mental representations* to be rules for interpreting language. Recognizing the need for interpretation in the application of a rule, the forever threatening infinite regress ensues when the interpretation itself consists in just another rule.

According to Kripke, Wittgenstein's work suggests that the skeptical problem involved here should be overcome by denying that there is some special fact about the individual, a fact about his grasp or understanding of the meaning of the command. Rather, there is a fact about the social practice related to that command.

> [I]f the individual in question no longer conforms to what the community would do in these circumstances, the community can no longer attribute the concept to him. Even though, when we play this game and attribute concepts to individuals, we depict no special 'state' of their minds, we do something of importance. We take them provisionally into the community, as long as further deviant behavior does not exclude them. In practice, such deviant behavior rarely occurs.[85]

Kripke's interpretation finds support in statements like the following from the *Brown Book*:

> [T]he last stage of the training is that the child is ordered to count a group of objects, well above 20, without the suggestive gesture being used to help the child over the numeral 20. If a child does not respond to the suggestive gesture, it is separated from the others and treated as a lunatic.[86]

On Kripke's account, not even God could look into my soul and know what it is that I understand when I apply the rule 'add 2', since there is no fact in my soul that consists in this understanding. On the other hand, He, like anyone else, could look at how my observable behavior conforms to or deviates from the social practice of the community in its use of the rule, and know that by those community standards I am sane or crazy—there is a fact of the matter here.

John McDowell disagrees with Kripke and suggests that this is not Wittgenstein's intent. Wittgenstein is not suggesting that we abandon the recognition that the meaningful use of language involves facts about the individuals who use that language. McDowell suggests that Wittgenstein is in effect displaying or picturing a complicated *reductio* of the original thesis that leads to the difficulty. This thesis McDowell calls "the Master Thesis" of representationalism, "the thesis that whatever a person has in her mind, it is only by virtue of being interpreted in one of various possible ways that it can impose a sorting of extra-mental items into those that accord with it and those that do not."[87]

Compare this Master Thesis with Hume's statement cited earlier that, "to form the idea of an object, and to form an idea simply is the same thing; the reference of the idea to an object being an extraneous denomination, of which in itself it bears no mark or character." Consider also Locke's statement in the *Essay* on nominal essences:

> 'Tis evident, that *Men make sorts of Things*. For it being different *Essences* alone, that make different *Species*, 'tis plain, that they who make those abstract *Ideas*, which are the nominal Essences, do thereby make the *Species*, or Sort.[88]

Of course Locke did not suggest that the mind could interpret "in one of various possible ways" the ideas present in it; but he did think that the nominal essences in the mind that determine the sorts or species of things are arbitrary combinations of ideas produced by the mind. An object is an X, where X is a species, because it exemplifies the various ideas constitutive of the complex species-idea in the mind, which just is the species X. The nominal definition of X provides necessary and sufficient conditions for *being* an X. In short, what counts as an X is determined by what is present in the complex idea that the mind forms for itself. To that extent he does seem to exemplify the latter part of the Master Thesis, that it "impose[s] a sorting of extra-mental items into those that accord with it and those that do not." Recall also the passage cited earlier in which Husserl spoke of "circles of representation," and wondered what it is that determines in a particular context a unity of sense in the representative function of an idea among these circles.

The Master Thesis is reflected in the memorable passage from the *Investigations*:

> A rule stands there like a sign-post.—Does the sign-post leave no doubt open about the way I have to go? Does it shew which direction I am to take when I have passed it; whether along the road or the footpath or cross-country? But where is it said which way I am to follow it; whether in the direction of its finger or (e.g.) in the opposite one?—And if there were, not a single sign-post, but a chain of adjacent ones or of chalk marks on the ground—is there only *one* way of interpreting them?—So I can say, the sign-post does after all leave no doubt. Or rather: it sometimes leaves room for doubt and sometimes not. And now this is no longer a philosophical proposition, but an empirical one.[89]

He makes a similar point in *The Brown Book*,

> How does one explain to a man how he should carry out the order, 'Go this way!' (pointing with an arrow the way he should go)? Couldn't this mean going the direction which we should call the opposite of that arrow? Isn't every explanation of how he should follow the arrow in the position of another arrow? What would you say to this explanation: A man says, "If I point this way (pointing with his right hand) I mean you to go like this" (pointing with his left hand the same way)? This just shows you the extremes between which the uses of signs vary.[90]

What this passage from *The Brown Book* adds to the one from the *Investigations* is that verbal instructions inherit the problem of the sign-post—

visual and spoken instructions are of a piece when it comes to the problems of interpretation, and what McDowell calls the Master Thesis.

One way of understanding Wittgenstein's point in these passages is to ask whether our ordinary practice of following a sign-post is determined by some intrinsic property of the sign-post itself. Consider the occasional practice in wartime of switching the direction of sign-posts in order to confuse the enemy. Knowing the practice of the ordinary person when confronted by such a sign, those who switch them count on the enemy *following* the direction of the arrow. But anyone who knows that this has been done, say the residents, will give someone directions to the town by saying "go in the direction opposite to that in which the arrow is pointing." But this latter "opposite to that in which the arrow is pointing" presumes the ordinary practice of taking it to point in the direction that proceeds *from* broad end *to* narrow end, the ordinary practice that they count on the enemy following. Knowing the wartime practice, should an enemy proceed according to ordinary custom, or according to wartime practice? Suppose the townsfolk anticipate the enemy's knowledge of the practice of switching signs, and so leave them as is. A skeptical problem quickly ensues on how to interpret the sign that cannot be resolved by an appeal to the sign itself, but only to the use of the community that placed it there. Without a knowledge of the community's use, one is faced with the prospect of flipping a coin or guessing how to interpret the sign. The instruction embodied in the sign has its character from the use of the community, not from an intrinsic property of the instrument employed to convey the instruction. So why not abandon the role of intrinsically representative properties altogether, as Kripke suggests is Wittgenstein's intent?

McDowell points out that comparing the mind's representations to sign-posts that just stand there involves denying that there is any intrinsic or normative relation between thought and world.

> [T]he master thesis implies that whatever I have in my mind on this occasion, it cannot be something to whose very identity that normative link to the objective world is essential. It is at most something that *can* be interpreted in a way that introduces that normative link, although it can also be interpreted differently ('I am free in the future to interpret it in different ways.') Considered in itself it has no relations of accord or conflict to matters outside my mind, but just 'stands there'. What I have in my mind is at most a potential vehicle for the significance in question, in the sort of way in which a sentence, con-

sidered as a phonetic or inscriptional item, is a vehicle for significance that it can be interpreted as bearing.[91]

McDowell uses as an example the thought, 'people are talking about me in the next room'. This thought is usually *interpreted* in such a way that it is true only if people are in fact talking about me in the next room; the real world fact is, to use John Searle's terminology, *the condition of satisfaction of the intentional thought*, having a "mind to world" direction of fit,[92] where the condition of satisfaction is intrinsic to the thought in the sense of being determined by the intrinsic content of the thought. However, the Master Thesis that McDowell describes denies this intrinsic representative character of the thought and holds that it is a matter of an arbitrary and imposed interpretation on the thought. Whatever the state of my mind is *in itself*, it simply happens to be interpreted in a way that requires that people are in fact talking about me in the next room; there is nothing intrinsic to the mental state that requires that this worldly fact, and not another, be a condition of satisfaction intrinsic to it, or in other words a principle of the thought's identity.

McDowell suggests that rather than accepting the Master Thesis and completely externalizing the grasp or understanding of meaning in social practices, as Kripke suggests is Wittgenstein's intent, Wittgenstein should be interpreted as providing reasons for us to abandon the Master Thesis. Where Kripke had placed significance on Wittgenstein's statement that a child will be treated as a "lunatic" if it does not learn to conform to community practices, McDowell places great significance on Wittgenstein's statement that "what this shews is that there is a way of grasping a rule that is *not an interpretation.*"[93] McDowell's suggestion is that for Wittgenstein to follow a sign is simply to do "what comes naturally to one in virtue of being initiated and trained in a custom in one's youth" and upbringing.[94] McDowell's use of "what comes naturally to one in virtue of being initiated and trained in a custom in one's youth," seems to be a foreshadowing of his attempt to retrieve "Aristotelian second nature" in his recent *John Locke Lectures: Mind and World*. There is a fact to be known about the individuals engaged in the meaningful use of language, the fact is the condition of their "second nature."

For McDowell no "philosophical" solution is called for where there is no difficulty. By "philosophical" solution, McDowell has in mind the tradition of providing solutions to skeptical doubts. He suggests that Wittgenstein is eliminating the doubts that give rise to the need for a "philosophical" solution. Presumably this custom into which one is initiated cannot involve an interpretation, at the risk of falling prey to the Master Thesis. Of course, this is only going to work for McDowell, as an alternative to Kripke, if being initiated and trained in a custom in one's youth and upbringing, "second nature" for short, results in some real world fact about the individual that is not simply a social-behavioral practice, or reducible to it, and that is intrinsically related to other elements of the world.[95]

Colin McGinn also disagrees with Kripke's interpretation of Wittgenstein. McGinn points out that in support of his interpretation Kripke quotes Wittgenstein as writing that "if everything can be made out to accord with the rule, then it can also be made out to conflict with it. And so there would be neither accord nor conflict here. (201)" For Kripke this passage serves as evidence for his interpretation of Wittgenstein, namely, that Wittgenstein believes that there are no facts about the individual language user that constitute the meaning of a term used by that person. But McGinn emphasizes that Kripke "fails to quote, or even to heed, what immediately follows." McGinn fills that gap, quoting Wittgenstein:

> It can be seen that there is a misunderstanding here from the mere fact that in the course of our argument we give one interpretation after another; as if each one contented us at least for a moment, until we thought of yet another standing behind it. What this shews is that there is a way of grasping a rule which is *not an interpretation*, but which is exhibited in what we call "obeying the rule" and "going against it" in actual cases.
>
> Hence there is an inclination to say: every action according to the rule is an interpretation. But we ought to restrict the term "interpretation" to the substitution of one expression of the rule for another. (201)

Then McGinn comments:

> Wittgenstein does *not* say that the paradox arises from the misunderstanding that ascriptions of rules state facts or have truth conditions, nor does he suggest that the underlying mistake is to consider the rule-follower in social isolation; what he is objecting to is the specific conception of understanding as a mental operation of translation.[96]

McGinn stresses that Wittgenstein asserts that the earlier part of the passage, the part quoted by Kripke, expresses a "misunderstanding," that resides in "giv[ing] one interpretation after another. . . . But we ought to restrict the term 'interpretation' to the substitution of one expression of the rule for another," and not apply it presumably to actions that may or may not accord with a particular interpretation.

According to McGinn, Wittgenstein is not posing a skeptical problem and solution. Instead, he is recommending that we entirely abandon the view that has objects appear before the mind that serve as rational foundations for the use of language. All such appearances, be they images, pictures, or formulae, are supposed to be rules for the rational application of language that must somehow be grasped prior to use in order for linguistic practice to be rationally grounded. But all such rules require an interpretation; and all rules that would appear before the mind to rationally instruct us how to interpret the one in question are simply other interpretations. All that takes place is the substitution of interpretations. Wittgenstein is not proposing a solution to this problem. McGinn, like McDowell, believes that Wittgenstein is giving us reasons to reject the picture that leads to the problem in the first place. There must be a way of grasping a rule that is not an interpretation.

For McGinn, Wittgenstein's positive view is one in which we grasp a rule not by interpreting it, but by the development of natural capacities for language use, natural capacities which are not themselves divorced from the natural capacities we exercise in other "forms of life." Wittgenstein "does offer an account of the sort of thing understanding is: it is mastery of a technique, possession of a capacity, participation in a custom."[97] The technique or ability that constitutes understanding is the "ability to use signs." Thus, Wittgenstein's emphasis on actions not being interpretations—presumably these actions proceed for good or for ill from the natural capacities that have been developed in one's upbringing. McGinn believes that Wittgenstein is rejecting the modern rationalistic foundationalism that would require in our use of language a constant looking to some rule by reference to which our use counts as rational. Wittgenstein is substituting for it a form of anti-foundationalist naturalism. In the exercise of one's developed capacity for bravery, one does not consult a rule first to determine what it is rational to do. One acts in accord with the virtue. Thus, "what this shews" that is, "this other way of grasping of a rule," consists in the exercise of one's natural human ability to use signs.

Conclusion

Before proceeding on to consider Putnam, it would be well to indicate a few themes that have begun to appear in the foundations of *representationalism*, and the criticism of it. First, the dichotomy looms large between what is *internal* and what is *external* to the mind. Second, the fundamental structure initially enunciated in Russell's criticism between mind, idea or concept and *external object or thing* plays a dominant role. Third, the relation, or lack thereof, between *mental representation* and *external object* plays a large part in the discussion. In particular, Berkeley raises the question of a necessary connection between representation and thing represented, and Wittgenstein exploits the problems associated with this question. Finally, the mental representation or idea is generally taken to be *what* the mind directly knows by a privileged capacity of introspection. This latter theme, as Russell indicated, gives rise to skepticism concerning our knowledge of the *external* world, and by extension the ability of language subordinated to thought to be about it.

Chapter 4

THE LANGUAGE OF THOUGHT
A Revival of Mental Representationalism

> *Homo sapiens* is, no doubt, uniquely the talking animal. But it is also, I suspect, uniquely the species that is born knowing its own mind.
>
> —Jerry A. Fodor

The Revival of Mental Representationalism

In this chapter we sample mental representationalism as it is represented in the work of Jerry Fodor. Fodor, by his own account, is trying to revive the tradition of British Empiricism concerning mental representations and the mind. To the extent that his work is representative of a broad segment of recent research within the Philosophy of Mind and the Philosophy of Language, mental representation and its relation to language survive the Wittgensteinian critique.

Fodor, certainly one of the leading expositors and advocates of contemporary *representationalism*, wishes to place his own discussion firmly within the tradition criticized by Husserl, Frege, and Wittgenstein. Introducing representationalist theories of mind, he writes:

> I shall be emphasizing repeatedly ... respects in which cognitive science is a recidivist account of the mind. In my view, much of cognitive science is the rediscovery of doctrines that were familiar—in fact, commonplace—in the tradition of classical epistemology that ran, say, from Descartes to Kant with intermediate stops at Locke, Berkeley and Hume.
>
> [C]ognitive science, with its emphasis upon the mental representation construct, is plausibly viewed as continuing a tradition of theorizing about the mind that has its roots in Classical epistemology. . . . I know of no better way to see how representational theories of the mind work than by considering the relatively simple versions that were developed by these early theorists. One can then understand the current theory as an elaboration of its predecessors, including various wrinkles adopted to avoid difficulties they were beset with.[1]

In other words, Fodor believes that, with respect to mental representations, the Empiricists got it fundamentally right; he wishes merely to fine tune their account. One of the wrinkles he wishes to avoid is identifying the referential or intentional character of internal representations with the character of being an "image" or "picture" or more broadly "resemblance" of *external* objects, taking it as a settled issue that these modes of representation render classical representationalism untenable.

Suggesting that there are really only two proposals that seem viable, *mental pictures* or *causal connection,* he negates the first disjunct and so must affirm the second. The structure of mental representation will have the syntax of an inner language with a causal connection to what it represents. With some reservations he suggests that "some version of the causal story must be true."[2] He says that the "naturalization problem for a propositional-attitude psychology" will be solved, "if we are able to say, in a nonintentional and nonsemantic idiom, what it is for a primitive symbol of Mentalese to have a certain interpretation in a certain context."[3] It is his belief that the causal relations to the world that primitive symbols of the Language of Thought enter into will provide him with the requisite nonintentional and nonsemantic idiom.

It is interesting, however, that writing after Wittgenstein Fodor's rejection of *pictorial* representation does not generally follow Wittgenstein's criticism. In several instances he is severely critical of Wittgenstein and those he thinks follow in his footsteps:

> the private language argument—at least as I have been construing it—isn't really any good for, as many philosophers have pointed out, the most that the argument shows is that unless there are public procedures for *telling* whether a term is coherently applied, there will be no way of *knowing* whether it is coherently applied. But it doesn't follow that there wouldn't in fact *be* a difference between applying the term coherently and applying it at random. A fortiori, it doesn't follow that there isn't any *sense* to claiming that there is a difference between applying the term coherently and applying it at random. These consequences would, perhaps, follow on the verificationist principle that an assertion can't be sensible unless there is some way of telling whether it is true, but *surely* there is nothing to be said for that principle.[4]

Fodor's rejection of the private language argument is particularly important, since one of the stages in his account of mental representation is to posit in each person a *language of thought,* which is in some sense private, though not in the traditional sense. But his quarrel with Wittgenstein goes further than simply criticizing one, though admittedly major, argument.

> I have argued elsewhere that confusing mentalism with dualism is the original sin of the Wittgensteinian tradition. Suffice it to remark here that one result of this confusion is the tendency to see the options of dualism and behaviorism as exhaustive in the philosophy of mind.... [But] there would seem to be nothing in the project of explaining behavior by reference to mental processes which requires a commitment to epistemological privacy in the traditional sense of that notion.[5]

Given his aversion to the Wittgensteinian tradition, it is easy to understand why Fodor sees only the need to modify and correct the Empiricist tradition, and why his correction shares many of the same presuppositions.

As he sees it, Empiricism failed for a number of reasons, but some of the problems do not bear directly upon the aspects of the tradition that prompt his modification. For example, as he understands it, the tradition of Empiricism went wrong by holding (a) that all empirical statements have a unique decomposition into sense datum statements, (b) that sense datum statements are somehow incorrigible, and (c) that each sense datum statement is logically independent of the rest.[6] Presumably he thinks the Empiricist tradition can, without loss, do without these assumptions, since they do not bear directly upon the modification he would make. The language of "sense datum," however, has more of the flavor of the Logical Positivist strain of the tradition in the twentieth century than of the classical British Empiricists.

More pertinent to his own modification of the tradition are ways in which it failed that are more clearly related to the Empiricists. Chief among these problems were the inability to account for the asymmetry of representation and the inability of "association" to provide an adequate account of the causal character of thought processes. The classical account held that representation is established by a relation of resemblance or similarity between the internal mental representation and the external thing it purports to represent. In the previous chapter we saw that because of the distinction between primary and secondary qualities, this sort of characterization is overly broad. Still, Fodor's criticism is that similarity is a symmetric relation, while representation is an asymmetric relation. If A is similar to B, then B is similar to A. But if A represents B, it does not follow that B represents A. For example, 'tree' as a piece of writing in a sentence may represent a tree outside my window, but it does not follow that the tree outside my window represents the written word. Thus, since one is a symmetric relation, while the other is asymmetric, the relations of similarity and representation cannot be the same.

Further, the psychology of association characteristic of Empiricism could not adequately account for the causal character of thought processes, in particular to the extent that they exhibit the normative character of argumentative thought processes. As Fodor characterizes them, the laws of association primarily involve spatio-temporal contiguity. But how could such laws account for the causal character of thought processes in which one step causes the next, and which when taken together constitute a rational argument? His favorite example is the sort of thought process that Sherlock Holmes recounts for Watson on how his grasp of certain salient facts led his mind on to make certain intermediary conclusions, which eventually led to solving the case.[7]

> This parallelism between causal powers and contents engenders what is, surely, one of the most striking facts about the cognitive mind as commonsense belief/desire psychology conceives it: the frequent similarity between trains of thought and *arguments*.[8]

Holmes is at one and the same time recounting the stages through which his thought progressed, one thought causing the next, and providing a rational argument from certain facts to conclusions reasonably inferred from those facts. As Fodor puts it, "the Empiricists had no good way of connecting the *contents* of a thought with the effects of entertaining it,"[9] and "what was wrong with Associationism . . . was that there proved to be no way to get a *rational* mental life to emerge from the sorts of causal relations among thoughts that the 'laws of association' recognized."[10]

The causal processes implicated in mental processes play a large role in the argument Fodor gives for positing mental representations. Since Fodor will end up advocating a causal theory of mental representation, it is important to distinguish *that* causal character from the causal character involved in mental processes. The causal character of mental processes refers to the ways in which my beliefs and desires cause other beliefs and desires, as well as actions, as the Sherlock Holmes example is supposed to elicit. So believing that whole milk increases my fat count, and believing that fat increases my chances of an early death, and desiring that I live long and prosper, and yet also desiring milk, I believe that skim milk is, all things considered, better for me. So, if my beliefs and desires are well ordered, I desire skim milk, and in appropriate circumstances act upon those beliefs and desires by drinking it. Some of my beliefs and desires cause others of my beliefs and desires, as well as my actions. This is what Fodor means by the causal character of thought or mental *processes*. The causal relations between beliefs and desires are all between mental states internal to the mind, that is, "intentional attitudes." The action that proceeds from my beliefs and desires may not be a mental state, but the sort of explanation that is given of it in commonsense belief/desire accounts involves the causal relations between these mental states. As we will see, the causal character of mental representation is a different question for Fodor. That has to do with how mental representations relate to the world, not how mental states causally interact with one another.

Why Mental Representations?

But why is there a need for mental representations at all? It is a constant theme of Fodor's work that any reasonably adequate cognitive science will have to account for the sorts of "causal" relations between the intentional attitudes involved in the commonsense account of why we act as we do.

> [M]ental symbols constitute domains over which *mental processes* are defined. If you think of a mental process—extensionally, as it were—

as a sequence of mental states each specified with reference to its intentional content, then mental representations provide a mechanism for the construction of these sequences; they allow you to get, in a mechanical way, from one such state to the next *by performing operations on the representations.*[11]

Take particular note of the point of the last phrase. Mental representations are subject to mental operations which manipulate them "in a mechanical way." This suggests that mental activities act directly upon these mental representations. Fodor notes that despite errors and self-deception, commonsense explanations in terms of beliefs and desires are vastly better confirmed than any rival accounts of human behavior that do not "quantify" over both intentional attitude states and the mental representations that individuate them. Behaviorism and its dispositions to act in certain observable ways come under particularly severe criticism. However, he does not want to rest the claim that there are mental representations on the "workability" of simple common sense, a foundation that might suggest a supposed conflict of our commonsense view with a scientific view.

He provides two basic justifications for a Representational Theory of Mind (RTM), both of which rely heavily on questions of scientific merit. First, for him, RTM involves a full-blown competition between scientific theories, and it rests upon the authority of science to tell us what really exists—"some version or other of RTM underlies practically all current psychological research on mentation, and our best science is ipso facto our best estimate of what there is and what it's made of."[12] Theoretical belief/desire psychology must be understood as a scientific refinement of commonsense belief/desire explanations, which shares in the benefits of the latter's well-confirmed status. In the appendix to *Psychosemantics* he argues that commitment to belief/desire psychology makes it "practically mandatory" to believe in a Language of Thought, and its associated mental representations. This line of argument, of course, opens him to the "eliminativist" attack, namely, that we are in the stages of developing a better cognitive science that does not quantify over intentional attitudes, but rather neural networks. When it is developed and tested it will be shown to be a better theory for cognitive science. But since it will not quantify over intentional states, we will have no reason for thinking beliefs and desires exist, and no need to investigate their nature. Nonetheless, Fodor thinks any theory that does not quantify over beliefs and desires will lack the degree of confirmation had by a psychology that does so quantify. His second justification for RTM is that current computer models of the mind involving the computational or syntactic manipulation of semantic representations provide a way of understanding how the causal processes involved in chains of thought are not mere "spatio-temporal associations," but rather have the normative character associated with chains of arguments. The causal relations of computer states serving as models of the mind are built precisely to mimic syntactic relations between symbols in a formal language.[13] The challenge of hardware design is to construct circuitry in which this kind of syntactic structure can be causally realized. In well-constructed formal systems, the semantic relations between symbols, the inferential relations between propositions in particular, can be "mimicked" by their syntactic relations:

> There must be mental symbols because, in a nutshell, only symbols have syntax, and our best available theory of mental processes—indeed, the *only* available theory of mental processes that isn't *known* to be false—needs the picture of the mind as a syntax-driven machine.[14]

The thought is that the causal character of mental processes reflects the normative character of arguments, because like a computer the hardware involved in mental processes, that is, the neural structure of the brain,[15] is "programmed" such that its causal states reflect syntactic structures. Presumably, mental processes would fall short of the normativity of arguments when there is something wrong with the hardware, that is, the brain circuitry. It is this sort of causal-syntactic structure that explains why Sherlock Holmes reasons the way he does, and why I desire skim milk.

According to Fodor, the best scientific-theoretical accounts of belief/desire psychology require an internal structure of mental representations, a "language of thought" upon which intentional attitudes can be directed, and which in fact are involved in the individuation of intentional attitudes.[16] Fodor contrasts his account of the individuation of intentional attitudes with what he calls the "standard story." On the standard story a belief is just a two-place relation between a person and a proposition. If one person has two distinct beliefs, presumably you can infer that he is related to two propositions. So for example, Oedipus believes that *he is marrying Jocasta,* but he does not believe that *he is marrying his mother.* Therefore, since the beliefs are distinct, the propositions involved must be as well. Similarly for the states of affairs involved; the following are separate states of affairs, namely 'Oedipus's marrying his mother' and 'Oedipus's marrying Jocasta'. Non-identity of intentional attitudes implies non-identity of intentional objects, for example, propositions.

Fodor disagrees with this "standard story." For him, an intentional attitude, in particular a belief, is a relation between a person, a proposition, a vehicle, and a functional role. What is of interest here is the "vehicle," which is the internal mental representation in the Language of Thought, or "Mentalese." It "mediates" the relation of the person to the proposition believed. The proposition determines the content of the belief or intentional attitude, but the vehicle mediates between the person and the proposition.

> What's essential to my story is that believing is never an *unmediated* relation between a person and a proposition. In particular, nobody "grasps" a proposition except insofar as he is appropriately related to a token of some vehicle that expresses the proposition.[17]

What is crucial for Fodor is that the very same proposition can be mediated by different vehicles. Thus concerning Oedipus, there is a single proposition that can be expressed by either vehicle, namely that *he marries Jocasta* or that *he marries his mother.* But Oedipus's beliefs differ because the very same proposition is mediated or *presented* (my term) to him by distinct vehicles or mental representations. *Mutatis mutandis,*[18] for Fodor, 'Oedipus's marrying Jocasta' and 'Oedipus's marrying his mother' are the same state of affairs in the world. But Oedipus's desire *to marry Jocasta* differs from his desire *not to marry his mother,* because of the way in which that same state of affairs is represented to him by different vehicles in the Language of Thought—whether it is represented to him as *a marrying of his mother* or represented to him as *a marrying of Jocasta.* So he can both dread and welcome the same external state of affairs, because it may be presented to him by different internal vehicles.

Thus, the vehicles or mental representations mediate the relation of intentional attitudes to the external states of affairs that are the objects of my desires or that render my beliefs true or false. Fodor recognizes that this solution is tendentious against the background of the complex argument in contemporary philosophy concerning the individuation of intentional states.[19] But all that is of interest here is the point that the internal vehicle or mental representation "mediates" between the person and the proposition or state of affairs.

Why a Language of Thought?

The structure of mental representations is a "language of thought" because it is straightforwardly syntactical, that is, the way in which the basic elements of a mental representation are arranged bear compositionally upon the content of the mental representation and the truth conditions of thoughts involving it.[20] Fodor thinks there are syntactical relations reflected in the causal processes *between* thoughts, but here we are interested in the structure of a single thought, say, *Oedipus loves Jocasta*, abstracting from its relations to others. As Fodor argues, the Language of Thought must be syntactical in order to account for its compositionality or combinatorial aspect. If mental representations did not have some sort of inner structure, we could not explain why someone who can understand the thought that *Oedipus loves Jocasta* can also understand the thought that *Jocasta loves Oedipus*.[21] Here we do have two distinct propositions, not one. We expect that any person who is capable of understanding one is capable of understanding the other, unless inhibited by some pathology. This is not the case with *Oedipus loves Jocasta* and *Oedipus loves carrots*, for example.

What is the difference? The difference is easily explained on the thesis that thoughts are intrinsically structured in a syntactic fashion, structured from more elemental representations. In the first pair we have the same elements *Oedipus*, *Jocasta*, and *loves* in both. In the second pair, there are these elements on the one hand, but on the other there is the additional element *carrots*. We expect that someone who understands one of the first pair can understand the other, since they differ only by a syntactical arrangement of the same elements. Further the difference of syntactical arrangement bears upon the semantic evaluation of the two, namely, what conditions in the world have to obtain in order for each to be true. The state of affairs that renders true the belief that *Oedipus loves Jocasta* is clearly different from the state of affairs that renders true the belief that *Jocasta loves Oedipus*. One state of affairs could certainly obtain without the other. In the second pair, namely, *Oedipus loves Jocasta* and *Oedipus loves carrots*, we do not expect that someone who understands one of the pair will understand the other, because they have at least one different elemental representation between them, and that difference seems to bear more upon the different truth conditions between the two than their syntax.

But suppose thoughts are not compositionally structured in this way. Suppose the thought that *Oedipus loves Jocasta* is completely unstructured and cannot be decomposed into simpler representative elements. Then, Fodor concludes, it is possible that someone could have the capacity to understand the thought that *Oedipus loves Jocasta*, but not be capable of understanding the thought that *Jocasta loves Oedipus*. If thoughts are unstructured, then there is no reason to think the two have anything in common at all. Given an understanding of the thought that *Oedipus loves Jocasta*, we have no better chance of understanding that *Jocasta loves Oedipus* than of understanding that *Oedipus loves carrots*, or that *William loves carrots*, or for that matter that *William hates carrots*. For Fodor, this is clearly unacceptable. Thus, because of the pervasive presence of syntactic considerations in the analysis of mental states, both between mental states and intrinsic to them, Fodor concludes that the internal system of mental representations must be a *Language of Thought*.

Fodor argues that this Language of Thought must be innate in human beings and that it lies behind the differences that are observed in natural languages, that is, languages that are actually spoken or written by various social communities. His argument is based upon what he believes are the conditions required for learning one's first natural language.

> [L]earning a language ... involves learning what the predicates of the language mean. Learning what the predicates of a language mean involves learning a determination of the extension of these predicates. Learning a determination of the extension of the predicates involves learning that they fall under certain rules (i.e., truth rules). But one cannot learn that P falls under R unless one has a language in which P and R can be represented. So one cannot learn a language unless one has a language.[22]

By "natural language" Fodor just means a language that one would find if one went out and observed social communities, for example, French, Spanish, Swahili, and so on.

To a large extent he places his own account against the background of Noam Chomsky's criticism of behaviorism and advocacy of universal grammar, though he does not think his account depends upon whether or not Chomsky is correct.[23] He does want to exploit the assumption that Chomsky and others make that

> there is an analogy between learning a second language on the basis of a first and learning a first language on the basis of an innate endowment. In either case, some previously available representation system must be exploited to formulate the generalizations that structure the system that is being learned. Out of nothing comes nothing.[24]

The notions of *translation, encoding,* and *decoding* play a large part in understanding why Fodor believes we are committed to an innate Language of Thought. For him, when we translate from one language to another, we provide rules that associate the true sentences of the one language with sentences of the other. For example, translating from French, we say, "'l'arbe est rouge' is true if and only if the tree is red." If this is a true biconditional, then according to Fodor, when we learn it, we learn the meaning of 'l'arbe est rouge' in French.

Notice that in the biconditional 'l'arbe est rouge' is mentioned, not used. It can only be used in its native language. But we do not need to know beforehand what it means in order to mention it in our language; we only need to be able to represent it, which we do by quotation. The quotation serves to represent the string, without requiring that its meaning be known.[25] On the other hand, 'the tree is red' is used, not mentioned. We must already know what it means in our native language. More generally, Fodor believes that when we learn a foreign language by learning how to translate it, among other things we learn such truth rules, that is, generally truth rules of the form "'$P(x)$' is true iff x is G is true for all substitution instances."[26] The predicate $P(x)$ is a predicate of the language to be translated, while the predicate $G(x)$ is a predicate of the native language. G in the native language has the same extension as P in the foreign language. Thus, two languages are necessary in learning a new language, the one to be learned and the unlearned one into which the former is translated.

What is significant then for the Language of Thought is extending this account to learning one's first natural language. By hypothesis, a child does not know any other natural language. But if learning one's first natural language is like learning a second natural language, then there must be some "innate" language, the Language of Thought, into which the first language is translated. However it is easy to see an infinite regress in the works here, where one has to learn the innate language, which requires a deeper innate language, and so on. Fodor anticipates this objection and cuts it off at the first stage. What he holds is that in order to learn a language one must already know a language. He does not hold that "you can't learn a language unless you've already *learned* one."[27] The Language of Thought is known and not learned! Thus, the semantic properties of natural languages are easily seen to be reduced to the semantic properties of the Language of Thought, since learning the first natural language is just learning the truth rules that relate the semantic elements of the new language to the already known, but not learned, semantic elements of the Language of Thought. It is important to note that if the fundamental mental representations within the Language of Thought are all or mostly innate, that is, neither derived from nor produced by experience, in this respect at least, Fodor's account appears less a form of Empiricism than a form of Rationalism.

Here again Fodor appeals to the computer model—"what avoids an infinite regression of compilers is the fact that the machine is *built* to use the machine language."[28]

> [O]n this view, what happens when a person understands a sentence must be a translation process basically analogous to what happens when a machine 'understands' (viz., compiles) a sentence in its programming language.... [W]hen you find a device using a language it was not built to use (e.g., a language it has *learned*), assume that the way it does it is by translating the formulae of that language into formulae which correspond directly to its computationally relevant physical states. This would apply, in particular, to the formulae of the natural languages that speaker/hearers learn, and the correlative assumption would be that the truth rules for predicates in the natural language function as part of the translation procedure.[29]

Note the explicit reference to "formulae which correspond directly to its [presumably the Language of Thought] computationally relevant physical states." At root the Language of Thought is a complex physical system. One must not lose sight of the fact that a 'cow' token in the Language of Thought is "really something else;" "in itself" it is just a neural state of the brain related in the appropriate way to cows.

What Is the Representative Character of Mental Representations?

This is all well and good, but in what does the representative character of a mental representation consist? Though he believes that the intentional character of mental representations, that is, their being *about* other things, is real, Fodor does not believe that this aboutness is a basic property of the world.

> If the semantic and the intentional are real properties of things, it must be in virtue of their identity with (or maybe of the supervenience on?) properties that are themselves *neither* intentional *nor* semantic. If aboutness is real, it must be really something else.[30]

Fodor seeks to "naturalize" semantics by finding this "something else." As he represents it, the aim of contemporary cognitive science is to provide, among other things, an account of the intentionality of mental symbols, or representations, in terms of their typical causes. "It was never in the cards that intentionality would prove to be a *fundamental* feature of the world (in the way that spin, or charge, or charm may turn out to be fundamental features of the world)." However, the further goal is then to provide an account of, and thus a reduction of the semantic properties of language to the intentional properties of mental symbols, which have been reduced to their causal and computational properties.

However, Fodor does not think that finding a naturalistic reduction of one thing to another shows that the reduced thing is not real or is "epiphenomenal" or is "eliminated." He does believe that intentional properties are real, they just aren't "fundamental." Fundamental properties are the properties of a completed physics. He gives the example of geological and meteorological properties. It may be the case that the causal properties of tornadoes are ideally reducible to properties of physics, but that is not to deny that tornadoes are real, and that tornadoes cause a lot of damage.[31] In that sense, even cognitive neurological states are real, but really something else, since they can presumably be reduced to the fundamental properties of physics. But they are more fundamental than intentional states since they are in some sense closer to the fundamental properties of physics. The intentional states have to be reduced to them first, before they can ultimately be reduced to fundamental physics.

Now consider a term in the Language of Thought like 'cow'—it *means* or *represents* cows in the world. How does it do this? In the first place, it does it because there is a "nomological" relation between properties—a covering law relating the following two properties, *being a cow* and *being a cause of 'cow' tokens or mental representations in the Language of Thought*. By a nomological relation, Fodor simply means a relation between properties that tells us *subjunctively* what instances of the properties "would do (or would have done) *if* the circumstances were (or had been) thus and so."[32] Strictly speaking, a causal relation is between events,[33] and events instantiate properties in virtue of which they enter into causal relations; the causal relations are "covered" by the "nomic" relations that obtain between the properties instantiated by the events.

At the risk of oversimplifying the complicated issues involved in accounts of causation, it is important to recognize what the nomic relation between properties buys for Fodor—a causal relation much more robust than mere *association* or *generalization*, and much more robust than the ambiguous relation of *similarity* characteristic of the Empiricist tradition. For example, suppose it is true that *Roosevelt's going to bed* always accompanies *Hirohito's riding his horse*. Is this true generalization a causal relation? Is the event of *Roosevelt's going to bed* also the cause of *Hirohito's riding*? Perhaps, but probably not. One does not generally expect the following counterfactual to be true, namely, *had Roosevelt gone to bed, Hirohito would have ridden.* All other things being equal, tomorrow Roosevelt might go to bed, and Hirohito might not ride. But all other things being equal, *if the temperature rises, so will the mercury in the thermometer.* All other things being equal, *had the pot been placed on the fire, the water would have boiled,* and *had the temperature risen, the water would have boiled.* It is a fact about this world that *the burning of a fire* is, in the appropriate circumstances, a cause of *the boiling of water.* Had the world been different enough, perhaps *the burning of a fire* would not be a cause of *the boiling of water.* But that is not the world in which we live. I may be wrong, but this robustness is what I believe Fodor intends to get, when he writes that "*ontologically* speaking, I'm inclined to believe that it's bedrock that the world contains properties and their nomic relations."[34] Some generalizations fall under nomic relations and some do not. Those that do are "causal" relations, while those that do not are mere generalizations of events that are otherwise unrelated.

So, for example, the many cows that exist in the world instantiate the property of *being a cow.* But, for Fodor, it is a brute fact of the world that *whatever is a cow is a cause of 'cow' tokens or mental representations in the Language of Thought.* It is simply a fact about this world that the properties *being a cow* and *being a cause of 'cow' tokens in the Language of Thought* are nomically related. This nomic relation covers the generalization that *cows cause 'cow' tokens in the Language of Thought,* "or instances of the property [being a cow] *would* cause tokenings of the symbol *were they to occur,* or both."[35] It supports counterfactual statements like *if there had been cows in the pasture he would have thought of them,* and so on. If the world were significantly different, perhaps this nomological relation would not hold. But Fodor is interested in what nomological relations hold in this world. So mental representations in the Language of Thought mean *cows* because of the nomic relation between the property of *being a cow* and the property of *being a cause of those mental representations.*

Recall that the specific criticism that Fodor directed against the classical Empiricists' account of representation was that resemblance is a symmetrical relation, while representation is asymmetrical. Because his own account rests on a causal account, and not resemblance, he avoids the weight of his own criticism. Causal relations are eminently asymmetrical. All Fodor needs to get the asymmetry of representation is the legitimacy of causal relations subsumed under nomological relations between properties.

There are complications, however. Suppose *being the cause of 'cow' tokens in the Language of Thought* is instantiated by things other than cows. For example, it might be a causal generalization that in certain circumstances, *large black cats cause 'cow' tokens in the Language of Thought*. The appropriate circumstances might be during moonless nights and at an appropriate distance. Why don't 'cow' tokens in the Language of Thought mean *cats*, or why don't they mean the disjunction *cows or cats*? Fodor introduces another asymmetry to deal with this problem.

> "[C]ow" means *cow* and not *cat* or *cow or cat* because *there being cat-caused "cow" tokens depends on there being cow-caused "cow" tokens, but not the other way around.* "Cow" means *cow* because . . . noncow-caused "cow" tokens are *asymmetrically dependent upon* cow-caused "cow" tokens. "Cow" means *cow* because *but that "cow" tokens carry information about cows, they wouldn't carry information about anything.*[36]

The very possibility of a 'cow' token in the Language of Thought falsely representing a cat from a distance on a moonless night depends upon 'cow' tokens truly representing cows. In somewhat strong language, Fodor writes that "false tokens are metaphysically dependent upon true ones."[37]

Introducing this asymmetry in the causal relations is, however, not simply an *ad hoc* means for Fodor to avoid these difficulties. It also helps him to distinguish causal relations that simply carry information from causal relations that carry information and support meaning.[38] Consequently, it also supports the difference between, on the one hand, mere causal relations in the "external world," and, on the other hand, the causal relations constitutive of meaning between the "internal" mental representations and the "external" objects meant. Fodor calls "pansemanticism" the thesis that informational content is just meaning content. A pansemanticist would hold that "meaning is just everywhere" that you can find a "reliable causal covariance." Fodor wants to deny this. For example, the length of a column of mercury in a thermometer indicates or carries information about ambient temperature, but the length doesn't *mean* the ambient temperature. Boiling water in appropriate circumstances indicates or carries information about a fire, but it doesn't *mean* fire. One might be tempted to object that "Well, doesn't smoke mean fire?" Fodor wants to say no. In this latter context we are using "mean" equivocally; we are really just using it as synonymous with "indicates." He presents an argument to demonstrate the equivocation, based upon the transitivity of meaning:

1) The mental representation 'smoke' means *smoke*,
2) *smoke* means *fire*,
3) therefore the mental representation 'smoke' means *fire*.

But obviously, the mental representation 'smoke' does not mean fire. 'Fire' means fire. If one does not buy his account of mental representation, one can still buy the argument by substituting "word" throughout for "mental representation."

In the Language of Thought representing cats as cows, on a moonless night at a distance, may well carry information about those cats. It may carry information because of a nomological relation between *being a cat* and *being the cause of 'cow' tokens in the Language of Thought*. But it does not support the thesis that *'cow' tokens mean cats*, because no other relation depends asymmetrically upon the latter nomological relation. 'Gold' tokens in the Language of Thought may not mean *iron pyrites*, but it does not follow that they do not carry information about *iron pyrites*. Not just anything enters into a nomological relation with the mental representation of gold. But the possibility of learning something about what does is metaphysically dependent upon a paradigm, namely gold itself.[39] It is just a basic "metaphysical" fact about the world that the causal relation between '*gold*' tokens in the Language of Thought and gold in the world stands out from all the other causal relations that '*gold*' tokens could enter into. If gold did not cause '*gold*' tokens in the Language of Thought, nothing else would. It will be important to remember the counterfactual character of this last statement when we consider Putnam's criticism of Fodor.

Narrow and Broad Content

Now difficulty arises for Fodor's account from two theses that he is committed to. First, there is the thesis that mental representations are ultimately reducible to physiological states of the human organism, in particular neurological states of the brain.

> [W]e have assumed a typology according to which the physiological identity of organisms guarantees the identity of their mental states (and, a fortiori, the identity of the contents of their mental states). All this is entailed by the principle—now taken to be operative—that the mental supervenes upon the physiological (together with the assumption—which I suppose to be untendentious—that mental

states have their contents essentially, so that typological identity of the former guarantees typological identity of the latter).[40]

Second, these physical states of the organism are causally related to things in the world, that is, they fall under nomological relations between properties, and some of those relations are metaphysically robust, while others are parasitic upon the former. Thus some neurological states are 'cow' tokens in the innate Language of Thought. But the causal generalizations and corresponding nomological covering laws are supposed to be "empirical laws." That is, they are not absolutely necessary. In other *possible worlds* "sufficiently" different from our own, they are false, that is, they are not true in all possible worlds, but are true in our world and worlds "sufficiently" like our world to support the true counterfactuals related to the nomological laws. I put scare quotes around "sufficiently" because depending upon the nomological law, the world in which it is false might be very much like ours, except for the falsity of this one law.

Now the Language of Thought is supposed to be innate, and in some sense reducible to physiology. It is universal and underlies translation between different natural languages because it already has an innate representational content. So at least in different parts of the globe, in our world in which the nomological generalizations are true everywhere on the globe, I am guaranteed that others have the same mental content or innate representations as I do, because of the identity of their physiology and the uniformity of the nomological laws governing our world. But what happens if the same physiology is put in a situation in which one or more of the nomological covering laws bearing on meaning is false?

Suppose there is a possible world in which (a) we can survive, that is, all the nomological biological laws necessary for our survival are true, but (b) the nomological covering law is false that obtains in our actual world between cows and the mental representations that "supervene" upon our physiology, all other things being equal. Suppose instead that there is the appropriate asymmetric nomological covering law between horses and those neurological states. Presumably this is possible because the nomological laws are not absolutely necessary, but merely nomologically necessary. Because of (a) I am supposing that physiologically we are identical in that possible world and in the actual world. Because we are physiologically identical, the mental content of the representations in the innate Language of Thought remains the same. That is the point of the quote above, that "physiological identity of organisms guarantees the identity of their mental states (and, a fortiori, the identity of the contents of their mental states)." Thus the mental states and their content, to the extent that they are really identical or supervene upon physiological states, have the sort of "in itself" character that I described in the third chapter. But because of (b), those mental representations in the innate language of thought are not causally related to cows, in the appropriate fashion necessary for them to *mean* cows. So those mental representations do not "mean" cows, according to the account of meaning given earlier. They "mean" horses. But since in the two worlds, the actual world and the possible but non-actual world, the mental content of the representation is the same, and its *meaning* is different, one can conclude that on Fodor's analysis meaning and mental content are different.

This is a bit of an embarrassment for Fodor's account, since it seems to suggest that one could *mean* two different kinds of things, horses versus cows, and yet the content of one's thought be the same.[41] It suggests that in another possible world one could be thinking the very same thing as one is in the actual world, that is, be in the same mental state, and yet be thinking *about* different things, for example, horses instead of cows.

One could eliminate the embarrassment by denying the supposition of the argument, that is, by denying that there is a possible world in which the biological laws necessary for our life obtain and in which the nomological relation between cows and the mental representation in the Language of Thought does not obtain. But that is a strong price to pay, since it would make it metaphysically necessary that *if human beings exist, then cows asymmetrically cause that mental representation*. Using that mental representation to think about cows would be as much an essential characteristic of human nature as breathing, supposing of course that breathing is an essential characteristic. But then is it an essential characteristic of human nature that cows exist, since on the proposed solution they must exist to asymmetrically cause the innate mental representation of cows?

Fodor is not interested in escaping the problem by this "metaphysical" solution. If the particular nomological relations of *mental representations* to *things* are as essential to the existence of human nature as are biological nomological laws, the idea that the former supervenes upon the latter, and that the former can be reduced to the latter, looks to be an arbitrary choice of biology over semantics. Surveying the essential characteristics of human nature, how does one decide which are more fundamental, the biological or semantic? The motive for a reduction would seem to fall away. On the other hand, if the biological laws are of broader extent, that is, are true in possible worlds where the semantic laws differ, the choice of the former as more fundamental, and the latter as supervenient seems more plausible. The semantic are still lawful, but they supervene upon the biological *in a context*. Semantic laws follow upon biological laws in a context. "If they are real they are really something else," *in a context*.

Instead Fodor distinguishes between *narrow* and *broad* content. *Narrow* content is associated with the mental representation of the Language of Thought considered apart from any particular possible situation or context in which a human organism might live. *Broad* content is narrow content relativized to a context in which a human person might live. It is easy to see that *broad* content is determined by the nomological laws that obtain in a situation, while narrow content prescinds from a consideration of the distinctive laws that obtain in any particular situation, though not from the laws that obtain across all the situations in which a person might live. "[Broad content is] what you get when you specify a narrow content *and fix a context*."[42] The context "anchors" the narrow content to the nomological laws that obtain in the world. The two possible worlds I considered above have identical narrow content, but different broad content.

Fodor makes the point that there is in fact never any "unanchored context":

> One wants, above all, to avoid a sort of fallacy of subtraction: 'Start with anchored content; take the anchoring conditions away, and you end up with a *new sort of content,* an unanchored content; a *narrow* content, as we say' (Compare: 'Start with a bachelor; take the unmarriedness away, and you end up with a *new sort of bachelor,* a married bachelor; a *narrow* bachelor, as we say.'[43]

Any actual *use* of a representation is always in a context. In a telling phrase, Fodor writes that "narrow content is radically inexpressible, because it's only content *potentially;* it's what gets to *be* content when—and only when—it gets to be anchored."[44] Compare this mention of the potential character of narrow content with McDowell's characterization of the Master Thesis,

> what I have in my mind is at most a potential vehicle for the significance in question, in the sort of way in which a sentence, considered as a phonetic or inscriptional item, is a vehicle for significance that it can be interpreted as bearing.

"In itself" the innate mental representation of the Language of Thought "just stands there" as a potential vehicle of thought. The nomological laws that obtain in a context or broadly possible world provide the interpretation, and thus "give life" to the mental representation. So even if the innate representations are neither derived from nor produced by experience of the environment, still the Empiricist emphasis on the importance of the environment actually determining or informing the *tabula rasa* remains.

According to Fodor, we may identify a *narrow* content, but we cannot express it. Any attempt to express it will require *using* the representation, but the representation can only be used in a context, and as such is anchored to that context. So *used*, the mental representation can only express the *broad* content. However, we can *mention* the narrow content, and thus identify it in that sense—"in absolute strictness, [we cannot] express narrow content; but as we've seen, there are ways of sneaking up on it."[45] We "sneak up" on it, by obliquely referring to it, that is, by mentioning it. For example, one might say, "The narrow content of the mental representation 'cows' is what is asymmetrically related to cows in the actual world, and what is asymmetrically related to horses in the possible world we considered."[46] Fodor seems to be saying that *broad content* is the only content that is ever expressed or used in a mental process; nonetheless, considered apart from the asymmetric nomological laws that in fact obtain, or would obtain if the world were different than it is, something *does not differ* among the possibilities, namely the *narrow* content of the innate Language of Thought. There is some continuity amidst the differences, between the use of the innate Language of Thought in the actual world and how it would be used in situations in which the asymmetrical laws of meaning differed.

It is important to recognize this difference in Fodor's account between *narrow* and *broad* content for a number of reasons. It bears directly upon the way in which he avoids Putnam's Twin-Earth puzzles. Further, and more fundamentally, recall that the problem arose for Fodor because of the conviction that if mental states are "real, they are really something else." They are really physiological states and supervene upon them. If they were not really physiological states, they would have to be really something else, and there is no other obvious "naturalistic" candidate for them to be. The distinction between *narrow* and *broad* content allows Fodor to hold that the same physiologically based system of mental representations can exist in quite different semantic contexts. But this also exhibits the lack of a necessary connection between the mental representation and what it actually represents. Vary the context a little, and it will continue to represent the same thing. Vary the context a lot, and it will represent something else. The representation has a certain "in itself" character upon which in an actual causal context the asymmetrical relations of meaning *this* rather than *that* are built.

Recall McDowell's statement of the Master Thesis of mental representation—mental representations in themselves just "stand there," and require an interpretation to be added on to them from outside themselves. This is no less true of Fodor's mental states identified with physiological states. The difference between Fodor and the classical Empiricists is that Fodor does not require a *homunculus* to interpret the representation. All he needs is the causal nexus of the world in which the physiology happens to find itself, in order for the mental states to have an interpretation. In that sense, we could say that the world interprets the innate mental representation. Nonetheless, he still has the fundamental structure of a mental item or entity that has a certain "in itself" character, upon which an interpretation is hung, an interpretation that is not ultimately necessary, and is only slightly less conventional than one is accustomed to thinking, less conventional in the sense of not being up to us.

Conclusion

Despite the substitution of a causal relation for a resemblance relation, there are clear similarities between Fodor's revision and the Empiricist tradition of relating language to mental representations. Fodor's mental representations are mental objects like the Empiricists' ideas. Indeed, psychological theories quantify over them, so in the sense that "to be is to be the value of a bound variable," they are mental entities; they exist *in* the mind. Though he does not say that the mind directly knows the mental representations, for Fodor, as for the Empiricists, the mind's mental activities bear directly upon them. Presumably, by bearing upon them, the mind is able to represent and thus come to know the world outside the mind. Though they have their actual representative function, that is, their *broad content*, because of the causal relations they bear to things in the world, nonetheless, in some sense they are individual things in themselves, namely, *narrow contents;* they are capable of becoming broad contents "anchored" to a context by entering into causal relations. In that sense, like the Empiricists, they "just stand there."

Chapter 5

Hilary Putnam's Criticism of Aristotelian Accounts of Language and Mental Representationalism

> 'meanings' just ain't in the *head!*
> —*Hilary Putnam*

Putnam and the "Tradition"

Hilary Putnam believes that in the relevant respects the tradition he intends to criticize is all of a piece.[1]

> Aristotle was the first to theorize in a systematic way about meaning and reference. In *De interpretatione* he laid out a scheme which has proved remarkably robust. According to this scheme, when we understand a word or any other "sign," we associate that word with a "concept." This concept determines what the word refers to. Two millennia later, one can find the same theory in John Stuart Mill's *Logic*, and in the present century one finds variants of this picture in the writings

of Bertrand Russell, Gottlob Frege, Rudolph Carnap, and many other philosophers.[2]

To the extent that he extends the blanket name 'Aristotelianism' to that tradition, and specifically cites Aristotle's *De interpretatione* as its foundation, his criticism purports to be fundamental and not weakened by individual differences between the philosophers that he considers to be constitutive of that tradition. They hold "the same" basic theory. His work serves as a particularly clear example of the criticism directed at "traditional philosophy" by those philosophers who have taken the *Linguistic Turn*.

By now Putnam's description is familiar, "[T]he picture is that there is something in the mind that picks out the objects in the environment that we talk about. When such a something (call it a "concept") is associated with a sign, it becomes the meaning of the sign."[3] For Putnam the central difficulty raised by the Fregean thesis that sense determines reference, once it is adapted to *representationalism*, is that the mental representation identified with the meaning of a word determines what the word refers to, if anything.[4] He makes clear that he equates "concepts" with "mental representations" when he writes, "instead of the word 'concept' I shall use the currently popular term 'mental representation,' because the idea that concepts are just that—*representations in the mind*—is itself an essential part of the picture."[5] In his now classic essay "The Meaning of 'Meaning'" he makes the point that

> most traditional philosophers thought of concepts as something *mental*. Thus the doctrine that the meaning of a term (the meaning 'in the sense of intension', that is) is a concept carried the implication that meanings are mental entities.

He believes, *mutatis mutandis,* that this is even true of Frege and others who deny that concepts are mental entities.

> Frege and more recently Carnap and his followers, however, rebelled against this 'psychologism', as they termed it. Feeling that meanings are *public* property—that the *same* meaning can be 'grasped' by more than one person and by persons at different times—they identified concepts (and hence 'intensions' or meanings) with abstract entities rather than mental entities. However, 'grasping' these abstract entities was still an individual psychological act. None of these philosophers doubted that understanding a word (knowing its intension) was just a matter of being in a certain psychological state.[6]

Somehow by a knowledge of the mental representation identified with the meaning of a term, the Aristotelian is supposed to come to know what the referent of that term is. It is this latter thesis, whether of the classical pictorial sort or the more recent causal account of Fodor, that Putnam criticizes as "fatally flawed" and untenable.

Putnam believes the Aristotelian picture of mental representationalism embodies a "cryptographer model" of language use. He is very explicit about this, using the cryptographer model to explicate what he takes to be "at issue" in the Aristotelian model of language and mental representation.[7]

> [T]he Cryptographer model—the model of sign understanding as "decoding" into an innate *lingua mentis*—postulates that at a deeper level there is an identity between sign and meaning (this is the fundamental idea of the model, in fact). The idea is that in the *lingua mentis* each sign has one and only one meaning. Two words in human spoken or written languages which have the same meaning are simply two different "codes" for the same item (the same "concept") in the *lingua mentis*.[8]

Thus Aristotle's claim that the passions of the soul are the same for all, despite the conventionality of vocal utterances, is supposed to be embodied in Putnam's implicit claim that the *lingua mentis* is the same for all, despite the surface variations of the vocal utterances into which it is encoded in a spoken language, or in different spoken languages. In Fodor this claim is explicit and fundamental. Putnam goes on to characterize the "cryptographer model":

> the mind thinks its thoughts in Mentalese, codes them in the local natural language, and then transmits them (say, by speaking them out loud) to the hearer. The hearer has a Cryptographer in his head too, of course, who thereupon proceeds to decode the "message." In this picture natural language, far from being essential to thought, is merely a vehicle for the communication of thought.

> [T]he *lingua mentis* is pictured as a kind of Ideal Language in which different signs always differ in meaning and in which different signs also differ in reference, not necessarily in the actual world, but at least in some possible world. If we succeed in decoding a message sent in our local natural language back into the *lingua mentis*, then by inspecting the resulting "translation" (in "clear," as cryptographers say) we shall see at once which words in the message have the same

meaning and which have different meanings, which words have the same reference in all possible worlds and which words differ in reference in at least some possible worlds.[9]

In the second chapter I raised the issue of the access the mind has, cognitive or otherwise, to the decoding mental symbols that Kretzmann's interpretation of Aristotle identifies with the passions of the soul. It is not at all surprising that Putnam uses "sign" where Kretzmann uses "symbol." That is a mere difference of inscription for the same thing. Kretzmann used the notions of *encoding* and *decoding* to explicate his interpretation of Aristotle on *signs* and *symbols*. For him, the speaker "encodes" passions of the soul into symbolic vocal utterances, while the hearer "decodes" the symbolic vocal utterances into corresponding passions of the soul, that is, into mental symbols. "In themselves" vocal utterances are mere *signs* or *symptoms* of passions of the soul.

Kretzmann does not suggest that the passions of the soul are themselves a *lingua mentis*. On the contrary, he writes that for Aristotle "writing, like speech, is a *linguistic* medium, as mind is not."[10] Presumably, for Aristotle a *lingua* is necessarily an instrument of communication between two or more individuals. However, it is difficult for Kretzmann's interpretation to maintain that the passions of the soul do not constitute a *lingua mentis*, once he has chosen to speak of encoding and decoding symbols, rather than mere symbolic representations as "the owl is of Athena."[11] If one encodes *into* a language, what is one encoding *from* but another language, and vice versa for decoding? The defenders of the *lingua mentis* view would simply argue that the mind speaks to itself its own language, that is, that it linguistically symbolizes the world to itself in a mental medium. In spoken language it simply wishes to communicate to another what it has already communicated to itself. Again, I am not claiming that Kretzmann's interpretation *logically* implies this; but the picture he presents of the mind encoding and decoding in a rule-governed way between the system of passions of the soul and the system of spoken language is consistent with, and highly suggestive of it.

As Kretzmann describes the passions of the soul, it never occurs to him to suggest that they are themselves subject to interpretation, keeping in mind again that he denies the passage has any semantic intent. But he does stress that Aristotle's mention of the relation of similitude that holds between passions of the soul and things is designed to be a foil against which to contrast the conventional relation that holds between vocal words as symbols and the passions they encode. Similarly, Putnam emphasizes this characteristic of the tradition when he writes that "there is an identity between sign and meaning (this is the fundamental idea of the model, in fact). The idea is that in the *lingua mentis* each sign has one and only one meaning." Presumably, the passions of the soul themselves are not subject to interpretation, because there is a perfect coincidence between sign and meaning, which is lacking in speech because of its conventionality. This supposed coincidence of sign and meaning is where Wittgenstein directed his attack, and Putnam will direct his.

Finally, in Kretzmann's interpretation there is nothing to exclude the possibility that a spoken sound may have many different meanings. So there is nothing about the sound "in itself" that requires that only one passion of the soul can be encoded into it by the speaker, and only one passion of the soul decoded from it by the listener. The problem naturally arises, then, of how the sound is invested with significance. This is, of course, a question for everyone, not just Aristotle. But Kretzmann's answer is telling, in light of Putnam's characterization of the Aristotelian tradition. A spoken word is a code for passions of the soul. The speaker encodes the passions he wishes to communicate, while the listener decodes them into passions in his own intellect. But then the *homunculus* seems to be merely waiting in the wings to come on stage, even if Kretzmann does not mention him. The very metaphors of "encoding" and "decoding" suggest this, and this is what Putnam exploits. The *homunculus* practically jumps out of Putnam's summary—

> then by inspecting the resulting "translation" (in "clear," as cryptographers say) we shall see at once which words in the message have the same meaning and which have different meanings, which words have the same reference in all possible worlds and which words differ in reference in at least some possible worlds.

In his very first characterization of the "cryptographer model," Putnam is explicit about a "cryptographer in [the] head."[12]

Putnam's emphasis on a *lingua mentis* with the characteristics he describes is certainly justified by the use of "encoding' and "decoding" to characterize the Aristotelian position, even as he shows no sign of being familiar with Kretzmann's interpretation in particular. However much Kretzmann may downplay the semantic character of the passage, in his interpretation a necessary condition for a sound to be a word, that is, for it to have semantic content, is that it be an encoding of a passion of the soul. Hence, if Kretzmann's interpretation is correct, Putnam does not seem far off base when he associates the *lingua mentis* account with Aristotle and "Aristotelians."

Putnam's Criticism of Mental Representationalism

It is important to note that the *cryptographer,* or *homunculus,* or *interpreter* is taken to be investing *spoken utterances* with significance, not the *lingua mentis*. In Fodor the spoken language is the speaker's second language; the first language is innate and has its representational content according to the nomological necessities that govern it. As Putnam characterizes the traditional view, there is no need for the *lingua mentis* to be invested with significance; "there is an identity between sign and meaning." However, it should be clear that one way of understanding Wittgenstein's criticism is that if one supposes a *lingua mentis* consisting of mental objects or entities upon which the mind operates, then it is just as much in need of an interpreter as any spoken *lingua*. This is also the aim of Putnam's criticism. He believes there are three key theses present in the "Aristotelian" cryptographer model.

> 1) Every word he uses is associated in the mind of the speaker with a certain mental representation [concept].
>
> 2) Two words are synonymous (have the same meaning) just in case they are associated with the *same* mental representation by the speakers who use those words.
>
> 3) The mental representation determines what the word refers to, if anything.[13]

1) simply seems to be a rephrasing of what Putnam said earlier, that when we understand a word or any other "sign," we associate that word with a "concept." I will rephrase this as:

> 1') understanding a word involves having a mental representation or being in a certain mental state.

Jointly the theses seem to constitute the cryptographer model, since presumably spoken words derive their meaning (2)), and thus their reference (3)), from the mental representation they are associated with (1)). However, Putnam says that these three theses are false. What he means is that "there cannot be any such things as 'mental representations which simultaneously satisfy all three of these conditions.'"[14]

So the substance of the criticism has its place in the interaction between 1'), 2), and 3). No mental representation according to Putnam can jointly satisfy those three. It is through 1') that language gets yoked to mental representations in the first place. How one thinks about 1), that is, how a speaker associates words and mental impressions or concepts, will depend upon how one thinks about 1') and about the joint satisfiability of 1'), 2), and 3). And it is to Putnam's criticism of that joint satisfiability that I now turn.

Putnam usually argues informally against these theses,[15] employing a number of strategies involving counterexamples and imaginative thought experiments, proceeding in Wittgensteinian fashion to argue that mental representations of whatever sort do not determine reference, and that there cannot be the identity of sign and meaning that the Aristotelians want. Perhaps his most famous counterexample is what he calls a "bit of science fiction," the idea of "Twin Earth," and it is that specific argument upon which I will concentrate. It surpasses by far his other puzzles in capturing the philosophical imagination of contemporary philosophers.[16] Though often repeated in other works, including *Representation and Reality*, the *locus classicus* of the Twin-Earth argument is in Putnam's article "The Meaning of 'Meaning'," which makes very plain the elements and presuppositions of his argument with *mental representationalism*.

The first point he makes in this article is what he calls the "traditional" ambiguity in the term 'to mean'. In one sense a term "means" the things it is truly said of. This is the sense we have in mind when we ask, "Did you *mean* this book, when you asked me to bring you the book on the table?" This sense of 'mean' is often called the extension of the term. So a particular rabbit may be *meant* by one instance of use of the term 'rabbit', while on another instance of use, all rabbits may be *meant*. In general the set of all rabbits is called *the extension* of the term 'rabbit', and in this sense of the term 'mean', 'rabbit' means this extension. The extension of a term is the set of all objects of which the term is truly predicated. Though Putnam does not do so, I will subscript this sense of 'mean' as 'mean$_2$'.

There is, however, another sense of 'mean'. This sense is what we have in mind when we say "'featherless biped' does not *mean* the same thing as 'rational animal'," or "'creature with a heart' does not *mean* the same thing as 'creature with kidneys'" or "'the star that rises in the morning at such and such location in the sky' does not *mean* 'the star that sets in the evening at such and such location'." Of course the examples are somewhat standard, and are chosen because it is plausible to suppose that in each pair the *extension* of the two phrases is identical.[17] Thus they would seem to mean$_2$ the same thing or things. And yet, in some sense, we have said that they do not mean the same thing. So this latter sense of 'mean' must be different from the former. It is traditionally called the *intension*, and I will subscript it as 'mean$_1$'. Putnam associates the *extension-intension* pair with the Fregean *Sinn-Bedeutung* pair, though he notes that it was present at least as far back as the Middle Ages. Perhaps the difference between 'mean$_2$' and 'mean$_1$' associated with extension and intension can be captured as the difference between these two questions, "Whom did you mean when you said the captain was sober this morning," and "What did you mean when you said the captain was sober this morning?"

Meaning$_1$ is what we seem to have in mind when we say that two terms are synonymous because they have the same meaning, and what we have in mind when we say that a good translation preserves as best it can the meaning of the original. It is this sense of meaning that Quine famously attacked as deeply misleading and subject to elimination. He wrote that

> the useful ways in which people ordinarily talk or seem to talk about meanings boil down to two: the *having* of meanings, which is significance, and *sameness* of meaning, or synonymy. What is called *giving* the meaning of an utterance is simply the uttering of a synonym, couched, ordinarily, in clearer language than the original.[18]

In his attack he claims that "the explanatory value of special and irreducible intermediary entities called meanings is surely illusory."[19] And of course his attack on the First Dogma of Empiricism, namely, that there is "some fundamental cleavage between truths which are *analytic*, or grounded in meanings independently of matters of fact, and truths which are *synthetic*, or grounded in fact," focuses its attention upon the obscurity of a realm of meanings, and criteria of synonymy. Presumably he was not attacking the idea that two terms may have identical extensions, that is, identical meaning$_2$. Indeed he distinctly separates the two:

> A felt need for meant entities may derive from an earlier failure to appreciate that meaning and reference are distinct. Once the theory of meaning is sharply separated from the theory of reference . . . meanings themselves, as obscure intermediary entities, may well be abandoned.[20]

For Quine, meanings$_1$ are the relics or "museum pieces" of a long-ago abandoned Aristotelian essentialism. "Meaning is what essence becomes when it is divorced from the object of reference and wedded to the word."[21]

Recall that Fodor's account of the Language of Thought relies heavily upon the notion of translation. Natural languages, first, second, third, and so on, are learned by learning how to translate them into the Language of Thought in the first instance, and into one another subsequently. To the extent that *sameness of meaning*$_1$ or *synonymy* is bound up with the notion of determinacy of translation, abandonment of synonymy as a criterion suggests the thesis of the indeterminacy of translation. Then the uses that Fodor and the "Aristotelian" make of the notion of translation or encoding to explain the determinacy of language cannot hope to appear anything but befuddled.

In sharp distinction to Quine, Putnam is not out to eliminate meaning$_1$. He just wants to show the difficulties encountered when it is identified with having mental representations. Having made the distinction between extension and intension as the two senses of 'mean', Putnam proceeds to make clear another element of the background of his criticism, the Fregean thesis that sense (*Sinn*) determines reference (*Bedeutung*). Now there are several different ways of putting this, including intension determines extension, and meaning$_1$ determines meaning$_2$. Putnam points out that traditionally everyone recognizes that two terms of different intension or meaning$_1$ can have the same extension or meaning$_2$. That is clear in the example of 'featherless biped' and 'rational animal', as well as the other examples. So *extension* does not determine *intension*. On the other hand, it is fairly clear that *intension* does determine *extension*. That is, if two terms have different extension, then they have different intension. Or contrapositively, if two terms have identical intension or meaning$_1$ then they have identical extension or meaning$_2$. So if someone tells me that the extension of 'A' is the set of all rabbits, and the extension of 'B' is the set of all men, then without actually knowing fully the intension of the terms, I know at the very least that they do not have identical intension or meaning$_1$. On the other hand, if someone tells me that 'A' and 'B' have identical intension or meaning$_1$ then even if I do not know their extension, I know at the very least that they must have the same extension or meaning$_2$.

In *Representation and Reality* Putnam's preferred term was 'mental state'. In "The Meaning of 'Meaning'" it is 'psychological state'. However it is fairly obvious from the context, as well as the Twin Earth puzzle, that he is using them synonymously, that is, he means$_1$ the same thing in both instances.

This seems to be reflected in the parallel between 1'), understanding a word involves having a mental representation or being in a certain mental state, and "knowing the meaning of a term is just a matter of being in a certain psychological state."[22] In "The Meaning of 'Meaning'" he clarifies what he intends by "mental" or "psychological state." 'State' is to be taken from its customary *scientific* sense.

> In science . . . it is customary to restrict the term state to properties which are defined in terms of the parameters of the individual which are fundamental from the point of view of the given science.[23]

For a state in physics he gives the example 'being five feet tall', in "mentalistic psychology" he gives 'being in a state of pain', and from cognitive psychology he says "knowing the alphabet might be a state," though he says "it is hard to say." He then describes what he says traditional philosophers have assumed about mental or psychological states. He calls it the assumption of "Methodological Solipsism":

> Methodological Solipsism: no psychological [mental] state, properly so called, presupposes the existence of any individual other than the subject to whom that state is ascribed.[24]

Notice that Fodor's mental representations considered as *narrow contents* meet this assumption. Recall that narrow contents enabled the possibility that while mental representations are innate, they could represent things other than the ones they do represent in the actual world. Narrow contents were said to be what remains abstractly the same, when the nomological contexts determining actual representation differ. They do not presuppose the existence of the individuals they *actually* represent. However, Fodor's mental representations taken as *broad contents* do not meet this assumption, since they cannot be taken as *broad contents* apart from the things they represent, namely, their causes in a particular context of nomological laws, which causes are other than the "subject to whom [they are] ascribed." In accord with Methodological Solipsism, if one ascribes a mental state to someone, it must be such that an account of what the state is *in itself* does not presuppose any other beings, in particular any beings in the environment of the individual to whom it is ascribed.

With these elements and presuppositions of the background in mind, the assumption of Methodological Solipsism in particular, consider again theses 1'), 2), and 3) above:

Hilary Putnam's Criticism of Aristotelian Accounts 145

1') Understanding a word involves having a mental representation or being in a certain mental state,

2) Two words are synonymous (have the same meaning) just in case they are associated with the *same* mental representation by the speakers who use those words.

3) The mental representation determines what the word refers to, if anything.

Why does Putnam believe that they are not jointly satisfiable?[25] These theses are supposed to be partially constitutive of the Aristotelian cryptographer model of language and mental representation. In 1') and 2) it is evident that the meaning that is at play there is meaning$_1$, *intension* or *sense*. The *meaning$_1$* or sense or *intension* of the term is either the mental representation itself or is determined by the mental representation. The proviso "determined by the mental representation" is added because, as I mentioned earlier, Putnam believes that Fregean attempts to make senses abstract do not escape the problem, since the mind will be in some "state" that will be its grasping of that *abstract sense*, which state Putnam would see as the Fregean analog of a mental representation.

> [I]f our interpretation of the traditional doctrine of intension and extension is fair to Frege and Carnap, then the whole psychologism/Platonism issue appears somewhat a tempest in a teapot, as far as meaning-theory is concerned. . . . [W]hether one takes the 'Platonic' entity or the psychological state as the 'meaning' would appear to be somewhat a matter of convention.[26]

1') says that someone who understands how to use a word has a mental representation. But 2) tells us that two or more speakers using a word with the same mental representation are using synonymous terms, that is, terms with the same meaning$_1$. Now if *intension* or *sense* is identified with *mental representation*, as the "Aristotelian" purports to do, and *intension* or *sense* determines *extension* or *reference*, then 3) is immediate—*mental representation* determines *extension* or *reference*.

So 1'), 2), and 3) together imply that two speakers using a word with the same mental representation (meaning$_1$) must refer to the same things; in other words, if the words they use are associated with the same mental representation, then the words must have the same extension. I will rephrase the last statement and state it as a thesis:

Thesis of Language–Mental Representationalism: two words cannot be associated with the same mental representation and have different extensions.

Putnam believes that this thesis is false. If he can show that it is false, it follows that the assumptions that imply it are not jointly satisfiable. If those assumptions are partially constitutive of the Aristotelian account of language and mental representationalism, then that account is false. It is the burden of the Twin Earth thought experiment to show that this thesis is false, and that therefore the Aristotelian account is false.

Twins and Twin Earths

The burden of the Twin Earth thought experiment is to indicate how two different persons or communities of persons, one on Earth and one on a fictitious Twin Earth, could have qualitatively identical mental representations associated with their words and twin-words, yet with those words refer to distinct natural kinds, where for simplicity the natural kind is identical with the set of individuals of that natural kind. Colin McGinn, considering the Twin Earth case, summarizes the point:

> The appearance (to us) of a natural kind, whether mineral or animal or vegetable, does not uniquely fix its identity. Therefore, if we hold the appearance constant, we do not necessarily hold the reality constant. But our words and thoughts are addressed to the natural kind itself—to its reality—and so they could vary in their reference despite constancy of appearance.[27]

The "reality" is varied by imaginatively considering two environments differentiated by the natural kinds present in them. The appearances are held constant by supposing that the different natural kinds can produce appearances that phenomenologically are qualitatively identical.

Suppose there is a Twin Earth. Suppose, for instance, it has all the land masses that we have, all the lakes, all the mountains, all the plants, all the animals, even all the people. There is a double of Luciano Pavarotti and a double of me, and of everyone else on Earth. Finally all the languages are the same, including their grammatical structure, as well as their verbal and written tokens. But suppose one difference. On Earth our lakes and oceans are predominantly filled with H_2O, which is what falls from the sky when it rains, and is pumped from wells in the ground. On the other hand, on Twin Earth their lakes and oceans are predominantly filled with a chemical compound whose formula is very long, but which will be abbreviated XYZ. XYZ falls from the sky when it rains, and is pumped from the ground. Suppose also that despite the fact that $H_2O \neq XYZ$, nonetheless, on Twin Earth the substance XYZ looks, smells, feels, and tastes just like our Earth H_2O.

Finally, suppose that modern chemistry has not been developed. Putnam sets the time on Earth and Twin Earth at about 1750. Then we on Earth do not know that what is in our lakes, rivers, and streams is H_2O, and the people on Twin Earth do not know that what is in their lakes, rivers, and streams is XYZ. All both communities know is that what is in their lakes, rivers, and streams is a colorless, odorless, potable liquid. Their experience of what is in their lakes, rivers, and streams is identical to our experience of what is in our lakes, rivers, and streams. Someone from Earth who is miraculously transported to Twin Earth, perhaps Ransom in Lewis's *Perelandra*, might well ask for a drink "from that stream over there," thirsty from his journey. Because what they know of what is in their lakes, rivers, and streams is identical with what we know of what is in our lakes, rivers, and streams, Putnam says "they [are] in the same psychological state," as we are. They have the same mental representation of what is in their lakes, rivers, and streams, as we do of what is in our lakes, rivers, and streams. That is why it was necessary to suppose that we don't have modern chemistry, so that the knowledge of the natural kind could not be built into the mental representation.

Now 'water' on Twin Earth is associated with the mental representation *colorless, odorless, potable liquid*. 'Water' on Earth is also associated with the very same mental representation, *colorless, odorless, potable liquid*. Each community knows how to use the terms on their respective planets. Thus, according to 1') the denizens of Earth have a certain mental state or representation associated with 'water' as they use it. The denizens of Twin Earth also have a certain mental state or representation associated with 'water' as they use it. But according to the situation as described, the mental representation on Twin Earth is the same as on Earth, namely a *colorless, odorless, potable liquid*. So according to 2) the uses of 'water' on Earth, and 'water' on Twin Earth are synonymous. Both communities mean$_1$ the same thing, when they use 'water' in their respective worlds. Thus it has the same meaning$_1$, sense, or intension for both Earthlings and Twin Earthlings.

Still, as a matter of fact, even though they do not know it, when our Twin Earth cousins use 'water' they are referring to XYZ, not to H_2O as we do when we use the term. This is why when they develop chemistry, and find out that water is XYZ, they are still talking about the same thing, even if their mental representation of it then differs. On the other hand, we are referring to H_2O, even though we do not know it. When we discover this through chemistry, we continue to talk about the same thing, even if our mental representation of it changes. The reference of a term is just what it is truly said of, and on Earth it is truly said of H_2O. If some XYZ were miraculously transported to Earth, it would not *be* water, since on Earth water *is* H_2O, even if no one knows it. Anyone who happened across it and called it water would be wrong, though he would not know his error. A more mundane example, without the science fiction, is simply the references of 'gold' and 'iron pyrites'. So the extension of 'water' on Earth is the natural kind H_2O, even though no one knows it in 1750. On the other hand, the extension of 'water' on Twin Earth is not the natural kind H_2O, but the natural kind XYZ. So 'water' on Earth and 'water' on Twin Earth have different extensions; they mean$_2$ different things.

Thus the Twin Earth thought experiment shows that two words, 'water' on Earth, and 'water' on Twin Earth, are associated with the same mental representation and have different extensions. But this is in direct contradiction to the Thesis of Language–Mental Representationalism, which says that two words cannot be associated with the same mental representation and have different extensions. So it looks like 1'), 2), and 3), which imply the Thesis of Language–Mental Representationalism, are not jointly satisfiable. If they are partially constitutive of the Aristotelian account of language and the mind, then that account ought to be given up.

Certainly Putnam thinks that they ought to be given up. Mental representations are not the meanings of terms; words do not have the same intension when they are associated with the same mental representation; and mental representations do not determine what words refer to. The point Putnam draws from the example is that not only the mental representations of the language users, if at all, but also their environment plays an essential part in determining reference or extension. The explanation for why 'water' refers to H_2O on Earth, but refers to XYZ on Twin Earth, is cashed out in terms of a difference in environment broadly construed. This leaves open the possibility that mental representations are *involved* in the referential use of language, but supports the view that reference is also determined by contextual factors external to the mind. This point is reminiscent of Wittgenstein's emphasis on "*context* of use." Vary the environment while holding the mental representations the same, and you vary the reference. Therefore, environment rather than mental representation is determining reference. Reference is achieved through an "indexical" element broadly construed—being in *this* environment rather than *that*.

Putnam provides variations on this example,[28] but what is common is that the qualitative identity of the mental representations is put in terms of the identity of descriptions predominantly involving sense experience, and the ability on the basis of sense experience to distinguish distinct external objects. But the sense experience so described is itself predominantly understood along the lines of appearances *internal to the mind*; we talk about water that exists *external* to the mind, but our experience of water consists of appearances that exist *internal* to the mind. This is why in "The Meaning of 'Meaning'" he concludes that "'meanings' just ain't in the *head*," because the suggestion had been that mental representations, things in the head, were meanings$_1$.

Putnam and Locke

The structure of the argument and the example of gold that Putnam uses elsewhere is reminiscent of Locke's treatment of nominal and real essences in Book Three of the *Essay*,[29] and it displays more clearly the Lockean background of the whole discussion. For Locke nominal essence determines membership in species. If all that one includes in one's nominal definition of gold is *extension* and a certain *color impression,* then one will be correct to conclude that any number of objects *are* gold that differ in their inner constitutions, that is, in their "real essences."[30] For Locke a thing belongs to the species of any nominal definition it exemplifies, irrespective of its "inner constitution." Iron pyrites *are* gold, where the nominal essence of gold is an *extended metallic thing giving rise to the sensation of this shade of yellow.*

This background helps one to understand the larger context of Putnam's work. Both he and Saul Kripke want to claim that scientific definitions (among others), like *water is H_2O*, go beyond Locke's nominal definitions.[31] A Lockean nominal definition of water, like *water is a colorless odorless potable liquid,* helps us to "pick out the reference" of our words. For Putnam, it may be part of a stereotype derived from "paradigmatic" cases, which stereotype functions in the context of an "indexical description" that picks out the reference of the term. So, for example, the "indexical description" associated with water might be "'stuff that behaves and has the same composition as *this,*' said by someone who is 'focusing' on a particular sample."[32] The indexical element is given by the "this" together with the "focusing," which latter is presumably achieved in virtue of the characteristics associated with the mental representation, perhaps being *colorless, odorless, and potable*. But what the indexical component attaches to is not the stereotype, but "the same composition," the nature of the thing. "We use the name *rigidly* to refer to whatever things share the *nature* that things satisfying the description normally possess."[33] Rigid designation is taken from Kripke. A proper name rigidly designates an object just in case it refers to it in all possible worlds in which it exists. One thesis of *Naming and Necessity* was that natural kind terms function something like proper names rigidly designating. But in Putnam's analysis this "indexical description" involving perhaps a mental representation, or Lockean nominal essence, does not give us a synonymous expression telling us what the word means, that is, its conceptual content. It is not "what we preserve in translation."

In addition, in Putnam's analysis such a description does not provide us with characteristics essential to *being* water. Some of the items in the world that we might pick out using the elements of a Lockean nominal definition may well be H$_2$O, but on the other hand some may not.[34] For Locke, *to be X* is to be a member of the species *X*. But to be a member of the species *X* is *not* to be something with a particular inner constitution. Now since an object is called *X* just in case it exemplifies the properties associated with the idea *X*, to be *called* X and to *be* X are equivalent. A nominal definition specifying the ideas associated with the term 'X' gives an analytic definition of its meaning, and hence necessary and sufficient conditions for *being X*.

Locke thought that it was senseless to try to refer with our words to what cannot be known, the *inner constitution* of things. However, Putnam and Kripke believe that we can refer to the inner constitution of things that stands behind the appearances of them.

> [I]f we *knew* the hidden structure we could frame a description in terms of *it*; but we don't at this point. . . . [T]he use of natural kind words reflects an important fact about our relation to the world: we know that there are kinds of things with common hidden structure, but we don't yet have the knowledge to describe all those hidden structures.[35]

Their hope is that we can come to know through theoretical investigation what that inner hidden constitution is. On their views, to call something *X* in virtue of certain properties that it exemplifies is not the same as specifying *what* that *X* is. "[I]f there is a hidden structure, then generally it determines what it is to be a member of the natural kind."[36] "If I agree that a liquid with the superficial properties of 'water' but a different microstructure *isn't really water,* then my ways of recognizing water . . . cannot be regarded as an analytical specification of *what it is to be water.*"[37] Something more is required for that, what Locke calls the "real essence" or "inner constitution." On Kripke's and Putnam's views, without knowing that "something more," we can still refer to it. In our ordinary use of a term we are making reference to the natural kind by means of standard stereotypes or descriptions of the phenomenological properties it has.

On Kripke's view, this tacit reference to the natural kind is the reason why the scientific discovery of the nature of a natural kind does not change the meaning of the term that refers to it. We were referring to it all the time. This tacit reference is why Kripke writes that the biologist corrects the layman when the latter calls a whale a fish. On Putnam's view what allows for reference in ordinary use to the unknown inner constitution is the environment of use. Thus, Putnam and Kripke are fundamentally denying and trying to overcome two Lockean theses, as well as the relationship between them: *that we cannot know the inner constitution of things and that we cannot refer to their inner constitution.* For Putnam, exemplifying a Lockean nominal definition may be sufficient for *calling* something *X,* but no particular Lockean nominal definition is necessary, since there might be other equally legitimate stereotypes. What is clear is that they provide neither necessary nor sufficient conditions for *being X.* This opens up a fissure between what it is *reasonable* to call something and what is *truly* said of something. Consider Ransom miraculously transported to Twin Earth. It is *reasonable* for him to say, "There is some water, bring it to me." However, recall that Ransom's referential use of the term is determined by, or tied to Earth; he intends to refer to stuff that is in its inner constitution identical with the stuff on Earth, which is as a matter of fact H_2O. Reference is tied to truth. So in his mouth, 'water' is reasonably but not truly said of the stuff on Twin Earth, just as I might reasonably but not truly call iron pyrites gold.

Fodor and Putnam

Putnam's criticism of Fodor varies as Fodor's view varies.[38] Fodor sets himself the task of providing an explanation of the intentionality of language and thought that does not involve intentional terms; if intentionality is real, "it is really something else." Putnam's general strategy is to show that the *explanans* always seems on reflection to involve the *explanandum*. He criticizes Fodor's account on two levels, both of which question the place of representations in causal accounts. First, can an account of the individuation of events be given that does not presuppose semantic elements? Since causal laws are between events, the question is whether the events that must be individuated in any causal account of semantic relations can be individuated outside of representational contexts. Here he is relying on Donald Davidson's puzzles about event individuation in "The Individuation of Events."[39] Second he wonders whether any account of the causal relation that Fodor relies upon can be sustained. In particular, will the sorts of asymmetries that he wants stand up? And can causal relations, that is, relations of efficient causation, be specified in non-interest, non-context bound ways that do not therefore involve representational elements? Causal laws involving counterfactuals will involve *ceteris paribus* clauses, the determination of which typically involves interest-bound contexts, which look like they will involve representational features. Putnam has the work of Hart and Honoré in mind in this criticism.[40]

Fodor recognizes that there are problems here, but believes his account can withstand Putnam's criticism. He believes they are problems for all the special sciences, not just cognitive psychology. This response is not quite fair, since the other special sciences are not trying to explain intentionality or representation, while presumably Fodor's cognitive science is. Putnam suggests that we do not have a robust enough relation between effect-events and cause-events as presently conceived; the causal account in the end suffers the same fate as the pictorial, because in the end a representation must be added on to the relation for it to be relevant to the account that is supposed to be given. Circularity of accounts always threatens. In a sentence that resonates for Aristotelians, Putnam writes that "efficient causation, as presently conceived, will not provide us with enough *form*."[41]

There is a sense in which Fodor basically grants Putnam the point. Fodor writes that it seems that "the Twin-Earth Problem *is* a problem, *because it breaks the connection between extensional identity and content identity*."[42] But he suggests that "*the Twin-Earth examples don't break the connection between content and extension; they just relativize it to context.*"[43] In the same sentence, Fodor seems to give Putnam the point and take it away. After all he recognizes that extension or reference is determined by the context, the situation or world. His account does not fall prey to Putnam's counterexample because environment does in fact play a role in determining extension. This is just what Putnam had said and Fodor grants this when he "relativizes it to a context." On the other hand, he takes it away when he says that in fact it is the content that is determining the extension, but it is broad content, not narrow content. Recall that the broad content of the mental representations was the narrow content anchored to a context, the context being set by the asymmetric nomological laws which obtain in the world.

Mental content is still determining extension because it is broad content, the only content that there really is. There is both an internal and external element in Fodor's account. The internal element is the narrow content, which remains the same no matter what the context. In this sense Fodor affirms Methodological Solipsism. Recall that the problem that gave rise to the distinction between broad and narrow content was that the physiological/neural states of the brain, upon which representation supervenes, were supposed to be specifiable outside the context of representation. They would remain physiologically identical even if the representation that supervenes upon them should happen to be different in different possible worlds. In that sense the structure is innate. The external element is determined by the broad content, the narrow content actually anchored to a context of representation. Because broad content involves its asymmetric causal relations to the world, mental content is still determining extension, but the mental content now is not internally individuated.

The external element in the individuation of broad content, however, differs from the role that it plays in Putnam's account. Putnam's account placed the determination of reference or extension in the context of *use* of the community, not how physiology is causally related to the environment. For Fodor, *use* plays no role in determining which world, among all the possible worlds, is actual, so it plays no role in determining which asymmetrical laws obtain between the things in the world, and the innate mental representations of the Language of Thought. Meaning for Fodor is just a "brute fact" in the actual world.

Summary of the Criticism

Putnam's use of the Twin Earth puzzle against the Aristotelian requires that on the Aristotelian's view, one can be talking about two distinct things or kinds of things, while the *internal* appearances that determine what one is talking about are qualitatively identical to one's introspection. So, in one's use of words, attending to these internal mental representations cannot be determining what one is talking about. The qualitative identity of the appearances embodies the assumption that the mental representation, in its representational aspect, is not intrinsically related to what it represents outside the mind. For if it were so related, one could not vary the environment in the manner necessary for the thought experiment and have the representation remain the same. This part of the puzzle is the embodiment of what Putnam called Methodological Solipsism.

Putnam believes the Aristotelian representationalist is committed to the dichotomy between *internal* appearances or concepts and *external* reality. Comparing Aristotelians with Platonists, he writes that their similarity consists in this, that

> even in . . . Platonistic versions . . . speakers are supposed to be able to direct their mental attention *to* concepts by means of something akin to perception, and, if A and B are different concepts, then attending to A and attending to B are different mental states. So even in these theories, the mental state of the speaker determines which concept he is attending to, and thereby determines what it is he refers to.[44]

In mentioning the "Platonistic versions," his intention is to describe what they share in common with the Aristotelian. In both instances, the Platonist and the Aristotelian, he believes there is a capacity, "something akin to perception," by which the speaker grasps concepts. For the Platonists these "'concepts' are not in the mind, but rather form a realm of abstract entities . . . independent both of the mind and of the world." Putnam believes that for the Aristotelian, by contrast, this capacity "akin to perception" directs itself upon "something in the mind that picks out the objects in the environment that we talk about."[45] This Aristotelian "something" Putnam calls the concept. "When such a something (call it a 'concept') is associated with a sign, it becomes the meaning of the sign." The Twin Earth counterexample shows Putnam's assumption that the Aristotelian pictures the language user as holding before his mind an *internal* appearance; by knowing this *internal* appearance the language user is supposed to know the *external* thing pictured by it, and so know the thing talked about. There is some "third thing" in addition to the language user and the external thing talked about. The third thing, the mental representation is an "intrusion of the idea between the mind and the object," as Russell says.

There are then three crucial elements to Putnam's argument by counter-example. First, there is a realm of *internal* appearances in addition to the mind and external things. Second, the appearances are not individuated by an intrinsic relation to the *external* things represented, but by facts entirely *internal* to the mind, or in Putnam's terminology solely by "theoretical and operational constraints." Third, we know what we are talking about, because of our *internal* and introspective knowledge of or relation to these appearances that represent *external* objects. It is important to recognize that this structure is also present in Fodor's account, despite his difference from what Putnam would take to be the traditional Aristotelian account. The effects that represent their causes in Fodor's account are conceived of as constituting a set of internal mental entities, a set of *third things* interposed between the mind and its activities on the one hand, and extra-mental reality on the other. The extra-mental things that cause these mental entities are present to the mind because the mental entities *qua* effects are present to the mind, and so causally represent their causes to the mind. Finally, though there is a nomologically necessary relation between mental representations taken as broad contents, there is no necessary or intrinsic relation between things in the world and the innate representations taken as narrow contents. The innate representations are things in themselves, upon which a relation of representation is added by the causal nexus of the world that happens to obtain. In this respect, Fodor also assumes Methodological Solipsism.

The Road Ahead

From this examination of Putnam's criticism of the "Aristotelian" account of language and the mind, I want to draw out three substantive theses that I believe can be seen at play in the discussion. They are not explicitly advocated by every figure that has been considered in the last three chapters, but they adequately represent certain crucial features of the background that Putnam assumes.

The first thesis is that there are things or objects in the mind which may be akin to pictures, images, appearances, effects, or some other mode of representation of things outside the mind. Putnam thinks that the underlying ontological class of these things consists of events or occurrences. Thus a *mental representation* is an event that occurs in the mind. So in addition to the mind and external things, there is a third realm of mental things. I will call this the *Third Thing Thesis*.

The second thesis is that the mind in its activity of thinking directs itself to these internal objects as *what* it primarily knows, or attends to, or is related to. Putnam says that mental representations at the level of conscious thought "are the only mental representations of whose existence we have any sure knowledge."[46] In Fodor, the emphasis was not so much on knowing or perceiving the mental representations, but rather upon the mind operating directly on them as objects, in various mental/computational processes. But for the purposes of the objection the mode of representation does not matter, what matters is the picture of the mind in cognition directly operating upon the representations in some way. Elsewhere Putnam characterizes a mental object as "something introspectible."[47] I will refer to this as the *Introspectibility Thesis*.

Finally, the last thesis is that there is no intrinsic or necessary relation between the so-called "mental representations" in the mind and the represented things outside it. Putnam writes:

> What makes it plausible that the mind . . . thinks . . . using representations is that all the thinking we know about uses representations. But none of the methods of representation that we know about—speech, writing, painting, carving in stone, etc.—has the magical property that there *cannot be* different representations with the same meaning. None of the methods of representation that we know about has the property that the representations *intrinsically* refer to whatever it is that they are used to refer to. All of the representations we know about have an association with their referent which is contingent, and capable of changing as the culture changes or as the world changes,[48]

and

> [e]ven a large and complex system of representations, both verbal and visual, still does not have an *intrinsic*, built-in . . . connection with what it represents—a connection independent of how it was caused and what the dispositions of the speaker or thinker are. And this is true whether the system of representations . . . is physically realized—the words are written or spoken, and the pictures are physical pictures—or only realized in the mind. Thought words and mental pictures do not *intrinsically* represent what they are about.[49]

For Putnam it betrays a "survival of magical thinking"[50] to suggest that there is some intrinsic relation between mental representations and what they represent, though it is clear from the passages that the "magical thinking" he has in mind cannot involve causal and dispositional elements.

In the third chapter, we saw Berkeley making the point that the partisans of matter did not hold that there is a necessary connection between a representative idea and what it purports to represent. For Locke, things differing in their "inner constitution" or "real essence" may well be represented by the very same idea in the mind. McDowell calls the suggestion that there is no such intrinsic connection the Master Thesis of mental representationalism—mental representations in themselves "just stand there" awaiting an interpreter and an interpretation. And Fodor makes a distinction between *narrow* and *broad* content, where the *narrow content* prescinds from the causal contexts that provide its meaningfulness or interpretation. Colin McGinn in *Mental Content* stresses the importance of this element of the background that Putnam assumes and wishes to criticize;[51] the internal things are thought to be individuated solely by facts *internal* to the mind.[52] John Searle makes a similar point in analyzing these cases, though it depends upon his analysis of intentional content. For Searle, identity of perceptual content does not imply identity of intentional content. So where *internalism* may be true of perceptual mental states, he asserts that it does not follow that it is true of intentional mental states, in particular conceptual meanings.[53] Finally, Putnam himself makes this thesis explicit, when in the "Meaning of 'Meaning'" he describes the assumption of Methodological Solipsism as

> when traditional philosophers talked about psychological states (or 'mental' states), they made an assumption which we may call the assumption of methodological solipsism. This assumption is the assumption that no psychological state, properly so called, presupposes the existence of any individual other than the subject to whom that state is ascribed.[54]

Following McGinn, I will call this the *Internalist Thesis,* though at times, I may also follow Putnam and call it "Methodological Solipsism."

In the next three chapters I examine the Thomistic-Aristotelian account I have been providing against the background of these three theses.

Chapter 6

THE THIRD THING THESIS

> Each previous period in the history of Western thought had a quite different idea of what such a term as *mind* or *soul* might stand for, and a correspondingly different idea of what the puzzles were that we should be trying to solve.
>
> —Hilary Putnam

The Internal-External Axis

In "Aristotle after Wittgenstein," Putnam suggests that the Aristotelian's "metaphysical form," as opposed to the "logical form" of the *Tractatus*, may be sufficiently robust to succeed in accounting for how language "hooks up" with the world. This seems to display a greater appreciation for how the notion of *form* plays a part in the Aristotelian account. In his even more recent Dewey Lectures, "Sense, Nonsense, and the Senses: An Inquiry into the Powers of the Human Mind," however, he suggests that after all he cannot follow the Aristotelian down this road. One of the difficulties that Putnam has is with the Aristotelian use of 'in':

> [T]he metaphor that Aristotle likes—that is, the metaphor according to which the form is "in" the object rather than "outside" it, or "apart" from it—is far from clear (which is why it is maddening when Aristotle's followers, up to the present day, simply repeat it as if it were self-explanatory).[1]

Putnam thinks that the Aristotelian has something very distinct in view when he uses such phrases as *in anima* and *res extra animam*, something the Aristotelian thinks is self-evident, but that he, Putnam, finds obscure.

In the "received view" of *mental representationalism*, these Latin phrases, or their rough English equivalents, are overwhelmingly construed in spatial-sensual terms—thus Locke's dark room letting in some light from *outside* and Hume's theater of perceptions. The mind is conceived of as a *locus* that is private and inaccessible, except for the person who possesses it. Events take place *in* it; appearances find their home *inside* the mind or theater. Consider against this background John McDowell's characterization of the dichotomy between the *inner* world of the mind and the *outer* world of ordinary objects, as he discusses Russell's theory of descriptions and direct acquaintance:

> In a fully Cartesian picture, the inner life takes place in an autonomous realm, transparent to the introspective awareness of its subject; the access of subjectivity to the rest of the world becomes correspondingly problematic. . . . [I]f we let there be quasi-Russellian singular propositions about, say, ordinary perceptible objects among the contents of *inner space*, we can no longer be regarding *inner space as a locus of configurations which are self-standing, not beholden to external conditions; and there is now no question of a gulf, which it might be the task of philosophy to try to bridge, or declare unbridgeable, between the realm of subjectivity and the world of ordinary objects.*[2]

Of course McDowell is considering ways of overcoming the Cartesian gulf that, as Walker Percy describes it, was generated when "Descartes ripped body loose from mind and turned the very soul into a ghost that haunts its own house."[3] The effort to overcome this gulf is set by the spatial connotations of "inner space" and *external objects*.

It would be unjust to hold Descartes, Locke, and Hume unyieldingly to a literal rendering of the spatial characteristics of their metaphors. But it is these spatial metaphors that set the context for construing the modality of representation along pictorial and imagistic lines. And the criticisms directed against this position by Wittgenstein and Putnam often trade on the spatial-sensual metaphors.

Some contemporary philosophers, on the other hand, do appear to intend something like a literal rendering. So John Searle writes:

> Some form of internalism must be right because there isn't anything else to do the job. The brain is all we have for the purpose of representing the world to ourselves and everything we can use must be inside the brain. Each of our beliefs must be possible for a being who is a brain in a vat because each of us is precisely a brain in a vat; the vat is a skull and the 'messages' coming in are coming in by way of impacts on the nervous system.[4]

The 'brain in the vat' is of course a reference to yet another of Putnam's thought experiments, in which a scientist keeps a brain alive in a vat, but prevents it from being able to know that it is a brain in a vat. Instead, the scientist provides it with all the qualitatively identical sensory inputs for it to conceive of itself as a full-blown human body living out in the world. It is in effect Descartes's evil demon thought experiment, updated for a more scientifically and perhaps less religiously oriented audience.[5] At the same time as Searle will insist upon "everything we can use [for representing the world to ourselves] must be inside the brain," he will forswear any association with the received view of *mental representationalism*. Still, if one changes "skull" to "dark room," and "impacts on the nervous system" to "light streaming through the window," the parallel with Locke does not seem too far-fetched. In addition, the boundary need not be drawn at the brain; it could be the body generally, as Donald Davidson suggests,

> we have been assuming, by connecting his beliefs to the world, confronting certain of his beliefs with the deliverances of the senses one by one, or perhaps confronting the totality of his beliefs with the tribunal of experience. No such confrontation makes sense, for *of course we can't get outside our skins to find out what is causing the internal happenings of which we are aware.*[6]

With Davidson in mind, among others, McDowell examines the effect of this dichotomy on epistemological and metaphysical questions. He describes the two spaces as an inner conceptual world of meaning, reasons, spontaneity, freedom, and responsibility, and an outer, disenchanted, meaningless, though intelligible world of law-governed deterministic mechanisms. He proceeds to criticize the oscillation this sets up between two poles. On the one hand, there are the philosophers who stress the spontaneous autonomy of the inner space of concepts and reasons, finally arriving at a coherentist account of truth, a position he calls a "frictionless spinning of reasons in the void." On the other hand, others, wishing to slow down this spinning in the void, cause friction by appealing to the "myth of the given." The given is supposed to provide rational justifications for our internal beliefs about the external world. But as McDowell points out, since "the given" is by supposition outside the inner conceptual space of concepts and reasons, it cannot provide justifications, but merely "exculpations"—namely, one cannot be justified in believing what one does on the basis of a non-conceptual unreasonable given; but one can be exculpated, for it was merely "given" to one, and as such one cannot but believe what it gives rise to.[7]

McDowell's own solution is to find *out there* in the world this realm of reasons, meaning, spontaneity, and freedom by finding it in "Aristotelian second nature," a set of properly human capacities that respond to the world of reasons and meanings through a proper upbringing; "the resulting habits of thought and action are second nature." This second nature has only the barest relation to that part of human being that is governed by the realm of law. McDowell cedes the realm of Aristotelian *first nature* to a certain understanding of modern natural science with its deterministic mechanisms; that battle has been won by modern naturalism with its emphasis upon the realm of deterministic laws.

> This should defuse the fear of supernaturalism. Second nature could not float free of potentialities that belong to a normal human organism. This gives human reason enough of a foothold in the realm of law to satisfy any proper respect for modern natural science.[8]

Any attempt to recover first nature would be nothing other than a return to pre-modern medieval modes of thought in which nature is seen to be "enchanted" and magically invested with meaning, "a return to mediaeval superstition."[9]

But modern naturalism is "forgetful of second nature." What is needed is a proper respect for the realm of law that does not seek to reduce the space of reasons to it. This is achieved by combining Aristotelian second nature with the Kantian insight into the autonomy of reason in its own domain. What results, however, is the barest of relations between human nature as it is subject to natural science, that is, as it is subject to the realm of law, and human nature as it responds to the space of reasons and meaning, the space of spontaneity and freedom. Describing Wittgenstein's sense of "our natural history," McDowell writes:

> [H]e must mean the natural history of creatures whose nature is largely second nature. Human life, our natural way of being, is already shaped by meaning. We need not connect this natural history to nature as the realm of law any more tightly than by simply affirming our right to the notion of second nature.[10]

Our toehold in the realm of law is, it seems, very tenuous indeed, consisting in our right to assert our freedom and spontaneity over against the determinism of first nature.

But what import does the term *second* have in 'Aristotelian second nature', if not through its relation to 'Aristotelian first nature', as a natural development of the latter? Fundamentally the dualism remains in the picture McDowell presents. *Out there* in the world there are still two distinct natures, the nature governed by natural science with its deterministic mechanisms, and the nature, second nature, that is not so governed, but is spontaneous, reasonable, and free. Rather than there being an organic and developmental unity between first and second nature, as in Aristotle's account, in McDowell's account they appear just as disconnected as when the one was "in the head." Though he has managed to project the mind *out* there into the world through one's capacities for proper socialization, calling one of the natures "second" does not eliminate the problem; it simply reasserts the dualism of what is *in* second nature, and what is *external* to it—what is *inside* the space of reasons and what is *outside* of it. Second nature appears to be an almost free-floating, though more sociable mind hovering over and reflecting upon the determinism of material reality, with a determination not to be reduced to it. I will have more to say about this aspect of McDowell's analysis in the concluding chapter.

For David Braine these dual worlds, whether of the classical sort or the contemporary, prevent even materialistic philosophers of mind from avoiding the traditional and fundamental problems that they themselves associate with mind-body dualism.

> For materialism to get going at all in its main contemporary form it is an absolute condition that one should have established a dualistic pattern of analysis of what goes on in human life. That is, before mental states and events can be identified with brain-states or events, or regarded as 'realized in the brain', these mental states and events have to be conceived in a way which makes them purely 'inner', logically segregated from the 'outer world' and the 'outer man' with his behavior in the way which is characteristic of dualism.[11]

The dualist principles are no longer the Cartesian *thinking immaterial thing–extended material thing* pair, but rather *thinking brain–everything else*, or *thinking body–everything else*. Thus in many quarters of current philosophy these classical metaphors retain their force. My point, however, is not to pin anyone down to a spatial rendering of these presuppositions. Perhaps when pressed, the metaphors will give way to the sense they have in Aristotle and St. Thomas.

Aristotle and St. Thomas on 'in' and 'extra'

When one looks at Aristotle and St. Thomas, it is certainly the case that the Latin prepositions 'in' ['ἐν' in Greek] and 'extra' figure prominently in the discussion of the relation between the things known and the mind or intellect. Aristotle himself provides us with the initial key for understanding the sense of 'en'. Early in the *Categories* he distinguishes two senses of being "present in a subject" [ἐν ὑποκειμένῳ]—first present as a part *in* a whole, and second present "not . . . as parts are in a whole, but being incapable of existence apart from the said subject."[12] He uses knowledge as an example of something that can be "in a subject" in the latter sense, not in the sense of "parts in a whole."

At *Physics* 210a14–24, discussing *place* in general, Aristotle expands the list from these two senses of 'in' to eight, including among them "as the finger is in the hand, and in general, as a part is in the whole," and "in the most important sense of all, as a thing is in a vessel, and in general, in a place," contrasting these spatial senses with "as health is in the hot and the cold, and in general, as the form is in the matter."[13] St. Thomas, commenting on these different senses, points out that 'form' in the latter sense can be taken as "substantial or accidental" and 'matter' as "matter or subject."[14] Parts do not *in-form* wholes as subjects, nor do things *in-form* their containers. The simple fact that Aristotle enunciates so many uses of 'in', and that the sense at issue here is not "the most important," should be enough to put aside the criticism that its use is supposed to be either "self-evident" or "metaphorical."

'Extra' counterpoised to 'in' will have corresponding spatial and non-spatial senses. Spatially it should be construed either *as a thing is not a part of a whole*, or *as a thing is not in a vessel*, or *as a thing is not in a place*. The last is ambiguous. It could be construed either *as a thing is not in this place*, or *as a thing is not in any place*. The other sense correlative to the sense of 'in' in the *Categories*, as well as the sense of 'form *in* subject' of the *Physics* will be *capable of existence apart from the subject of which it is said to be 'extra', and not in-forming that subject.*

As used by St. Thomas, when 'in anima' and 'extra animam' are said of the concept and the res of which it is a natural similitude, respectively, they are to be understood not in the spatial senses but in the formal-subject sense—"in it just as a form, as it were, in matter."[15] The *species* "in" the intellect is the formal principle of intellectual operation.[16] "In" the intellect the intelligible species is an accident of intellect, existing "in" it;[17] the *intelligible character* that *in-forms* the intellect as a subject, resulting in a concept, is a principle of knowing.[18] The thing known usually[19] *exists* independently of the knower,[20] remaining the same whether known or not.[21] A principle of being *extra animam* can be a principle of knowing *in anima*, but not every principle of knowing is a principle of being of the thing known.[22]

Here again, the central importance of the *De ente et essentia* appears. When St. Thomas distinguished two modes of *being* for a nature, *in the soul* and *in singular things*, the use of 'in' is clearly an existential sense. Natures only exist through singular things or through the intellect. When we say that Roscelin's human nature is *in* him, we do not intend to place it spatially. We intend to claim that it only exists through his existence. Similarly, if we say that human nature is *in* Abelard's intellect because he knows human nature, we do not intend to "locate" or "place" it. The fault line between what is *in anima* and what is *in singularibus,* which singular things are *res*, distinguishes modes of existence, not modes of location.

The spatial sense of "in" for "classical representationalism," whether taken metaphorically or literally, is simply not at play in St. Thomas's discussion of intellect and world, and the reading of Aristotle in that light. Once one understands the existential character of 'in' and 'extra' in St. Thomas's thought, it is more appropriate to ask which beings actively depend upon me for their existence, and which do not?[23] But this question does not lend itself to imagining two "spaces," the inner space of the mind, and the outer space of the world, as well as a gulf between them.

Concepts Are Not "Third Things"

Now that it is clear how 'in' and 'extra' (beyond) are to be understood within this Thomistic account, it is appropriate to consider whether the Third Thing Thesis applies to it, that is, whether the concept should be taken to be a thing within the knower, distinct from the thing known and distinct from the knowing intellect. The method will be to consider how Aquinas uses 'res' (thing) and how it applies to the concept.

The Third Thing Thesis asserts that there is some internal mental thing (*res*) that is interposed between the mind and the world. John Haldane, in summarizing the difficulty, indicates that there is a

> common tendency to assume that the existence of a process implies the existence of a product distinct from it. More particularly, to suppose that talk of forming a thought, or of making a representation, implies the manufacture of a *third thing* to which consciousness is then directed.[24]

Is there such a *tertium quid* manufactured and interposed between intellect and world? Just this way of asking the question betrays the influence of the spatial metaphors in the received view of mental representationalism. In the existential sense of the Aristotelian account, what sense does it make any longer to ask whether there is some *third thing* "interposed between" intellect and world? And strictly speaking, for St. Thomas, it is not the intellect itself that knows, but rather the intellectual substance by *means* of his or her intellect.[25] But perhaps I can ask the question in a way that avoids the spatial connotations—is the concept a thing (*res*) other than the intellect of the knower, and other than the worldly thing (*res*) that is known? Robert Sokolowski, for the most part confining his criticism to modern trends, thinks that an affirmative answer to this question is "philosophically" naive; 'thing' is not properly applied to concepts. The answer to the question, how many *things* are essentially involved, ought to be two (knowing thing and known thing), not three.

Consider the following text from St. Thomas that may be taken to suggest just such a third thing:

> [There are two] operations found in the intellect. For in the first place the passivity of the possible intellect may be considered inasmuch as it is informed by the intelligible species. So informed, it forms in the second place either a definition or a division or a composition, which is signified through an articulated sound. Hence, the *ratio* which the name signifies is a definition, and the enunciation signifies a composition or division of the intellect. Therefore, articulated sounds do not signify the intelligible species, but rather that which the intellect forms for itself to judge of exterior things.[26]

This passage comes from the *Summa* and is a reply to an objection that explicitly quotes the *De interpretatione* passage, arguing that the intelligible species is what is actually understood by the intellect. The definitions that the passage mentions pertain to the first act of intellect, while the divisions and compositions to the second act. These are acts of the possible intellect. What the passage makes clear is that prior to these acts of the possible intellect, the possible intellect must be rendered capable of such acts by being informed by the *intelligible species*. This informing of the possible intellect is the act of the agent intellect, as St. Thomas argues in the article immediately preceding the one in which this text is found. The informing of the possible intellect by the agent intellect is the act of abstraction, according to St. Thomas.

What is interesting here is the last line. The passage seems to suggest that the possible intellect forms some *being*, some *thing* when it forms definitions, combinations, and divisions. Thus the first and second acts of the [possible] intellect do not bear upon *res extra animam*, but rather are productive acts that produce entities *in anima*; they produce concepts that are entities distinct from the productive acts, entities produced in order to judge of external things.

However, this is a misreading of what is taking place in the passage. When St. Thomas discusses the soul, its powers, and acts, he is engaged in a reflective analytic study. He is laying before the reader those aspects or features characteristic of being human. Such an analytic study by its very nature distinguishes for separate consideration features that in reality may not be separate, much like the Aristotelian natural philosopher will analytically distinguish form from matter, though they are one being in reality, and cannot exist apart. A constant refrain of St. Thomas concerns what he calls "the error of the Platonists," namely, that they mistook the mode of knowing for the thing known. Taken generally, this is nothing other than a warning against mapping the abstract and analytic characteristics of our human mode of knowing back onto the reality known. *Rationality* and *animality*, for example, are considered separately by the inquiring intellect, yet St. Thomas is committed to their formal identity in man.[27]

Synthesis, by contrast, is not just the mirror image of analysis, as if preserving *extra animam* the distinctions wrought by analysis. Indeed, according to St. Thomas's account of truth, a judgment is true to the extent that the intellect achieves through it a unity between what the intellect has first distinguished, the features or aspects signified by subject and predicate separately, a unity in the judgment that adequately signifies the actual unity in *res extra animam* of those known aspects or features. If the real unity signified is *per accidens,* the truth is contingent; if the real unity signified is *per se,* the truth is necessary. Thus, the intellect achieves truth when it unites what it has first distinguished.

In the case of acts generally, and concepts in particular, there is a sense in which making them the objects of a reflective analytic study robs them of their life. It *objectifies* them so that they can be rationally dissected. While one must keep in mind the differences between St. Thomas and Frege on the ontological character of concepts, the latter recognized something like a parallel of what I am suggesting when he wrote that "the concept as such cannot play [the part of what is meant by a grammatical subject], in view of its predicative nature; it must first be converted into an object, or, more precisely, an object must go proxy for it."[28] Our talk about concepts deforms them in a sense; it is *as if* some object or entity or third thing must go proxy for them. The primary being of concepts is to express our understanding, not to be the subjects of it. By analogy, it is one thing to analyze a walk, and another to walk.

Recall that in St. Thomas a *concept* is the informed activity of the intellect as it grasps *res extra animam.* Of course the intellect is simply the human substance's capacity or power to engage in just that act; thus the concept is to the intellect as act to potency. As Haldane notes there is a process-product ambiguity when describing this informed activity of the intellect, just as there is in the case of many of our activities. Often in describing an act, we are correct in our estimation that, in addition to the act, something other than it is produced. So for instance, we normally think that the product of the pitcher's *act of pitching* results in the ball having a certain trajectory, which we call the pitch.[29]

On the other hand, there are activities we engage in that involve no such additional product, though the language we use might suggest otherwise. We might say "Sebastian strolled in Bologna" or we might say, "Sebastian took a stroll in Bologna."[30] Strictly speaking, we take *things;* taking is the act and something other than the act is *taken.* Do we suggest by the phrase "took a stroll in Bologna" that there is some additional *thing* or *object* that Sebastian took to accompany him while strolling? If we do, how does it relate to other predicates with modifying phrases that we normally find appropriate in descriptions of *things* taken hold of? Did he take the stroll off the shelf and so on? No. Or suppose we say "Marat took a fateful bath." Here there *are* other *things* involved — the tub, the soap, the water, and so on. But none of these is the *thing* that presumably we would have in mind when we say "Marat took a fateful bath." What we mean to say is that he bathed.

'To take a stroll' or 'to take a bath' are elliptical ways of saying 'to stroll', or 'to bathe', which by their forms do not indicate additional *things* that are "taken." Nominalization of verbs into substantives is a way of reflectively talking about and analyzing our activities, not a way of recognizing another realm of *things* in addition to our activities. "We are up against one of the great sources of philosophical bewilderment: a substantive makes us look for a thing that corresponds to it."[31]

The difference for St. Thomas is between a transient act and an immanent act.[32] In a transient act, the act has its termination in some being other than the agent, completing or perfecting the other being. In an immanent act, the act has its termination in some sense within the agent, completing or perfecting it. Unfortunately a term like 'grasp' in its literal sense suggests a transient act—we grasp a pen and actually move or change it. But thinking about or understanding some being in the world does not actually change it. Rather the knower is changed or perfected. This asymmetry in thought or knowledge is why St. Thomas writes that the relation of knower to known is a real relation, while the relation of known to knower is a merely logical relation. Thought and understanding are immanent acts. When we metaphorically apply 'grasp' to the intellect, we run the risk of being misled by its grammar into conceiving of thought as a transient act. When we employ words like 'grasp', 'apprehend', 'employ', 'produce,', 'develop', and so on, as in 'he grasped the concept', or 'by speaking with them, we develop in children the concepts necessary to function linguistically', our use of such words should be analyzed along the lines of 'he took a stroll'. 'Concept' is a nominalized form of talking about our act of conceiving, not a way of referring to an additional class or category of objects or things in addition to our acts. Similarly, this is the case for 'definition' and defining, 'enunciation' and enunciating, and so on. Strictly speaking, if we retain the tactile word *grasp* we do not grasp concepts. We conceive, and in our conceiving we grasp things other than our conceiving. Our conceiving is informed by the *forms* of things, and the activity differs thereby, just as Sebastian's bodily movements may have the form of a stroll, or the form of a run, or may be informed more or less by speed or grace.

Consider another analogy. If I grasp the pen in my hand, then I can certainly say that my grasp is of a pen, and not of a ball. In the antecedent of this conditional, 'grasp' functions as a verb, while in the consequent 'grasp' functions as a noun. Do I need, then, to posit a *third thing* between the pen and my grasping hand—the grasp itself—a *third thing* named by the use of 'grasp' as a noun? There is no *third thing* that exists between hand and pen—or in the non-spatial mode of making the point, there is no *third thing* that is other than the pen and my grasping hand. Similarly, there is no third thing other than the conceiving intellect and the *res extra animam*.

St. Thomas often repeats that the intelligible in act is the intellect in act.[33] This thesis may appear to be yet another assertion on his part that the actually understood thing is a mental thing. It appears that the *res extra animam* is actually understood when it is converted into a *res in anima* that is directly related to the act of understanding. The *res in anima* bears a relation of formal identity to the *res extra animam* that is potentially known, which *res extra animam* is thus known *through* the *res in anima* because of its formal identity with it. So this *res in anima* appears to be a third thing between knower and *res extra animam* known through it.

But this is a misunderstanding of what St. Thomas means. A *res extra animam* considered in itself is not actually known or understood, but it can be. Thus it is not *actually* intelligible, but only *potentially* so. To be rendered actually intelligible, it must come to be known by some intellect. Thus for *it* to be actually intelligible is for *it* to be actually known.

> [T]he sensible in act is the sense in act, and the intelligible in act is the intellect in act. For we actually sense something or understand it when our intellect or sense is in act as informed by the sensible or intelligible species.[34]

"The intelligible in act" is not a mental entity distinct from the act of intellect, but is rather the act of intellect itself informed by the *intelligible species*. Thus, a *res extra animam* is rendered actually intelligible insofar as its form actually informs the act of intellect; it is not that a mental entity has been fashioned by the act of intellect, a mental entity with which the *res extra animam* is formally identical. In another place, he puts it this way, "[The intellect] actually understands a thing when the species of the thing is made the form of the possible intellect. This is why we *say* that the intellect in act is the understood in act."[35] Here he is explaining to the reader what he *means* by a phrase like "the intellect in act is the understood in act." The species of the thing known is made the form of the act of intellect by the abstraction which is the proper effect of the agent intellect. The agent intellect makes the possible intellect capable of actually understanding some *res extra animam*. St. Thomas is saying that the intellect in act is formally identical to the *res extra animam* that is actually understood through that *act*; the act is *formally* the *res extra animam* that it actually understands. That formal identity of the act constitutes the *res extra animam* as actually understood.

St. Thomas writes, "[S]omeone has a form in act inasmuch as he is able to execute an operation of that form."[36] The operation or act in question here is the act of understanding some thing beyond the soul. "The intellect in act is the intelligible [form] in act."[37] For the form of the thing known to be an intelligible in act is for the intellect to execute an operation informed by that form, that is, "of that form." Thus some *res extra animam* is actually known when its form informs the act or operation of intellect by which it is known; taken as informing that act it is called the *intelligible species*. The act is a knowing, not a producing; and the acts of understanding differ according to the forms that inform them. If I grasp with my hand a ball instead of a pen, the form of my grasp will be different than when I grasp a pen, and that difference of form in my grasp is determined by the difference of form between the actual ball and pen. Now my hand, grasping either a ball or a pen, looks like neither a ball nor a pen; nonetheless in its relation to the pen that it grasps, it is perfectly appropriate to call its form *the form of a pen-grasping hand*, rather than *the form of a ball-grasping hand*. I grasp differently according as the form of my grasping must differ for the diversely in-formed things I wish to grasp.

With this analysis in hand, I see no reason in the passage quoted earlier from the *Summa*, or in general, why St. Thomas must be interpreted as holding that acts of intellect produce mental entities, third things, when he speaks of the intellect "forming" definitions or combinations and divisions. What ontological chasm separates the definition from the defining, or the play from the playing? None.

Elizabeth Anscombe uses a similar example in an interpretation of Wittgenstein.

> [S]urgeons may order the manufacture of instruments adapted to catch hold of different items. Catching-hold-of is in every case the same kind of thing, but the objects caught hold of vary in shape and so they may need instruments the business parts of which are differently shaped. Consider now the difference between naming a number, naming a particular man, and naming a kind of fruit. We might conceive it on the analogy of the surgeon's instruments, and, while this would suggest that naming, or "using a word for a ____" was always the same kind of thing, still it would also give us the idea of the analogue of a "difference of shape" in the catching-hold part of the instrument. But what is in question is a difference of *logical* shape.[38]

She is discussing Wittgenstein on his sense of "grammatical" and "logical form." Anscombe adds almost immediately that "strange to say, Wittgenstein's conception of the grammatical is far closer to the Platonic-Aristotelian tradition than that of the linguistics which seems to hold the field at the present day."

Of course comparing the soul to a grasping hand goes back to Aristotle. Anscombe's surgeons' instruments are not that different, which is why I find the parallel interesting. The metaphor can, of course, be misleading if it is taken literally; a surgeon's instruments do mediate as entities or objects between grasping hand and the physical organs operated upon.

Once one recognizes that a concept is related to the intellect as act to potency, it makes very little sense to treat intellect and concept as if the intellect is *one thing* and a concept *another thing*. In another context, St. Thomas provides interesting insight on the more general question of how act and potency are one and not diverse. Discussing various erroneous views about what makes soul and body one, he sets them in a more general setting concerning the unity of act and potency.

> [T]hey held such [erroneous views] because they inquired about what makes a potency and an act to be one thing, and they sought the differences of these things, as if it were necessary to collect them as through some single medium, just like things which are diverse according to act. But as has been said, the ultimate matter, which is appropriate to a form, and the form itself are the same thing (*idem*). For one of these is as potency, while the other is as act. . . . [P]otency and act are in a certain respect one. For that which is in potency is made to be in act. And so it is not necessary for them to be united through some bond, like those things which are entirely diverse. So no cause makes those things which are composed from matter and form to be one, except that which moves a potency to act.[39]

Although St. Thomas is pursuing the union of soul and body, I do not believe I am taking the passage out of context, for he discusses that point by bringing to bear the larger context of how act and potency are not *diverse things*, but *one thing*.

In his analysis of knowledge St. Thomas extends by analogy the pairs form-matter and act-potency from their original use in the discussion of change and motion, just as he extends it to soul and body. The intellect as such is taken as matter to the form that in-forms it. If one recognizes that a concept is compared to the intellect as act to potency, then the general point that St. Thomas makes in the passage above applies straightforwardly. Indeed, he confirms this point in an early text:

> In the human intellect the likeness of the understood thing is other than the substance of the intellect, and it is as a form of it; hence from the intellect and the likeness of the thing is made one complete thing, which is the intellect understanding in act; and the likeness of it is received from the thing.

Later in the same work, he writes:

> And it is necessary that this species, which is the intellect in act, should perfect the intellect in potency: from the union of which one perfect thing is brought about, which is the intellect in act, just as from soul and body one thing is brought about, which is a man having human operations. Hence just as the soul is not other than the man, so the *intellect in act* is not other than the intellect actually understanding, but the same thing.[40]

Thus, the intelligible in act is *the same thing* as the intellect actually understanding, not a *res in anima* distinct from it. The concept is the intellect's act. Extending a little what St. Thomas writes, we should not be looking for something to relate concepts to the intellect, but for the cause that moves the intellect to its act.

Finally, in his commentary on the *Metaphysics*, St. Thomas finds in Aristotle the distinction between transient and immanent acts, where a transient act proceeds to some external thing, perfecting it, while an immanent act remains within the agent, perfecting it.

> [W]hen there is no product of the one acting other than the act of the power, then the act exists in the agent both as the perfection of him and does not proceed to some exterior thing as a perfection of it; just as sight is in the one seeing as his perfection, and contemplation in the one contemplating.[41]

Of course one could object that this only denies that something *exterior* is produced in seeing and contemplating, contemplation being an act of intellect. It does not explicitly deny that something internal is produced in seeing or contemplating, something internal that is a being distinct from the act of seeing or the act of contemplating. But, at this point such a response appears excessively *ad hoc* in order to save the Third Thing Thesis. The passage certainly does assert that in these instances nothing *other than* the seeing or the contemplating is produced. And the general point of the passage pertains to "when there is no product . . . *other than the act of the power.*" Thus, something *in anima* is produced, but it is nothing *other than* the acts of seeing or contemplating. This is how we should think about what the intellect does in defining and judging.

An Objection: Concepts Are Had, Not Done

Now, someone might object that one *has* a concept; it is not an act or doing. And a thing had is distinct from the act of having it. Thus the analysis above makes no sense to the extent that it identifies concepts with acts of intellect. I do not deny the ordinary language connotation of 'concept' in English, in which one speaks of "having a concept," and its importance to the philosophical discussion, but it shouldn't dictate finally the substance of the philosophical discussion. As the quotation from Wittgenstein about looking for something to correspond to a substantive noun suggests, ordinary use can mislead. 'Concept' does have a cognate verb form 'to conceive', and that presumably is something we do. The difficulty here is akin to that of 'thought'; we *have* thoughts, and we *do* think.

There are two senses in which I am using 'concept' here. First, for the act, which is the primary sense we have been discussing. That is the substance of the claim that 'concept' is a nominalized form of 'to conceive' used to talk about our conceiving. The second sense of 'concept' indicates the stable capacity or developed potency to engage in just that act, whether or not the act is in fact exercised. Having concepts, we do conceive, think, and speak. The question is how are we supposed to think about the union between the having and the doing in St. Thomas. They are compared, as Aristotle himself remarks, as "knowledge to the exercise of knowledge."[42] Think of this in terms of the union between a habit (a developed potency) and its episodic exercise (a fully actualized potency). According to St. Thomas,

> it ought to be said that sometimes the intelligible species is in the intellect only potentially, and then the intellect is said to be in potency. Sometimes, however, it is in the intellect according to the ultimate completion of the act, and then the intellect is actually understanding. Sometimes it is in an intermediary mode between potency and

act, and then the intellect is called habitual. And in this manner the intellect conserves the species, even when it is not actually understanding.[43]

St. Thomas is simply saying that one can understand habitually without actually exercising that understanding. It is certainly the case that one's knowledge or understanding, in one sense, is not an episodic doing. I may know, in one sense, that the square root of two is an irrational number, even as I am not presently exercising that knowledge. It is a bit of knowledge that I "have." But this example just points up the fact that knowledge is not simply a having. It is also a bit of knowledge that I can exercise. It is a having that is ordered to a doing. Without the doing, it is in a sense incomplete.

Habits for an Aristotelian are stable, developed dispositions that stand somewhere between an undeveloped natural capacity to act and the actual exercise of that capacity. They are developed and strengthened through the exercise, and facilitate the subsequent exercise of the capacity. Habits are had, but the nature of what is had is ordered toward doing. Bravery is something we have *and* something we do, or more properly, ordered toward something we do. Roger Bannister had great running speed, and ran swiftly, because he developed his skill through exercise. Thus, I attribute the ordinary language connotation of 'having concepts' to the structured *habits* of the power of intellect. 'To conceive' I attribute to the *acts* that proceed from the power of intellect, often but not always structured by those habits. In the next two chapters I will have more to say about powers and their exercise. In the end, if one gets too hung up on *having* but not *doing* concepts, simply substitute in the appropriate contexts 'first act of intellect' and 'second act of intellect' where I have written 'concept'.

A Further Objection

Here I want to consider an objection that directly contradicts my account on historical grounds. In *Theories of Cognition in the Later Middle Ages* Robert Pasnau examines St. Thomas's discussion of the *intelligible species*. He argues that the intelligible species is indeed a *third thing* interposed between the act of understanding and the *extra-mental object* understood. This way of putting it is slightly different from the way in which I have been considering it. I have been speaking of the concept, and whether it is or is not a third thing interposed between the act of intellect and what is understood. I have argued that the concept is the act of understanding. Pasnau speaks of the species itself as the third thing. But the opposition to my view is clear, since I hold that the *species* is the form of the act of understanding and thus cannot be some thing in addition to it. Comparing St. Thomas's account of *species* with the parsimony of Ockham's account that denies them, Pasnau writes:

> On [Aquinas's] species theory, there seem to be three elements in an ordinary cognition: a cognitive power or faculty, a species of the right sort, and an object to produce that species. . . . But on Ockham's theory, only two things are required . . . a cognitive power and the thing cognized (which is an external object). So . . . it appears that Ockham's theory can do with two things what Aquinas's needs three to do.[44]

According to Pasnau, Ockham rejected species as *efficient* causes that precede cognition. So, by taking the *species* to be an efficient cause, Pasnau is able to maintain that it is an entity distinct from the cognitive power in St. Thomas's account. My argument is that it is not distinct because it is the *formal* cause of the cognitive power's act.

Pasnau's interpretation is based primarily upon two pieces of evidence. First, he finds two "functions" for the intelligible species in St. Thomas's account. One function consists in the intelligible *species* providing the content of the act of understanding, while the other function consists in the intelligible *species* being the agent cause of the act of understanding. He bases his account of these two "functions" upon a text in which St. Thomas writes:

> Every cognition occurs in virtue of some form that is the source of cognition in the cognizer. But a form of this sort can be considered in two ways—in one way in terms of the existence it has in the cognizer, in another way in terms of the relation it bears to the thing of which it is a likeness. In virtue of the first relation, it makes the cognizer actually cognize; in virtue of the second, it determines the cognition to some determinate cognizable thing.[45]

According to Pasnau, the content-determining function is based upon the second consideration of the form. The function of being an agent cause of the act of intellect is based upon the first consideration. Pasnau interprets St. Thomas's use of 'source' (*principium*) and 'makes' (*facit*) as the evidence for taking this function to be agent or efficient causation. Throughout the work, whether referring explicitly or implicitly back to this text, he stresses the causal role of the species as making cognition happen, as bringing about cognition, as producing it, as the agent of cognition, as its source.[46]

The second piece of evidence for the agency of *species*, sensible or intelligible, comes from Pasnau's interpretation of the medieval Aristotelian discussion of *species in medio*. St. Thomas, like many medievals, did not think that material beings could act at a distance. In order to account for the transmission of the form of the object known to the intellect of the knower when there is some distance between them, Aquinas speaks of *species in medio*, that is, the *sensible* and *intelligible* forms of the object existing "intentionally" in the medium, usually air, between the knower and the known. The object acts upon the medium causing its sensible and intelligible species to exist within it, and the medium transmits them to the cognitive power of the knower. We might think of the way sound is transmitted from a locomotive through the rails to someone's ear miles down the track. Significantly Ockham denied the need for such mediation and the attendant *species in medio*, holding instead the thesis of immediate action at a distance of the object upon the cognitive faculties.

When Pasnau considers the *species in medio*, he holds that in St. Thomas's account the species are agent causes. He quotes St. Thomas's writing, "Air altered by color makes the pupil be of this sort (i.e., makes it have a certain quality) impressing on it a species of the color."[47] By way of explanation, Pasnau paraphrases St. Thomas as follows:

> The agent in question is the intentionally existing species, transmitted from sensible object to sense organ. This species *in medio*, as an agent [sufficient for bringing its form into the patient], is completely sufficient for producing a sensation. In the case of sight, for instance, colors produce an effect in the air—a species *in medio*—which is transmitted to the visual sense, where the species produces vision.[48]

Notice the explicit identification of the *species* as the agent, rather than the air that is the intervening medium. In the case of intellectual cognition the *species* is not "completely sufficient" as an agent cause, but is rather "another kind of agent that of itself suffices for bringing its form into a patient only if another agent intervenes."[49] The "other agent" is the agent intellect in St. Thomas's account.

Later Pasnau will distinguish between the *species in medio* and the "final species" that actually informs the cognitive faculties. This "final species" "informs the cognitive faculty and actualizes it—that is produces a cognitive action. This species will be either a sensible species, if the faculty is one of the senses, or an intelligible species, if the faculty is intellect."[50] Thus, it is a general characteristic of *species*, whether in the cognitive faculties or *in medio*, that they are agent causes that "produce a cognitive action."

It is easy to see that if Ockham conceived of the *species* as an efficient or agent cause, he was compelled to conceive of it as a distinct entity. In the Aristotelian scheme of things nothing can be the agent cause of itself. Thus, if some power is brought to actuality from potency, it cannot be the agent of that alteration. Something distinct from it must be. If the *species* is the agent cause, either wholly in sensation or partially in understanding, then it must be distinct from the act of cognition. But, if St. Thomas did not hold that *species* are agent or efficient causes of cognition, either Ockham did not understand St. Thomas's view or he was not arguing against it. However, Pasnau does not recognize this dilemma because he chooses to read Aquinas through Ockham's eyes,[51] among others, and Pasnau's account shows no sign of recognizing the difference between agent and formal causes.[52] He thinks of formal causality as a subclass of agent causality. Early in the book he clarifies for the reader that the use of 'species' in his discussion is not taken in the logical sense as a "class within a genus," but in the Latin sense of "form or appearance."[53] In the context I am now examining, 'species' just means form. Consider then, in reverse order, the two pieces of evidence for Pasnau's account.

First, take the discussion of the *species in medio*. Look again at the quotation from St. Thomas, which reads in Pasnau's own translation "air altered by color makes the pupil be of this sort (i.e., makes it have a certain quality) impressing on it a species of the color." While in his paraphrase of this text Pasnau had explicitly identified the *species* as the agent, the plain meaning of the text is that air is the agent cause of sensation— "air altered by color makes the pupil be of this sort." Clearly the passage does not identify the agent with the *species* or form of color within it. It is also a mistake to take the phrase 'air altered by color' to indicate that *color* is an agent cause of the altered character of the air. Color is the intentional alteration of the air. The air is altered in that way by a colored *thing*. So, color is the accidental form of the thing that alters, and the *intentional* form of the air that is altered. It is in that sense that we should understand "air altered by color."

This confusion of formal causality for agent causality shows up when Pasnau cites a passage in support of the "complet[e] sufficien[cy]" of the sensible *species* as an agent cause. St. Thomas writes "things beyond the soul . . . are related to the exterior senses as sufficient agents. . . ."[54] This passage only states that things (*res*) beyond the soul are the sufficient agents in sensation. Paraphrasing the passage, Pasnau goes on to identify those *things* with *species in medio*, that is, forms. But the most straightforward way to understand what is meant by 'things' here is the sensible material objects that human beings encounter in the world, not simply the sensible forms of those objects as if the latter were free-floating agents. Pasnau cites no text in which St. Thomas himself refers to the *species in medio* as an agent cause of cognition, whether sensible or intelligible. Rather his paraphrases of St. Thomas's texts need to interpret him in that way, in order to make him subject to Ockham's objections.

So consider now the first piece of evidence for Pasnau's interpretation that attributes agency to the *species* as such, not *in medio*.

> Every cognition occurs in virtue of some form that is the source of cognition in the cognizer. But a form of this sort can be considered in two ways—in one way in terms of the existence it has in the cognizer, in another way in terms of the relation it bears to the thing of which it is a likeness. In virtue of the first relation, it makes the cognizer actually cognize; in virtue of the second, it determines the cognition to some determinate cognizable thing.

The text is explicitly about the *form* in virtue of which cognition takes place. Yet Pasnau finds in the term 'source'[55] and the second to last clause which says that "it makes the cognizer actually cognize" the evidence for the form's *agency*. But consider 'source'. Even in English the term *source* does not necessarily indicate an efficient cause. Just consider, 'the source of the Nile is Lake Victoria', or the 'source of the leak is a hole in the pipeline', or 'the source of my text is the *Opera Omnia*'. None of these indicate an agent or efficient cause of that for which they function as a source. 'Source' is Pasnau's translation of the Latin *principium* which has a number of different senses, like 'beginning', 'commencement', 'origin', 'foundation', 'principle'.[56] Deferrari associates it with Aristotle's ἀρχή, which has roughly the same range of senses. In the fifth book of the *Metaphysics*, describing the meaning of his terms, Aristotle begins, appropriately enough, with 'ἀρχή'. The common meaning he attributes to it is *the first thing from which a thing either is, comes to be, or is known*, and he distinguishes between an intrinsic αρχή and an extrinsic αρχή. When St. Thomas comments upon this text it becomes in Latin a discussion of *principium*. He clearly associates agent causality with *extrinsic* principles, while he associates formal causality with *intrinsic* principles.[57] In his own *De Principiis Naturae* (Concerning the Principles of Nature) St. Thomas makes this same association of formal and agent causes with intrinsic and extrinisic principles respectively. The *species* as a form must be an intrinsic principle of cognition, and cannot, therefore, be an agent cause of cognition, an extrinsic principle of cognition, as through Ockham's eyes Pasnau would have it.

Now consider St. Thomas's use of 'makes the cognizer actually cognize'. 'Makes' is Pasnau's translation of 'facit'. Pasnau takes this to be more evidence of agent causation. But once again, even in English it does not necessarily have that connotation. Consider, for example, the question "What makes the bronze a sphere instead of a cube?" One answer might be, "The sculptor makes it a sphere rather than a cube," an answer that gives the agent cause. Another thoroughly legitimate answer is "Its shape or form makes it a sphere. The bronze could have been a cube, but what actually makes it a sphere rather than a cube is the form of sphericity intrinsic to it." Pasnau's reduction of formal causality to efficient would, it seems, lead him to say that the shape or form is an agent cause operating upon the bronze to make it a sphere; neither the sculptor nor the sphericity is a "sufficient agent" of the bronze sphere, but acting together they are. But this misunderstands the relationship between agent and formal causality in Aristotelianism. In generation an agent cause does not act *with* a formal cause, but *brings about* a formal modification of an object. The generation is complete when the formal cause has been brought about by the agent.

The point about how to properly understand the relationship between agent causality and formal causality is not restricted solely to the generation or coming to be of some object, but also to the subsequent changes the object may undergo. A physicist may give at least two answers to the question, "What *makes* this bronze sphere begin to roll?" One answer would be in terms of some external force acting upon it. Another quite different but equally legitimate answer would be that it is spherical. Were it cubicle it would slide but not roll given that force. Without an analysis of how the shape of an object determines its *moment of inertia* and its subsequent *angular momentum*, as well as its instantaneous pivot points, the physicist cannot give an adequate account of what "makes" the motion of some object to be what it is, in this case a roll rather than a slide. But the causal role of the shape, *in making* the motion be what it is, is quite different from the causal role of the forces acting upon the object; it is not another agent acting internally to the bronze.

Aristotle's insight was to see that even if one exhaustively accounted for all of the agent causes of some change, one would still need to give an account of why the object of the change responds in the way it does, why in my example the bronze rolls rather than slides. *Ex hypothesi*, the further explanation would not be just another agent cause. It would specify *what* the character of the change consists in. On the point at issue, even if one exhaustively accounted for all of the agent causes of cognition, what is lacking in an account like Ockham's that denies *species*, sensible or intelligible, is why diverse cognitive processes and acts have the character they do. Why this process of *mediated* cognition leads to an act of sight rather than an act of smell. Why this act of understanding is an act of understanding a tree rather than a dog. *Ex hypothesi*, the explanation that is lacking is not just another agent cause of cognition, as Ockham seems to have believed. In St. Thomas it is the formal cause of the processes and acts, the sensible and intelligible *species*.

We have seen that for St. Thomas an efficient cause of some being is a principle extrinsic to it, while a form is an intrinsic principle. So whatever St. Thomas means by saying that the *species*, that is, *form* "makes" the cognizer actually cognize, he does not mean "makes" as an efficient or agent cause. The fact is that 'makes', Pasnau's translation of 'facit', is a common locution for St. Thomas in the context of formal causality. He writes throughout his work that "the form makes a thing to be" (*forma facit esse*)[58] or "the form makes a being actual and a *this something*" (*facit enim ens actu et hoc aliquid*).[59] Most often the contexts in which he speaks of the form "making" something be or be actual have to do with the substantial union of soul and body, but it is by no means restricted to substantial formality. Thus in the *Summa*:

> A substantial and an accidental form are partially alike and partially different. They are partially alike because each is act, and according to each something is in some way in act. They differ, however, . . . because a substantial form makes a thing to be absolutely, and its subject is a being in potency as such. An accidental form, however, does not make a thing to be absolutely, but to be in a certain respect, or so great, or in some relative manner, for its subject is a being in act.[60]

Thus 'makes to be' is far from being an extraordinary way for St. Thomas to characterize formal causality.

However, lest one conclude from that *use* of 'makes to be' that formal causality is a subclass of agent causality, three things should be kept in mind. First, formal causality is an intrinsic principle of the being for which it is a cause, while agent causality is an extrinsic principle. Second, St. Thomas explicitly distinguishes them in a number of places, where he indicates that formal causality is not agent causality, but is rather brought about by it.[61] Finally, despite the fact that the soul as form *makes* the body *to be actually* alive, St. Thomas writes explicitly that

> when the soul comes to the body, it does not make (*facit*) the body exist as an efficient cause, but only as a formal cause. However, the efficient cause that makes (*facit*) the body exist is that which gives the form to the body as perfecting it.[62]

Here St. Thomas uses 'makes' to characterize both formal and efficient causality, while he explicitly denies that the formal *making* is an efficient *making*. This distinction is not based upon the special case of the soul, but is the application to the soul of a general account of the difference between agent and formal causality, insofar as they are both said to "make," that is, "facit" some being. Thus, it is clear that *species* as formal causes, "making" the intellect to be in act, are not efficient causes; any argument to a "third thing" based upon their efficient causality is unsound.

Res and Ens in St. Thomas

I have argued thus far that concepts are not "third things" interposed between the act of intellect and objects in the world. But is it altogether inappropriate to call concepts *things* in any sense? From many of the critics it would seem so. Many believe, in Sokolowski's terms, that we must once and for all "exorcise them." Against the background of St. Thomas's metaphysical analysis of *being* (*ens*) and *thing* (*res*), in the next section I will explain the sense in which concepts can be called *things*, and why we need not exorcise them. J. L. Austin writes in a different context, "there's the bit where you say it and the bit where you take it back."[63] Having said that concepts are not "third things," this is the bit where I take it back, though not entirely.

It is important not to completely abandon the use of 'thing' ('*res*') in this Thomistic account of concepts. As used by St. Thomas, '*res*', like '*ens*', is an analogous term; indeed it is very much like '*ens*' in that it expresses a mode of *ens*, "which is not expressed by the name of *ens* itself." Though it does not "add" to the signification of '*ens*' by expressing a "restricted manner of *ens*," as each of the categories do, it "adds" to it by expressing "a manner that generally follows every being."[64] So of everything of which '*ens*' is said, '*res*' is also said.

When analyzing the ways in which '*ens*' is said, St. Thomas initially divides the senses into two classes. According to him, in one sense it is said of anything of which a true affirmative proposition can be formed. In this sense, '*ens*' is said to indicate the truth of a proposition. In this latter sense we may say that defects, lacks, absences, fictional creatures, and so on, are *entia* (beings). In this sense, '*ens*' may be said of blindness or unicorns, since we might say truly "blindness is a pathology in beings who ought to have sight," or "unicorns are sung about in Irish songs."[65] In this sense, '*ens*' may also be said of dogs, men, quantities, colors, and acts, since one may form true affirmative propositions about them. So in this sense at least, one can say that concepts are *ens* because one can truly say "concepts are written about by those who wish to deny their existence," just as one might truly say that "the present king of France is written about by those who wish to deny his existence."

Here, under threat of contradiction, '*ens*' cannot be synonymous with 'what has existence', where 'existence' has the sense that it has in "the present king of France is written about by those who wish to deny his existence." Since the present king does not have existence in the sense that it is truly denied of him, to say that he is an *ens* in that sense would be to assert that he both exists and does not exist. So even if one says that dogs, colors, and concepts in particular are *ens* in this sense, one is not thereby committed to their existence. I will use ens_p for this sense of *propositional being*—the being proper to something taken as the subject of a true affirmative proposition.

However, St. Thomas suggests another sense of *ens* that does seem to be synonymous with *what has existence*, in the sense of *existence* at play in "the present king of France is written about by those who wish to deny his existence." In this sense, it is divided among the categories, and this is to take *ens in re*. Notice the use of the preposition *in* here in an existential sense. I will use ens_r for this sense of *ens in re*. Here it is true to say that dogs, quantities, colors, and acts are $entia_r$, while blindness, unicorns, and the present king of France are not.[66] However, care must be taken to recognize that the use of $entia_r$ in this last statement has analogous senses included in it, since, as divided among the categories, ens_r is not applied univocally.[67] So when I write "dogs, quantities, colors, and acts are *entia*,"

I do not intend that each is an *ens$_r$* in the same way. Of course I could employ superscripts to distinguish all these senses, but I think the point is made without actually doing so. The intelligible character of *ens$_r$* differs for different categories, though they are all related to the fundamental category of substance.[68]

As in the examples of blindness, unicorns, and the present king of France, some *entia* are *entia$_p$* only, that is, only in the sense which indicates the truth of a proposition; on the other hand, as in the examples of dogs, colors, and acts generally, some *entia* are both *entia$_p$* and *entia$_r$*—that is, in the sense that indicates the truth of a proposition *and* the sense that indicates one or another of the categories *in re*. *Concepts* are acts of intellectual substances, namely human beings, and so as acts they have the sort of *ens* that acts have, that is, they are both *entia$_p$* and *entia$_r$*, just like a dog as a substance, in a more fundamental but analogous sense, is both.

How then does the use of '*res*' relate to these uses of '*ens*'? In the first disputed question of *De veritate*, St. Thomas relies upon Avicenna to indicate how *res* differs from *ens*—what it adds in its conception to "that which the intellect first conceives."[69] In article 1, while discussing *ens, unum* (one), *aliquid* (something), *verum* (true), and *bonum* (good), he writes that *ens* is taken from the act of being (*actus essendi*), while *res* expresses quiddity or essence of an *ens* (*quidditatem sive essentiam entis*). The distinction is not made in Question 21, article 1 when St. Thomas discusses what *bonum* adds to *ens*,[70] though he repeats much of the discussion of Question 1, article 1 on what *unum* and *verum* add. On the other hand, he does make the distinction again in the *Commentary on the Metaphysics*.[71] This latter text suggests that where *human nature* is considered in a way that abstracts from this or that particular human being, for example, Socrates or Plato, *res* involves a further abstraction from this or that kind of quiddity or nature, for example, human nature or equine nature.

These distinctions are useful in metaphysics, as one might say "every *ens* is a determinate *res*, that is, has a *what it is* or *essence*" (*quod quid est*), in other words, "every being is a determinate thing." Contrast this with "every determinate *res* is an *ens*, that is, has an act of being, or is actual"; in other words, "every determinate thing is a being." The apparent triviality of the latter statement, in comparison with the former, indicates how the intelligible character of *res* or *thing* includes within it *ens*, but not vice versa. In fact, the first statement, "every *ens* is a determinate *res*," suggests that *res* might also be a way of specifying the sense of '*ens*' at play in a specific context, for example when discussing the *properties* of *ens*, say *unum, bonum, verum, aliud,* and so on. It seems trivial to say, "every *res* is an *ens*." However, given the two senses of *ens* that St. Thomas distinguishes, ens_p and ens_r, it is at best ambiguous, and therefore not trivial to say, "every *ens* is a determinate *res*." What sense of *ens* is at play in such a statement? Is it ens_p, or ens_r, or both? Consider its contradictory, *some ens is not a determinate res*, that is, *some being is not a determinate thing*. Is this nonsense? It seems not for St. Thomas, if there is mere ens_p that is not also ens_r.

But not everything that is an ens_p is an ens_r, *blindness, the present king of France, the golden mountain*, and *unicorns* being prime examples. Now generally an ens_p that is not an ens_r does not have a quiddity or essence in the strict sense.[72] So it seems such an ens_p is not a *res*, because *res* is taken from the quiddity of an *ens*. According to the *Commentary on the Posterior Analytics*, I can give an account of the meaning of the term 'unicorn', namely *a horse with a single horn proceeding from its head*.[73] Such an account would be of the form "'unicorn' means *a horse with a single horn on its head*." However, this account of the name is not the sort of proposition that *establishes* unicorns as $entia_p$, since 'unicorn' is only mentioned in such an account, not used. So what we have is merely a true affirmative proposition about 'unicorn', and a term is an ens_r. By contrast, we know that unicorns are $entia_p$ because Irishmen sing about them, not because we can give an account of the common meaning of 'unicorn'. Yet, given our present state of knowledge, we are relatively certain that they are not $entia_r$, and consequently do not have *quiddities* or *essences* in the strict sense. Unicorn is my example. Here it might be of historical interest to some that the unicorn is not St. Thomas's favorite example of a mythological creature; the phoenix most likely is. Indeed, St. Thomas does not seem to think that the unicorn is a mythological creature, unless the rhinoceros is.[74]

In any case, the *account of the name* given for 'unicorn' is not an account of, or about unicorns, since unicorns are not terms used in speech. *A fortiori*, the account of the name is not an account of the *quiddity* or *essence* of unicorns, not an account of *what it is for unicorns to be*. Talking about *A* or *X*s, even when *A* or *X*s do not exist, is different from specifying the meaning of '*A*' or of '*X*'. To take another perhaps more famous example, to give an account of 'God' as *that than which none greater can be thought to exist* provides an account of the name or term used in speech, but it does not provide an account of the *essence* or *quiddity* of God, and it is not about God. Thus, St. Thomas asserts that a proof for the existence of God, based solely upon the account of 'God' and no other knowledge of some ens_r, is an invalid proof. Proofs *for* existence start *from* existence. To demonstrate existence, one must start with something that exists, even if the existence to be demonstrated is that of God and the existence that forms the basis of the demonstration is that of a creature; names alone will not do.

Although *entia*$_p$ do not have essences or quiddities in the "strict sense," there are contexts in which St. Thomas is willing to use 'essence' or 'quiddity' in speaking of *entia*$_p$ that he does not yet know exist, and which may turn out not to be *entia*$_r$.[75] As an example, he considers knowing the "essence" or "quiddity" of a phoenix or a man, without knowing whether there is a phoenix or a man. Joseph Bobik points out that in this example St. Thomas is not claiming ignorance of the existence of men, since surely in giving it, he knows that there are some men.[76] The fact that one of the disjuncts is an instance of *ens*$_p$ which he knows to be *ens*$_r$, suggests by contrast that he is certain that the other is not an *ens*$_r$, namely, there is no phoenix. The juxtaposition of the two suggests that the point he is going to make does not depend upon their difference, but is that by knowing *essentia*, one does not know, by that very fact, the *esse* of some being having that *essentia*. The contrast between human nature and the nature of the phoenix suggests that knowledge of essence provides no more knowledge of existence in the case of *entia*$_r$, than it does in the case of mere *entia*$_p$, namely, none. Attributing an *essentia* to a phoenix involves speaking loosely, but it is done in order to further the point being made, namely that knowing *esse* in an *ens*$_r$ is "other than" knowing *essentia* or *quiddity* in that *ens*$_r$. In this instance, St. Thomas is willing to use 'essence' and 'quiddity', though loosely. But of course this is not the context of the first chapter of the *De ente*, where he says that '*essentia*' applies only in instances of *ens*$_r$. When one *knows* existence, '*essentia*' and 'quiddity' properly apply, while when one *knows* or is at least certain of non-existence, they do not. So, strictly speaking, only *entia*$_r$ are said to have *quiddities* or *essences*. But this suggests, since *res* is taken from *essentia* or *quidditas*, that only *entia*$_r$ are *res*. *Entia*$_p$, which are not at the same time *entia*$_r$, are not *res*. In "every *ens* is a *res*," the sense of '*ens*' at play must be *ens*$_r$; in "some *ens* is not a *res*," the sense of '*ens*' must be *ens*$_p$. So there is no dual sense of '*res*', paralleling the dual sense of '*ens*', as if there were *res*$_p$ and *res*$_r$. There is simply *res* which follows upon *ens*$_r$.

This reservation of '*res*' to *ens*$_r$ has clear implications in a number of contexts. When considering the transcendental properties of *ens*, since *res* is one of them, one must keep in mind that the sense at play is *ens*$_r$, not *ens*$_p$. When, as in the *Quaestiones disputatae* and *In Commentarium Metaphysicam*, St. Thomas discusses the properties of *ens* qua *ens* like *res*, *unum*, *aliquid*, *verum*, *bonum*, he is discussing the properties of *ens*$_r$, not *ens*$_p$. So for instance in the *Commentary on the Metaphysics*, he writes:

> [F]or it ought to be understood that this name 'man', is imposed from the quiddity or nature of man; and this name 'thing' is imposed from the quiddity merely; indeed this name 'being' is imposed from the act of being: and this name 'one' from order or indivision. For one [thing] is an undivided being. But it is the same [thing] that has an essence and quiddity through that essence, and that is undivided in itself. Whence these three, 'thing', 'being', and 'one' signify altogether the same, but according to diverse intelligible characteristics.[77]

Here he identifies the sense of '*ens*' ('being') as that which is taken from the *actus essendi* ('act of being'), which sense I have been calling ens_r. He writes earlier in the same paragraph that "they [unity and being] are the same thing (*res*), differing only in intelligible character."

In the very next lesson, while discussing in what sense *unum* is a negation, St. Thomas writes, "negation, which is included in the intelligible character of *one*, is a negation in a subject (otherwise a non-being could be called *one*)."[78] For our purposes what is interesting is in the parenthetical remark. In the same passage he had earlier used a "chimera" as an example of a non-being. In that passage he is analyzing why the kind of negation involved in the *unum* that is coextensive with *ens* is more akin to a privation and a negation in a subject, than to simple negation. He says simple negation like *non-seeing* can be said of non-beings as much as of subjects for which sight is a natural perfection, as opposed to *blindness* which can only be a negation in a subject that ought to have sight.

> [H]ence absolute negation can be verified as much of a non-being which is not of such a nature as to be the subject of an affirmation, as of a being which is of such a nature as to have something affirmed of it, though [as a matter of fact] it does not have it. For non-seeing can be said of a chimaera as much as of a stone and as much as of a man. But for a privation there is a definite nature or substance of which the privation is said: for not everything that does not see can be called blind, but only what is of such a nature as to have sight.[79]

It is then that he writes that the negation that is included within the intelligible character of *one* is said like a privation of a subject, "otherwise a non-being could be called *one*." St. Thomas takes it for granted that it is illegitimate for a non-being, that is, a mere ens_p, to be called *one*. It seems to follow then that the remaining transcendentals, namely, '*res*', '*aliud*', '*verum*', and '*bonum*' are also illegitimately said of ens_p.

Now consider St. Thomas's position on what the "intellect first conceives as most known":

> [B]ut that which the intellect first conceives as most known, and into which it resolves all concepts, is *being*, as Avicenna says in the beginning of his *Metaphysics*.[80]

The fact that in this context St. Thomas is telling us what 'res', 'unum', 'aliquid', 'verum', and 'bonum' add in their signification to 'ens' suggests that he does not have in mind the ambiguous use of 'ens', prior to the division into 'ens$_p$' and 'ens$_r$'. He is writing of ens$_r$ as that which the "intellect first conceives as most known." He is not writing of some sort of prior common conception of *ens*, common to both ens$_p$ and ens$_r$. In each of the two acts of intellect, understanding of indivisibles and combining and dividing, there is something "first." In particular the first operation conceives the *quod quid est*, or quiddity, and

> indeed in the first operation something is first, which falls within the conception of the intellect, namely what I call *ens*; nor is something able to be conceived by the mind by this operation, unless *ens* is understood.

The sense of 'ens' here, presupposed to the conception of *quiddity*, must be ens$_r$. When he writes that "nor is something able to be conceived by the mind by this operation, unless *ens* is understood, this also suggests that ens$_p$ is only derivatively, and secondarily, or *secundum quid* conceived by the intellect, depending upon the intellect's primary and *per se* conception of ens$_r$. Ens$_p$ and ens$_r$ do not have a common intelligibility, a common intelligibility prior to *actus essendi*. Whatever intelligibility ens$_p$ has is derivatively and analogously had from the prior intelligibility of ens$_r$.[81] In other words, the intelligibility of propositional being (ens$_p$) is dependent for its very conceivability upon the prior intelligibility of the being of things (ens$_r$). So, in addition, the division in the *De ente* of the ways in which 'ens' is said, into 'ens$_p$' and 'ens$_r$', is not an *a priori* division based upon the very notion of *ens*, as if it were engaged in a Porphyrian division of *being*. It is *prima facie* a division based upon how 'ens' is "said," and how it is said, depends on how *ens* is known. "We name as we know." It is known as existing or *thing-like* (*res*), before it is known as propositional.

A mere ens$_p$ is not a *res*, nor does it have any of the other properties of ens$_r$ qua ens$_r$. If the reservation of *res* to ens$_r$ is correctly seen against this rich metaphysical structure, it will not be surprising if in addition it proves useful for sorting out the difficulties of the Third Thing Thesis. 'Res' indicates an ens_r having a quiddity or essence. Mere $entia_p$ are not *res*. Blindness and other pathologies, unicorns, Santa Claus, and Hamlet, all *ens* in some sense, namely ens_p, nonetheless are in no sense *res*; 'res' simply does not apply to mere ens_p.[82]

Here it is important to note the breakdown of the translation of the Latin 'res' into the English 'thing', at least the metaphysically charged sense of 'res' that St. Thomas discusses. In English, in certain contexts, we do not have a difficulty with saying that beings of fiction or imagination are things; we just deny that they are *real* things. We do not shudder at the locution 'unicorns are not real things, but imaginary things', or 'The Golden Mountain is not a real thing, but an imaginary thing that plays a prominent role in philosophical discussions of Logic'.' In these contexts, if we were to translate the Latin 'res' it seems appropriate to translate it with 'real thing', rather than 'thing' simply. If 'real' adds non-redundantly to 'thing' in 'real thing', then 'thing' as it stands cannot already have the sense of 'real' contained within it. In English, 'thing' appears to be synonymous with 'being'. In this way, when we say in English that *some thing is not real*, we are merely saying the equivalent in Latin of '*ens non est res*', with the sense of ens_p, not ens_r.

Consequently, when philosophers suggest that a concept is not a *thing*, there are any number of ambiguities that arise. In St. Thomas's terminology, it cannot be the claim that concepts are not *entia* in any sense, for they have at least the propositional ens_p involved in the true affirmative proposition *concepts are talked about by those philosophers who wish to deny their existence*. The denial must be construed along the more robust and significant lines that concepts are *real things* or *real beings*, and this does correspond in St. Thomas's terminology to a denial that they are *res*. The denial would hold that concepts are merely $entia_p$, and not $entia_r$. Concepts are like unicorns sung about by Irishmen, no more real than fairies and ghosts. Why we should then exorcise them is not clear, since we do not exorcise delusions of demons, but counsel those suffering from them, and perhaps medicate them. Exorcism is reserved for real possession by real demons.

At the very beginning of the discussion, I classified concepts as ens_r, because they are in the category of acts. Now, just as 'ens,' has analogous uses across the categories, so will '*res*'. In its particular application to a category, '*res*' will "add" to the sense of 'ens_r' appropriate to that category the note of *quiddity* or *essence* appropriate to that category. Consequently, its own sense will vary as ens_r varies. St. Thomas writes:

> But because *ens*$_{[r]}$ is said absolutely and primarily of substances, and secondarily and as qualified of accidents, so it is that *essentia* is properly and truly in substances, but in accidents in a certain way, and as qualified.[83]

Like the fundamental sense of '*ens*,', the fundamental sense of '*res*', taken as it is from *quidditas*, will apply to the category of substance. However, in relation to this fundamental sense, '*res*' will also be applied to the other categories, conceived broadly as accidents of substance.[84] Thus quantities are *res* in their relation to substance, places are *res* in relation to substance, and so on. Indicating quiddity or essence in an *ens*$_r$, '*res*' will vary just as the notion of definition, which expresses quiddity, varies across the categories.

In particular, acts are *res* in their relation to substance. An act of conceiving some *res extra animam*, whether that *res extra animam* is a substance or accident of substance, is itself a *res*, namely a *res in anima*. But a *res in anima* is just as much an *ens in re*[85] as is any *res extra animam*, such as a walk in the park. This is in sharp contrast to the more modern dichotomy between what is "in reality" and what is "in the mind," where it often seems that something is not "in reality" if it is "in the mind." Reading back into the Latin terms, it seems that the more modern fundamental distinction is between *ens in re* and *ens in anima*. But this opposition of *in anima* and *in re* is not in St. Thomas. The distinction in him is between *ens*$_r$ and *ens*$_p$. And *res in anima* are just as much *ens*$_r$ as are *res extra animam*, recalling again the existential character of the prepositions *in* and *extra*.

One might think that St. Thomas's remarks in Book IV, lesson 4, passage 574 of the *Commentary on the Metaphysics* contravenes the analysis I have been giving. The English text from Regnery Press says that "there are two kinds of beings: beings of reason and real beings." Since it speaks of "two kinds of being," it seems straightforward to identify the "beings of reason" with *ens*$_p$, and the "real beings" with *ens*$_r$. As the translation has it:

> [T]he expression 'being of reason' is applied properly to those notions which reason derives from the objects it considers, for example notions of genus, species and the like, *which are not found in reality but are a natural result of the consideration of reason*. (Emphasis added)

And he says that such "beings of reason" form the proper subject matter of Logic, suggesting that *entia in anima* are not "real beings," but rather *beings of reason*. It seems that genera and species are akin to unicorns and chimera, and the study of Logic akin to literary criticism.

Where the translator has "being of reason," the Latin has "ens rationis." Here the translation appears appropriate. But where the translator has "real beings" and "not found in reality, but are a natural result of the consideration of reason," the Latin has "ens naturae," and "non inveniuntur in rerum natura, sed considerationem rationis consequuntur," respectively. Here I am less sanguine about the translation. "Ens naturae" can be straightforwardly translated "[a] being of nature," or more gracefully "a natural being." The only reason for translating it as "real being" seems to be the appearance of "rerum" in the other passage, namely "in rerum natura." But that passage seems better translated either as "not discovered in natural things, but following the consideration of reason," or "not discovered in the nature of things, but following the consideration of reason." In the first possible translation, "natural things" would have its sense from the contrast with "follows the consideration of reason." This is the sense of a phrase like "art imitates nature," or the sense in which we contrast a "natural" occurrence with something that occurs as a product of deliberation. Art and acts that occur as a product of deliberation "follow the consideration of reason." In neither suggested translation does it seem appropriate to say that what is not "natural" is not "real." Deliberative acts and artistic productions are no less real for following upon the consideration of reason.

The first hint that something is amiss in identifying *ens rationis* with unreal mere ens_p is that it would include beings like unicorns, evil, and blindness. But unicorns do not "follow upon the consideration of reason," at least not in the straightforward sense that genera and species do. Second, the passage brings to mind the *De ente*, where Aquinas argues that such intentions as *genera*, *difference*, and *species* are not found in the nature of things "absolutely considered," but accrue to a nature following its mode of existence in an intellect, that is, accrue to a nature following upon the activity of the intellect conceiving it.[86] This parallel is striking, and suggestive of how 'in rerum natura' ought to be understood in the passage from the *Commentary on the Metaphysics*—not as 'in reality', but as 'in the absolute consideration of the nature of things,' or taking 'absolute consideration' as understood, more simply 'in the nature of things'. This interpretation is strengthened when St. Thomas writes later that the dialectician proceeds by considering the "*intentionibus rationis, quae sunt extranea a natura rerum.*" The translator has "conceptions of reason, which are extrinsic to reality," but it is more felicitous as "the intentions of reason, which are extraneous to the nature of things." But being *extraneous to the nature of things* does not exclude being real, since there are any number of *real things* (*ens$_r$*) that are extraneous to the *nature of things*, namely, accidents, as for instance the color of my skin is extraneous to my human nature, but no less "real." This again is reminiscent of the nature "absolutely considered" in the *De ente*.

The division between *ens rationis* and *ens naturae* being discussed here is a division between *being* that in some sense follows upon the activity of the intellect and characteristics of beings found in their natures absolutely considered. If there are texts in which fictions like unicorns and chimera are called *ens rationis*, it is not *because* they are unreal beings, though they are in fact unreal beings; it is because whatever characteristics they have as *ens$_p$* are not found together as *one* in the natures, absolutely considered, of any *entia$_r$*.[87] The unity of *horse, horn,* and *head* in unicorn proceeds solely from the consideration of reason; it is not an essential unity, a unity of elements discovered in the nature, absolutely considered, of an *ens$_r$*. However, from the proposition that *some beings that follow upon the consideration of reason are unreal beings*, one cannot conclude the proposition that *everything that in some way follows upon the consideration of reason is an unreal being*.

The analysis I provided above allows me to put some sense into the phrase 'unreal being'. *Unreal being is mere ens$_p$*. Now, it is the case that every *mere ens$_p$* is an *ens rationis*. But, from *some entia rationis are mere entia$_p$*, it does not follow that *every ens rationis is a mere ens$_p$*. Entia rationis that are mere *entia$_p$*, unicorns for example, seem to differ from *entia rationis* that are *entia$_r$*, genera and species for example, because the former do not accrue to a nature as an accident of its mode of existence in the intellect, like the latter do. Further, *entia rationis* that are mere *entia$_p$*, like blindness, evil, and other pathologies, seem to differ from mere *entia$_p$*, like unicorns, because the former are related logically, as negations, to real accidents, that is, types of *entia$_r$* that *ought* to accrue to a nature in its mode of existence *extra animam*. Fictions as *entia rationis* seem to be unique in that they are neither natures absolutely considered, nor accidents, *extra animam* or *in anima*, that accrue to a nature in its modes of existence, nor the negation of such accidents. Their being (*ens$_p$*) and unity (*unum*) has no foundation in a thing (*res*), no *fundamentum in re*.

There is an unfortunate tendency to translate '*fundamentum in re*' as 'foundation in reality'. This tends to suggest that if one says that some *being of reason* has a *fundamentum in re*, a foundation in reality, the *being of reason* is itself unreal by contrast. So for example, if one says that *species* and *genera*, which are *beings of reason*, have a *fundamentum in re*, one is tempted to think that by contrast they are *unreal beings with a "foundation in reality."* By now it should be clear why it is better translated 'foundation in a thing', where the thing (*res*) in question has an essence or quiddity that constitutes the foundation—the *fundamentum in re*. Thus, beings of reason like *species* and *genera* have foundations in a thing or things that are not *beings of reason*. But this way of translating it in no way suggests that *species* and *genera* are *unreal* beings. Fictional beings like *unicorns*, by contrast, have no *fundamentum in re*, no foundation in a thing.

So *ens rationis*, at least in this particular text concerning *genera* and *species*, is not to be contrasted with *ens in re*, as if *ens rationis* like *genera* and *species* were *mere ens$_p$*, and not *ens$_r$*. And *ens in anima* should not be identified with "beings not found in reality," as if reality was exhausted by the realm of *ens extra animam*. It is also a mistake to identify mere *entia$_p$* with beings in the soul or mind, along the lines of the modern theory of ideas. Unicorns, blindness, the present king of France, and so on are not "in the head." Not existing, they are nowhere.

Something may follow the consideration of reason, that is, be an *ens rationis*, without being a *res in anima*. But consider the science of Logic, which does not study natures absolutely considered. It studies the formal relations or intentions that accrue to natures *as known*, that is, that accrue to natures in the mode of existence that they have *in anima*. Genus, species, and difference are *ens rationis* that are *res in anima*. The recognition that concepts are *res in anima*, that is, are quidditative, opens up the vista of so-called "second intentions" to a speculative investigation. Studying them is not akin to studying fictions like unicorns. A study of the essential characteristics involved in a nature *as known* will reveal the formal relations or intentions between such concepts as species, genera, and difference, as well as the formal relations or intentions between more complex concepts as are formed by the intellect's act of combining and dividing, and proceeding from one thing known to another. But because these are essential formal relations between *res in anima*, one can begin to see a justification for the normative claim that our thoughts, if they are not to be defective, ought to embody these formal relations, without on the other hand lapsing into the sort of *psychologism* that appeared attractive to many in the nineteenth century.[88] Further, one can now more fully appreciate the assumption of unity that St. Thomas makes about the order of Logic based upon the acts of reason.

In modern English 'thing' and 'being' may be synonymous, that is, without the difference of intelligible character that St. Thomas attributes to *'ens,'* and *'res'* in Latin. Even if this is so, against the background of *analogy* in St. Thomas, it is perfectly legitimate to say that a concept is a *res*. As a *res* it has a quidditative character, which is *to be knowledge* of some *res*. However as a *res* itself, it is also an *ens*, which means it has an *actus essendi* (an act of being). In its *actus essendi*, rather than its character as knowledge, a concept is an act of the substance that knows. In the sense of *accident* that covers all the non-substantial categories, a concept is a *res* in the way in which accidents are *res*. But the result is that it is not a thing or *res* in the sense that pertains to the Third Thing Thesis; it is not a thing distinct from the act of understanding.

The Analogy of 'Being' and 'Thing'

This analogous use of *'res'* applied to concepts provides a response to what Putnam generally thinks is missing in contemporary accounts, namely a recognition that such terms as 'being', 'existence', 'object', and 'thing' are not univocal terms. In *The Many Faces of Realism* Putnam sees himself as overcoming the sorts of worries motivated by various dualisms, including that between the *objective* and the *subjective*—"the idea of a 'point at which' subjectivity ceases and Objectivity-with-a-capital-O begins has proved chimerical."[89] In particular, he stresses how in our talk about *objects* existing in the world, what counts as an *object* depends upon the conceptual categories we bring to the discussion. He is concerned to battle what he calls "metaphysical realism," one element of which, as he characterizes it, is the thesis that the world comes to us already "cut up" into a determinate number of objects, the cutting up of which is wholly independent of our choice of conceptual schemes. Because the world comes to us already cut up into objects, there is one true theory, consisting of all the truths about that determinate set of objects. In such a theory the truths are independent of our conceptual activities and our task is merely to discover them. What Putnam is objecting to is the thesis that objects in the world are "things in themselves" apart from the classifications of our conceptual scheme.

He asks us to consider the question "How many objects exist in a world?" For the metaphysical realist, this question should have a determinate answer, an answer that does not depend upon any "subjective" conceptual states. So consider a world consisting of three marbles. Initially the metaphysical realist is inclined to answer that such a world consists of three objects or individuals, and no more. But this is only because he tacitly assumes the background concept *marble* as the criterion by which something will be judged to exist or to be an object. Suppose he had tacitly used a different conceptual apparatus for counting, say that of the Polish logician Lezniewski, where an object may be a logical atom or a sum of logical atoms. Then the world consists of seven objects, $marble_1$, $marble_2$, $marble_3$, $marble_1 + marble_2$, $marble_1 + marble_3$, $marble_2 + marble_3$, and $marble_1 + marble_2 + marble_3$; these are the objects that the Polish logician existentially quantifies over. A less esoteric example of a conceptual apparatus than the *mereological sum of logical atoms* above might be that of *molecule*. Had he tacitly used *molecule*, instead of *marble*, he would not have counted the three marbles as objects, that is, as existing. He would have counted the number of molecules, but not the marbles, passing them by as not truly objects—as not truly existing.

Suppose then he decides to combine the two concepts into a disjunctive concept. *An object is a molecule or a marble.* Then he will count both molecules and marbles. But what about atoms? Now suppose we ask how many objects exist, not in this imaginary marble world, but in the real world of the metaphysical realist? What sort of disjunctive concept will he need in order to count all the objects in a room, let alone the whole world? Nothing will count as a thing or object, except against some background conceptual apparatus that allows for the counting. But that is to admit that what counts as an object is relative to our conceptual functioning; it is to make the status of objects subjective. If we take concepts to be ideas, we are likely to say with Locke that "'tis evident, that *Men make sorts of Things*," while if we reject ideas and find concepts in language, we might say with others that "everything is a text." Against the background of my analysis of *ens rationis* and *mere ens$_p$*, it appears that every object or thing other than the mind is at least an *ens rationis*, and in the extreme perhaps every object or thing is even a *mere ens$_p$*.

Putnam draws the conclusion that among other things the existential quantifier is not univocal.

> [I]t is no accident that metaphysical realism cannot really recognize the phenomenon of conceptual relativity—for that phenomenon turns on the fact that *the logical primitives themselves, and in particular the notions of object and existence, have a multitude of different uses rather than one absolute 'meaning'*.[90]

Quantification and existence questions are always answered against a background conceptual scheme; there are no pure objects, only conceptually mediated objects; there is no pure existence, only conceptually mediated existence. This is why he writes that the "trail of the human serpent is over all." Though relative to our conceptual apparatus, Putnam does not think that existence or objectivity is thus somehow completely conventional or subjective, "once we make clear how we are using 'object' (or 'exist'), the question 'How many objects exist?' has an answer that is not at all a matter of 'convention'. . . ." Conventionality is *across* conceptual schemes, not *within* them. He urges us to give up the *object-subject* axis—to give up the illusion that there are things the existence of which does not depend upon our conscious activity, that is, *objective things*, and things the existence of which does depend upon our conscious activity, that is, *subjective things*.

Putnam's attack is directed at those philosophers who hold that concepts are human capacities for certain kinds of activity, in particular capacities for linguistic expression, or scientific understanding, who at the same time believe that the truths about what exists are wholly independent of those capacities. If they are human capacities, since what counts as an object is relative to a concept, what counts as an object is relative to a human capacity. In that case, the metaphysical realist is wrong for thinking that the world is already parsed into objects and truths about them, prior to any consideration of human conceptual capacities and activities. Against a background in which 'object' is synonymous with 'thing' ('*res*' in Latin), and is an ontological term having to do with "what exists," Putnam objects that there are no such mind independent things, that is, that there are no objects or things in themselves. What counts as existing, as a thing, as an object is determined not by the world itself, but by our conceptual activity. Terms like 'thing', 'object', 'being', 'exists', and so on each have a variety of different uses determined, presumably, by our conceptual choices. Once the choice of conceptual scheme is made, the web of truth unfolds from within.

Notice, however, that Putnam does not tell the reader whether and how the multitude of uses might be related. Indeed, it doesn't seem that he can provide such an account of how they are related. For him, the multiplicity of uses marks a multiplicity of conceptual schemes. Reacting to the position that the world comes already parsed into a determinate set of objects, where 'object' is used univocally, Putnam believes that a term like 'object' has non-univocal uses. This recognition of the non-univocal use of these terms is, for the Thomistically oriented philosopher, all well and good. Putnam does not go on to say, however, what kind of non-univocal uses they have. His account in terms of distinct conceptual schemes that are "chosen" suggests pure equivocation, since to the extent that the conceptual schemes are truly distinct, the terms would then be *prima facie* unrelated. However, if they are analogous uses, they must be related in some way, which would suggest that they are elements of the same conceptual scheme, rather than distinct conceptual schemes.

Now consider again terms like 'being', 'existence', 'object,' and 'thing' that Putnam believes have non-univocal uses. In St. Thomas's appropriation of Aristotle these terms have the variety of uses that Putnam is looking for; they are analogous terms. In particular, they have this analogous variety in their application to the categories. Indeed, Putnam tends to treat 'being', 'object', and 'thing' as if they are synonymous, and then to argue that these synonymous terms have non-univocal uses. From my analysis, it is clear that the terms are not even synonymous, as 'being' ('ens') has a different sense than 'thing' ('res'), even as they both have analogous uses across the categories.

This recognition of analogy does not alone overcome the metaphysical conundrum dreamed up by Putnam. As long as he maintains that we *determine* conceptual content by "choosing" our conceptual schemes, the conundrum appears to remain. He thinks the diverse uses of these terms put one in diverse conceptual schemes. But, since diverse conceptual schemes determine what counts as *the world*, these uses put one in diverse *worlds*. But if he were to recognize analogy, he could still hold this, since there might well be different conceptual schemes to "choose" among, with different self-contained structures of analogous use. While his "metaphysical realist" comes with a "readymade world," Putnam's "internal realist" comes with a repetoire of "readymade" conceptual schemes. But if conceptual content is determined for us by our active engagement with the world, that is, our active response as cognitive beings in the world to the structures of being that we encounter in the world, as a Thomist thinks it is determined, then the conundrum does appear to fade away. If we think of conceptual content as our developmental and specifically human response to being in the world, it is not particularly obvious that we "choose" conceptual schemes any more than we choose the shape of our toes, even if our conceptual response is in some sense voluntary while the shape of our toes is not. It appears more likely that our specifically human development is or is not adequate to our being in the world, that is, that it is true or false, good or bad.

Of course what we already know or claim to know typically informs our interests in such a way as to determine what questions we find interesting and what directions we wish to pursue in our investigations, and we adopt partial, inadequate, and even erroneous views that need development and correction where necessary. Our passions are involved in our "development" as rational animals, and we can choose to act more or less rationally. Nevertheless we are the sort of beings for whom it is a natural part of our development to be informed by the structures of the material being amidst which we live, and for that "information" to inform in general and particular ways our activities as rational animals. Thomistic Realism is not committed to the position that acquiring knowledge of the world is a purely passive, automatic, and mechanistically determined process on the part of the knower, as if she is simply being stamped with the structure of things. Neither, however, is it committed to a purely active process, as Putnam's talk of "choice" of conceptual schemes and "determining" conceptual content by those choices suggests. Here a more adequate understanding of abstraction is necessary, which I will attempt to sketch in the next chapter.

Chapter 7

THE INTROSPECTIBILITY THESIS

> It is manifest that the object of the intellect is not an intelligible species, but rather the quiddity of a thing understood.
> —*St. Thomas*

St. Thomas held that we know corporeal things. A cursory look at Questions 84–86 of the first part of the *Summa* shows that he never seriously considers the possibility that we do not know corporeal things. He makes plain in the introduction to Question 84 that the order of procedure in considering the knowledge of the incarnate soul proceeds from what is below it, to itself and what is in it, to what is above it. In particular, concerning corporeal things he writes that "three things ought to be considered: first, through what do we cognize *them*, second, in what manner and order, and third, what do we cognize in them."[1] It never occurs to him to ask, "Do we cognize corporeal things?" It is also quite clear that he held that the first concept of the intellect is *being*. After our consideration of the Third Thing Thesis in the last chapter, we see that this concept of *being* is the concept of *existing being, ens,*. Finally, he held that the proper *obiectum* of the human intellect is the nature of material beings:

> Thus it appears from this that the Philosopher says that the *proper obiectum* of the intellect is the quiddity of a thing, which is not separate

from things, as the Platonists held. Hence that which is the *obiectum* of our intellect is not something existing apart from sensible things, as the Platonists held, but something existing in sensible things.²

Notice the stress on the fact that the *obiectum* is not something "existing apart from the sensible things;" it is "something existing in sensible things." This conclusion follows for St. Thomas because the human intellect, contrasted with the angelic and the Divine, is a power of a form that is the substantial form of a material substance.

But these positions of St. Thomas are not completely sufficient to address the Introspectibility Thesis. Putnam, comparing "Aristotelians" with "Platonists," writes that

> even in . . . Platonistic versions . . . speakers are supposed to be able to direct their mental attention *to* concepts by means of something akin to perception, and, if A and B are different concepts, then attending to A and attending to B are different mental states. So even in these theories, the mental state of the speaker determines which concept he is attending to, and thereby determines what it is he refers to.³

In mentioning the "Platonistic versions," his intention is to describe what they share in common with the Aristotelian versions. In both instances, the Platonist and the Aristotelian, he believes there is a capacity, "something akin to perception," by which the speaker grasps concepts. For the Platonists these "'concepts' are not in the mind, but rather form a realm of abstract entities . . . independent both of the mind and of the world." Putnam believes that for the Aristotelian, by contrast, this capacity directs itself upon "something in the mind that picks out the objects in the environment that we talk about."⁴ This Aristotelian "something" Putnam calls the concept. Thus, he is under the impression that the concept is grasped by a mental act "akin to perception."

We have seen the difficulties with this picture of the intellect grasping concepts, when we considered the Third Thing Thesis. Still, once the concept as an *act* is recognized, as well as the senses in which it is and is not a third thing, the Introspectibility Thesis is not automatically overcome. An act of understanding need not take as its object a being other than itself. The prime example when it does not is when the knowing being is God. St. Thomas argues that in God's knowledge of Himself, "that which is understood, and the intelligible species [by which it is understood], and the act of understanding itself are altogether one and the same thing."[5] I do not want to go into St. Thomas's argument for this position, but those familiar with Aristotle might usefully think of "self-thinking thought thinking itself." This example shows that for St. Thomas it is not automatic that an act of understanding take as its object a being other than itself. Even if the concept is not an entity distinct from the act of understanding, it might still be the case that it is "self-introspectible," and by being known *first* or *primarily* makes other things known subsequently. Thus, in the case of intellectual creatures, it is necessary to ask whether the thing primarily known is identical with the act by which it is known. In the case of human beings, St. Thomas's answer is no, but it is necessary to see why.

When Putnam contrasts the Platonist with the Aristotelian, he says the difference is a matter of whether the mind grasps a thing in an abstract realm apart from the mind, and presumably apart from sensible things, or an immanent mental thing equally apart from sensible things. When St. Thomas contrasts the Platonists with Aristotle, he agrees roughly on the account of the Platonists, but there is no suggestion that the other option is an immanent mental thing. Rather, for him, the Aristotelian option is the quiddity or essence of sensible things existing in those sensible things. However, even as he says this, one can see the possibility of misunderstanding. In the same sentence in which he writes that the proper *obiectum* is something existing in sensible things, he also seems to concede something else.

> [T]he [proper] *obiectum* of our intellect is . . . something in sensible things, allowing that the intellect apprehends the quiddities of things in a manner that differs from the way they are in sensible things: for it does not apprehend them with the individual conditions to which they are adjoined in sensible things.[6]

The concession at the end seems to suggest something like Putnam's view. Essences as they exist in sensible things have particular conditions accruing to them. This is clear from the discussion of the *De ente,* among other places. But here he says that the intellect "apprehends the quiddity of things" without the individuating conditions adjoined to the essence in sensible things. Doesn't this suggest that it must apprehend the essence as it exists apart from sensible things, not in the Platonist way, but in the way suggested by Putnam—as an immanent mental thing? It is also clear from the discussion of the *De ente* that the essence in intellect exists without the individuating conditions of material things. Wouldn't

Putnam's reading of this text be plausible—that the intellect grasps the essence as it exists in the mind without individuating conditions, and then, because this essence in the mind is formally identical with the essence as it exists in material things, it grasps the material things? One can agree that the proper *obiectum* of the intellect is the nature of material things, but it only attains its proper *obiectum* by first grasping the natures as they exist in the intellect. This issue was raised in the previous chapter while we considered what is meant by the thesis that *the intelligible in act is the intellect in act*.

The plausibility of Putnam's account of the Aristotelian is bolstered by considering one translation of St. Thomas's earlier passage in the *De anima* commentary. The passage in Latin is:

> Ipsa autem natura cui advenit intentio universalitatis, puta natura hominis, habet duplex esse: unum quidem materiale secundum quod est in materia naturali; aliud autem immateriale secundum quod est in intellectu. Secundum igitur quod habet esse in materia naturali, non potest ei advenire intentio universalitatis, quia per materiam individuatur; advenit igitur ei universalitatis intentio secundum quod abstrahitur a materia individuali.[7]

This is another statement of the theses from the *De ente*. Natures can be considered with respect to two modes of existence, in the intellect immaterially, and in material things. Universality is an intention that accrues to a nature as it exists immaterially in intellect, not to a nature as it exists materially. But consider this translation of the sentence beginning "secundum igitur":

> As in the material mode of existence it cannot be represented in a universal notion, for in that mode it is individuated by its matter; this notion only applies to it, therefore, as abstracted from individuating matter.[8]

By attributing to the *Commentary* the view that a nature existing in things "cannot be represented in a universal notion," the translator seems to be suggesting that a nature can only be "represented in a universal notion" once it is "abstracted from individuating matter." Abstraction produces something, a nature apart from individuating conditions; this something that abstraction produces will, of course, be a mental something. This mental something can then be "represented in a universal notion."

This suggests two mental *somethings*—the mental something that is represented in a universal notion, and the representing universal notion itself—the represented and the representation.

However, a more literal translation would be:

> Therefore according as it has being in material things, the intention of universality cannot accrue to it, because it is individuated by matter; therefore the intention of universality accrues to it as it is abstracted from individual matter.

Nothing about this passage suggests that because universality does not accrue to a nature in material existence that that nature cannot be *represented* in a universal notion. The other translation introduces a thesis about the structure of knowledge and representation into the translation. But *prima facie* the passage is not about representation.

It is important to emphasize what St. Thomas has to say about the proper *obiectum* of the intellect because he considered something very much like the Introspectibility Thesis and what he has to say about it fundamentally undercuts Putnam's characterization of the Aristotelian. Examining what he says about this thesis will provide insight into how to think about our knowledge of our knowledge.

St. Thomas on *Obiectum*

In the preceding paragraphs, I left *obiectum* untranslated in some instances. There are troubling ambiguities involved with the English 'object', and the Latin *obiectum*, both of which, respectively, feature prominently in present and medieval discussions of knowledge and language.

In many contexts, in particular Putnam's, the English term 'object' is roughly synonymous with 'being', or 'thing', the elements of the set over which the quantifiers range in an "objectual" interpretation of the first order predicate logic.[9] Indeed, this is one of the terms Putnam says needs to be recognized as non-univocal, along with 'exists', 'being,' and 'thing', when he discusses *Metaphysical Realism*. In the previous chapter 'object' was used as roughly synonymous with these terms. In the discussions of representationalism, 'object' answers to a variety of questions: "What do we think about? An object"; "What do we know? An object"; "What do we speak about? An object"; "What is the reference of our words? Objects"; "What do we sense? Objects"; "Are the objects we know, sense, think, or talk about internal or external to the mind?" "If internal to the mind are these objects related to objects external to the mind?"

Frege, though not giving a definition of object (*Gegenstand*), characterized it as "something that can never be the whole reference of a predicate, but can be the reference of a subject."[10] A proper name is a sign of an object and has the object as its reference,[11] whether the proper name is a term like 'Aristotle' or a definite description like 'the student of Plato and the teacher of Alexander the Great'.[12] A predicate, on the other hand, has as its reference a concept,[13] and "it is of the reference of the name of which the predicate is affirmed or denied."[14] Objects "fall under" the concepts that are truly predicated of them, and such concepts as are truly predicated of an object are that object's "properties."[15] Thus an object, the reference of a "proper name" in an assertion, is what we talk about, what we make true or false assertions about, the properties of which we recognize and truly predicate, what our thoughts and knowledge are about, whether Aristotle, Fido, Excalibur, or Rosebud. But more fundamentally, an object is what exists, whether we refer to it or not, think about it or not, know about it or not, Putnam's objections to *Metaphysical Realism* notwithstanding. A set which is the extension of a concept consists of all the objects of which the concept is truly predicated.

Barry Smith credits Bernard Bolzano, as well as Frege, with introducing the contemporary sense of 'object'.

> The modern notion of object, encompassing not only external objects of the real world but also the abstract objects of, for example, mathematics, was introduced into philosophy with the invention by Bolzano and Frege of existential quantification. Thus the notion has established its position in philosophy hand in hand with an approach to ontology, defended most persuasively by Quine, according to which object-status is relativised to membership in the domain of quantification of a formalised scientific theory.[16]

Dummett also credits Frege with providing the dominant influence on the use of 'object' in contemporary philosophy. He characterizes it as

> that sense, namely, in which a distinction is drawn, among objects, between concrete and abstract ones, or in which the problem of ontology is taken to be that of saying what kinds of objects there are, or ought to be posited. [17]

And Quine writes:

> [W]e are prone to talk and think of objects. Physical objects are the obvious illustration when the illustrative mood is on us, but there are also all the abstract objects, or so there purport to be: the states and qualities, numbers, attributes, classes. We persist in breaking reality down somehow into a multiplicity of identifiable and discriminable objects, to be referred to by singular and general terms. We talk so inveterately of objects that to say we do so seems almost to say nothing at all; for how else is there to talk?[18]

Dummett and Quine are stressing that the sense of 'object' involved is distinguished by its ontological character: what exists. Do abstract objects exist? Are sets among the objects that exist? Are qualities? Are attributes? 'Object' in this sense is roughly synonymous with 'thing' (the Latin *res*). Though by no means agreeing upon what should be considered an *object*, nor on the issues of metaphysical and epistemological realism with respect to *objects*, this sense is at play in the work of many recent philosophers, including Wittgenstein, Kripke, Quine, and Putnam.[19] 'Object' in this sense is the being that Quine is trying to get at when he writes "to be is to be the value of a bound variable."

However, it would be seriously misleading to suggest that this is the only sense of 'object' in English generally, and in philosophical contexts particularly. Consider Wittgenstein's statement that

> the object of the training in the use of tables . . . may be not only to teach the use of one particular table, but it may be to enable the pupil to use or construct for himself tables with new co-ordinations of written signs and pictures.[20]

Here object does not mean *existent* or *thing*, but rather *goal*.

In Book II, lesson 6 of the *De anima* commentary, St. Thomas provides an account of how one should understand *obiectum*. He is commenting on Aristotle's efforts to define the different kinds of soul. Aristotle had said that in order to understand each kind of soul, it is necessary to understand the functions characteristic of each kind of soul, that is, the faculties or powers of each kind. So the question arises of how to define faculties or powers. St. Thomas provides what amounts to a concise summary of the *De anima* discussion at the beginning of the response to Question 77, article 3 of the first part of the *Summa*:

> It ought to be said that a power, inasmuch as it is a power, is ordered to an act. Hence it is necessary that the *ratio* (intelligible character) of

the power be taken from the act to which it is ordered; and consequently, it is necessary that the *ratio* of a power is diversified as the *ratio* of [its] act is diversified. But the *ratio* of an act is diversified according to the diverse *ratio* of [its] object. For every action is either active or passive. But the object is related to the act of a passive power as a principle and moving cause.... And the act of an active power is compared to the object as a term and end.

The example he gives of an *obiectum* of a passive power is color with reference to vision, while of an active power it is "perfect quantity" as the *obiectum* of the power of growth. St. Thomas summarizes a constant theme, powers are defined by their acts, which are in turn defined by their *obiecta*.[21] Thus *obiecta* are not things (*res*) simply. They are things taken as either moving a power to act or terminating the act of a power. It is the sense of term or end that is present in the quote from Wittgenstein above.

But because cognitive powers, in particular the intellect, are passive powers,[22] I am interested in the first sense—things taken as moving cognitive powers, in particular the intellect, to their acts. Presumably, since something is knowable only inasmuch as it is in act,[23] something will be a sensible *obiectum* because it possesses an act, that is, a particular form in virtue of which it is capable of effecting an alteration in the sense power.[24] Something will be an intelligible *obiectum* because it possesses an act, that is, a form in virtue of which it is capable of effecting an alteration in the intellect.[25] The form of a thing capable of effecting an alteration in a sense power is a "sensible form," while the form of a thing capable of effecting an alteration in the intellect is an "intelligible form."[26]

But keep in mind that it is the *thing*, not the form, that effects the alteration. Typically when St. Thomas uses the singular *obiectum* he is specifying the *formal* aspect, in virtue of which a thing moves a cognitive power. One can more generally say that the *obiectum* of a cognitive power is the *form* in virtue of which a thing can move the power to its act. So color, generally speaking, is the *obiectum* of vision, while odor is the *obiectum* of smell, and the quiddity of material things the *obiectum* of intellect. On the other hand, St. Thomas's typical use of the plural brings one back to a healthy sense of reality, for it emphasizes that it is things (*res*) which are sensed, and things (*res*) which are known, not color apart from things (*res*), or material natures apart from things (*res*). So the *obiecta* of vision are colored things, the *obiecta* of hearing are audible things, and the *obiecta* of intellect are intelligible material things. In other words, it is not color itself but colored things that move vision to its act, and it is not intelligible forms, but things with intelligible forms that move intellect to its act. This latter point about colored, audible, and intelligible *things* is the sort of point that St. Thomas thinks is so obvious as not to need explicit treatment. In the last chapter I used it to respond in part to the objection from Robert Pasnau's work. St. Thomas does, however, make the point incidentally in other instances. For example, he uses it as the basis of an analogy for understanding the statement "the first of created things is being," where he writes "it is similar to the mode of speaking, as if it were said that *the first visible thing is color,* though that which properly speaking is seen is a *colored thing.*"[27] And more importantly for the issue we are pursuing here, he makes another analogy in the De anima commentary, "what is seen is color, which is in a body."[28]

A few comparisons might be in order. In the modern sense of 'object', it would not be at all odd to say that the object I see is the very same object that I hear, namely Socrates. Or the object that I know is the very same object that I taste, water. Further, if one were to suggest that there would be no objects if there were no cognitive powers, one would risk being accused of some form of Metaphysical Idealism. But one can see how the sense of *obiecta* differs for St. Thomas in his discussions of cognition. In St. Thomas if there were no cognitive powers, there would be no *obiecta*. It does not follow that there would be no *res*. There would indeed be *res*. And those *res* would possess the very same formal characteristics in virtue of which those *res* are capable of effecting alterations in cognitive powers should there happen to be any cognitive powers.

On St. Thomas's account, it cannot be said that the *obiectum* of vision is the *obiectum* of smell, or the *obiectum* of taste the *obiectum* of intellect. A thing *qua* moving vision to see is not the very same thing *qua* moving hearing to hear. A thing *qua* moving hearing to hear is not the very same thing *qua* moving intellect to understand. So using the English, the visible object is not the audible object, and neither of them are the intelligible object, except "incidentally."[29]

Without this last qualification, and if one stressed the contemporary "entitative" sense of 'object', one might be tempted to think that there are distinct realms of things, visible things which are not audible things, neither of which are intelligible things. This is precisely why it is important to stress St. Thomas's sense of *obiectum*. A being, for example Socrates, may be visible, audible, and intelligible, but he is not three things, a visible thing, an audible thing, and an intelligible thing. Still, with respect to Socrates there are several *obiecta* of human cognitive powers, the visible, the audible, the intelligible, and so on. Responding to an objection that according to the analysis of the diversity of powers, the same object could not pertain to different powers, St. Thomas substitutes 'subject' for the objector's 'object' and writes "[N]othing prohibits that which is the same subject from being diverse according to *ratio*. And so it can pertain to diverse powers of the soul."[30] The very same "subject" may be cognized as visible, audible, and intelligible, because of the different forms that inform it. But the very same "object" cannot be. As the objector uses 'object' it seems to be akin to the modern sense I have discussed. It is significant then that in his response St. Thomas substitutes 'subject' for 'object', signaling thereby that the objector is not using 'object' in his, St. Thomas's sense.

The Introspectibility Thesis

Throughout the rest of the chapter, I will use 'object' in St. Thomas's sense. The Introspectibility Thesis essentially involves the issue of *what* the intellect first or primarily knows in its act of knowing, or more generally what it is primarily related to in its act of knowing. Does the intellect first know, or consider, or perceive, or hold before its conscious attention (a) the *passions of the soul* that are *similitudines* or *likenesses*, or does it first know (b) the *res extra animam* of which the *passions of the soul* are *likenesses*? Seen in this light, the anti-representationalists inspired by Wittgenstein assert that the representationalists, those who in this context would hold to (a), are fundamentally misguided. Consequently, examining the problem of what the intellect first knows shows the way to addressing the Introspectibility Thesis.

St. Thomas explicitly poses something very much like this problem in both the *Summa*, at I, Question 85, article 2, and in the *De anima* commentary, at Book III, lesson 8, passage 718, when he asks whether the "intelligible species . . . is related to our intellect as that which is understood?" The treatments of this question in the two works are very similar, but there is an interesting difference. The treatment in the commentary, which usually provides more expansive discussion than parallel treatments in the *Summa*, is in this instance much more concise, while the *Summa* supplies the details.

The Response: Part 1

St. Thomas's response to this question is divided into two parts. In the first part, he rejects as unsound an argument that moves from the premise that *the intellect only understands the impressions made upon it* to the conclusion that *the intelligible species are related to our intellect as that which is understood*. The premise of this argument is a position that he thinks some philosophers have actually held. Editors typically take it to be a reference to Protagoras as described by Aristotle in *Metaphysics* IX, and St. Thomas's own commentary on the *Metaphysics*. "According to this position, the intellect understands nothing except its own passion [of the soul], that is the intelligible species received within it. And according to this, a species of this kind is that which is understood."[31] St. Thomas's statement of the argument that he judges unsound is highly compressed, but it goes something like this:

P1) Our intellect only understands the impressions made upon it.
(Protagorean Assumption)

P2) Understanding a thing is understanding its form.
(Implicit premise)

C1) Our intellect only understands the form of the impressions made upon it.
(Conclusion from P1 and P2)

P3) The intelligible species is the form of the impression made upon the intellect.
(Stipulation)

C2) Our intellect only understands the intelligible species of the impressions made upon it.
(Conclusion from C1 and P3)

P4) If something is the *only* thing understood by our intellect, then it is related to our intellect as *that which is understood*.
(Implicit premise)

C3) The intelligible species is related to our intellect as that which is understood.
(Conclusion from C2 and P4, *modus ponens*)

Thus, the position he is considering in the first part of his response is that the *intelligible species* that informs the concept (the passion of the soul) is that which the intellect understands (C3), *because* it is the *only* thing that the intellect understands (C2).

St. Thomas does not reject the conclusion of this argument as *false*. It will turn out that he does think the intelligible species is related to our intellect as that which is understood, but not primarily. He rejects the argument as *unsound*, because according to him P1 "appears manifestly false." In the Summa, he cites two reasons for denying the truth of P1. First, he appeals to the subject matter of the sciences, "the things which we understand are the same as the things of which the sciences treat." He simply takes it for granted that the things of which the sciences treat are *res extra animam*, not the intellect's own impressions or *intelligible species in anima*.

It is significant that at this point he makes reference to the Platonists "according to [whom] all sciences are about ideas which they held are the actually understood." The Platonists are brought up in order to provide a contrasting alternative to the thesis that the soul only knows the impressions within it. But in the context, they do not simply provide that possible alternative, but bring to mind St. Thomas's earlier rejection of it in Question 84, article 1. Reminding the reader of his rejection of the Platonist alternative paves the way for his own solution. His response in Question 85, article 2 is similar to his earlier one where he argued against what he takes to be Plato's position that we know material realities by knowing the Ideal Forms in which the material realities participate. There he also asserted that this "appears false." He wrote, "[I]t seems ridiculous (*derisibile videtur*) that when we inquire into the knowledge of things which are manifest to us, that we should put other beings in between which cannot be the substance of [the manifest] things, since they differ in being (*esse*) from them." In the earlier discussion, the manifest things in question are material bodies beyond the soul (*res extra animam*). The "other beings" differing in being are the Platonic forms. In this discussion, by mentioning the Platonists again, St. Thomas is not making the mistake of identifying the Platonic Ideal Forms with mental entities. He is fully aware that they are supposed to be abstract eternal realities. By bringing up the Platonic Ideas in the discussion of the intelligible species, he is not trying to *identify* the two. Rather he makes the general point that it makes no difference whether these "other beings" are abstract eternal beings or abstract mental beings. We pursue knowledge of material things; why then would we go in search of something else? The reason for rejecting the mental impressions as the *only* things we truly know is the same fundamental reason St. Thomas had earlier used to reject the Platonic Ideas as the *only* things we truly know. Increased knowledge of what is manifest will not be advanced by appealing to what is not.

This discussion in the *Summa* is essentially a recapitulation of St. Thomas's account of Aristotle's criticism of Plato in his commentary on the *Metaphysics*.[32] And it is very important to understand this rejection of Plato against the larger discussion of these issues in that commentary, in particular the discussion of participation and the knowledge of sensible things by knowledge of the Platonic Forms. St. Thomas does not completely deny that one thing can be known by knowing another thing. In particular, when an effect is known through its cause *qua* cause, this is genuine knowledge. This is what Aristotelian science strives after, demonstration *propter quid,* that explains the effect in virtue of its cause. But in this explanatory structure, the effect is not known *only* through the cause. Rather, the knowledge of it that one already has is enhanced when it is known *qua* effect of some cause, and why. But this presupposes that one can know and give a substantive account of the causal relation. Lacking such knowledge, one does not have an enhanced knowledge of the effect by knowing the cause.[33]

When in Book I of the *Metaphysics* Aristotle turns to consider Plato and the relation of participation that the latter used to account for the relation between sensible things and the Ideal Forms, among and in addition to his more substantive arguments against the Forms, his passing judgment upon it is harsh—"but what the participation or the imitation of the Forms could be they left an open question,"[34] "to say that they are patterns and the other things share them is to use empty words and poetical metaphors."[35] The key for St. Thomas's understanding of Aristotle's criticism is that "the cognition of anything whatsoever is achieved through a cognition of its substance, and not through the cognition of certain substances extraneous to it."[36] The emphasis on substance is driven by the emphasis of the *Metaphysics* on developing a science of substance, but the general discussion throughout of the Ideal Forms of accidents and relations makes it clear that the point applies *mutatis mutandis* across the categories of being. In Aristotle's judgment participation does not provide a causal explanation sufficient for knowing material things through Ideal Forms.

By parity of reasoning, in Question 85, article 2, if the *only* things we understood were *intelligible species in anima*, then the sciences would not treat of the *res extra animam* that they do. If the sciences are going to be about *res extra animam*, it will not be because we *only* know the intelligible species *in anima*. St. Thomas also gives this argument in the *De anima* commentary, citing as an exception the "science of reason," that is, Logic. Thus, if the only things that are known are the *intelligible species* of *res extra animam*, St. Thomas is committed to denying that *res extra animam* can be known. But since *res extra animam* can in fact be known, by *modus tollens* C2 is false, which leads St. Thomas to conclude P1 is false. Consequently, the argument that follows from it is unsound.

Now let us consider a possibility that St. Thomas does not himself consider, but I would argue he is implicitly committed to denying as well. Notice the limited character of St. Thomas's commitment at this point. It is still possible that *res extra animam* are known by knowing the intelligible species *and something else* besides the intelligible species. If this is a real possibility, what might this something else besides the intelligible species be? Here the importance of his rejection of Plato's position is clear. The rejection is based upon his claim that *X is not known by knowing something other than X*. So if *res extra animam* are known by knowing the intelligible species *and something else* besides the intelligible species, this *something else* can only be the *res extra animam*. But that renders knowing the intelligible species otiose. Thus, once St. Thomas has rejected the position that *intelligible species* are the *only* things we know, and in conjunction with the presuppositions of his earlier rejection of Platonism, *knowing* the intelligible species plays no part at all in *knowing* the *res extra animam*.

The parallel between the rejection of Platonism and the rejection of intelligible species as the only things known by the intellect is important. In the next chapter, we will consider the formal identity between knower and known, formal identity in virtue of the concept. Despite this formal identity, the concept and the *res extra animam* are distinct in being. The *De ente* discussion shows that the nature *in singularibus* and the nature *in anima* have distinct modes of being, despite their identity "absolutely considered," that is, they are the natures of distinct beings. Putnam recognized a parity between what he understood to be the Platonistic and the Aristotelian versions of mental representationalism. For Putnam, the difference between Plato and Aristotle on where these intermediary beings exist, *outside* the cave or *inside* it, makes no difference, for the truly objectionable part is that only by cognitively attending to these entities do we succeed in knowing some *other* entities, the entities that we think we cognitively attend to, trees, dogs, other people, and so on. Here I believe St. Thomas would agree with Putnam. His own reference to his earlier rejection of the Platonic forms indicates that he precedes Putnam in this criticism. It makes no difference to St. Thomas whether the intermediary entity is a Platonic form *extra animam* or an intelligible species *in anima;* if these are the only things we know, then we cannot have the knowledge of material reality that we in fact do have. In both questions

St. Thomas appeals to our knowledge of material reality to deny both types of cognitive intermediaries.

This result is sufficient for denying the Introspectibility Thesis, since it denies that we know *res extra animam* by knowing the concept *in anima*, but it is worthwhile to go on and consider the rest of St. Thomas's response. In the first part of the response, he provides a second reason for denying P1. In an argument that does not appear in the *De anima* commentary, he appeals to the experience of error. If our knowledge only grasps the internal appearances of things, then error would be impossible, for everyone would always judge truly of how things appear to him or her. Thus echoing Aristotle, he writes that

> the error of the ancients would follow, who said that everything is true as it appears . . . and so . . . contradictories would at the same time be true . . . and so it would follow that every opinion will be equally true.[37]

Far from leading to skepticism in the sense of "nothing can be known to be true," the thesis leads in the other direction, to a sort of hyper-anti-skepticism—nothing at all can be known to be false; every judgment is true.

It might be tempting to take the two arguments in this first part of the response to be St. Thomas's answer to skepticism, his effort to get outside of the mind. But St. Thomas is not trying to argue that we can and even do know material beings beyond the soul; he presupposes that we do. That presupposition allows him to answer a technical question that comes up in his account of the knowledge of material things beyond the soul—what is the mode and order of our knowledge of those things. Skepticism is not the problem, and he is not trying to solve it. It is *presumed* false, not concluded false. Since St. Thomas is in pursuit of a philosophical account of the soul, he is not trying to cross the chasm between what appears to him "in his skull" and what he hopes is outside of it.[38] The reason he denies that the intelligible species constitute the only thing that the intellect knows is not a fear of skepticism, but because he believes it would provide a false account of the role that intelligible species play in the "mode and order of understanding."

The Response: Part 2

The second part of St. Thomas's answer to the question whether "the intelligible species are related to our intellect as that which is understood" provides an answer to how and in what order intelligible species are themselves known.

He begins by stating what he intends to show, "and so, it must be said that the intelligible species is related to the intellect as that *by which* the intellect understands." St. Thomas wants to concentrate upon *what* the intelligible species is, in order to determine *how* it is related to the intellect. At this stage he intends to show that it is a *means* or *that by which* (*quo*) the act of intellect proceeds. St. Thomas intends to show that the intelligible species is the form of the act of intellect, not so much in the sense of a demonstration, but as rather a recapitulation of how the intelligible species functions as likeness and form in cognition.

He brings up the distinction between immanent activity that "remains within the agent, as for example to see and to understand" (*manet in agente, ut videre et intelligere*) and transient activity that "proceeds to an exterior thing, as to heat and to cut" (*transit in rem exteriorem, ut calefacere et secare*). Here he presupposes his discussion in Question 77, article 3 of "whether the powers are distinguished by their acts and objects." As we saw in the previous chapter, the distinction between immanent and transient activity is a distinction he finds in the *Metaphysics*, a citation which editors attribute to Book VIII, chap. 8, 1050a23–b2, where Aristotle distinguishes between acts like "seeing and contemplating" in which nothing is produced other than the act, and other activities like building where something other than the act is produced.

In both types of action, "each is done according to some form."[39] St. Thomas writes that in the case of transient activity, the form of the act is a likeness of the object of the act (*obiecti actionis*), that is, the thing produced by the act. This is the sense of 'object' that refers to the goal or end that is produced by an active power. The example he provides is the fairly mundane one of "heat in the heater is a likeness of the heated." In the case of an immanent act the form of the act is again a likeness of the object of the act, but here the object of the act is whatever moves it to its act. This is the sense of 'object' for a passive power that I discussed above. Thus, the form by which an immanent act takes place is a likeness of whatever moves it to act. In the case of sight, a visible thing *extra animam* moves vision to its act, in which a likeness of that visible thing is the form of that act. In the case of intellect, a potentially intelligible thing *extra animam*, rendered actually intelligible by the agent intellect, moves the possible intellect to its act, in which a likeness of that intelligible thing is the form of that act. But then, "a likeness of the visible thing is that *by which* (*secundum quam*) sight sees, and a likeness of the thing understood, which is the intelligible species, is the form *by which* (*forma secundum quam*) the intellect understands."⁴⁰

Of course 'likeness' plays a prominent part in this discussion, and we will take that up later. Here, however, St. Thomas is simply trying to make the point that in terms of "what" the intelligible species is, it is related to the intellect as that likeness "by which," as a form, the intellect engages in its act of understanding. The intelligible species, that is, the form informing the intellect's activity, is a *means* of knowing something else. By itself this emphasis upon "means" is not sufficient to overcome the Introspectibility Thesis, which holds something like this, that internal things, inasmuch as they are known, become *means* for knowing whatever external things they represent. Instead, St. Thomas avoids the Introspectibility Thesis by his denial that these acts and the species that inform them function as objects of knowledge, the knowledge of which provides knowledge of *res extra animam*. He is *implicitly* committed to that denial by the principles he uses to reject Platonism, but he also rejects it explicitly and forcefully. Commenting on the *De anima*, he writes that

> it is clear that the intelligible species, by which the intellect is made to be in act, are not the object of the intellect. For they are not related to the intellect as what is understood, but as that by which it understands.⁴¹

It is here that he makes the analogy to sight:

> just as the species which is in sight is not what is seen, but that by which sight sees, and what is seen is a color which is in a body; similarly what intellect understands is a quiddity which is in things, and not an intelligible species.⁴²

There are differences between vision and intellect, which St. Thomas discusses in his commentary, but the question of the exteriority of their objects is not one of those places where they differ. *Res extra animam* insofar as they possess certain forms are capable of moving the power of sight to its act. Similarly, *res extra animam* insofar as they possess certain forms are capable of moving the intellect to its act. The intelligible species is the form of the act so moved.

This is why when the intelligible species is called "that by which" the intellect understands, it differs from the representations in the Introspectibility Thesis, which are fundamentally related to the intellect as "that which it understands," or "knows," or "grasps," or "perceives," or operates upon in some fashion or another, and subsequently they are "that by which" the intellect understands *res extra animam*. In St. Thomas the direction is reversed.

> [The first object of the intellect is] something extrinsic, namely the nature of a material thing. And so that which is first cognized by the human intellect is an object of this kind; and secondly it cognizes the act itself by which it cognizes the object.[43]

St. Thomas's use of "by which" to indicate a principle of action has been a constant of this discussion. In general a form is a principle of act or action; it is that "by which" a *res* is *what it is* and that "by which" an agent acts. The intelligible species received is a principle that gives determinate form to the intellect's act, and so is a principle "by which" that act takes place.

Considering the question of whether the intellect knows material things *in* the Eternal Types, St. Thomas makes a distinction about "knowing something in something else" that may be useful here. He says in a sense, yes, and in a sense, no.

> [S]omething is said to be cognized in something else in two ways. In one way, as in a cognized object (*cognito obiecto*); just as someone sees in a mirror those things of which the images are reflected in the mirror.... In another way something is said to be cognized in something as in a principle of cognition; as if we were to say *the things which are seen by the sun are seen in the sun*.[44]

Of course the distinction as it is used in the article allows St. Thomas to say that material things are known "in" the Eternal Types as principles of knowledge, not that the Eternal Types are objects of knowledge. But as he presents it, it is a perfectly general point and can be applied to intelligible species as formal principles of knowledge, as he does in an earlier parallel passage.

> It ought to be considered that something is cognized in two ways: in one way, in itself; in another way, in another. Something is cognized in itself when it is cognized through a proper species adequate to the cognized thing itself, as when the eye sees a man through the species of the man. A thing is seen in another, however, through the species of that which contains it, as when a part is seen in a whole through

the species of the whole, or when a man is seen in a mirror through the species of the mirror, or in any other manner in which it may happen that something is seen in another.[45]

Here the "principle" of knowledge under consideration is not the Eternal Types, but "species"; still the parallel is clear. Notice, however, the difference between the two passages. The Eternal Types as principles of knowledge count as cognizing in "something else," though not as cognized objects. In this second passage, the species as a principle of knowledge counts as cognizing the "thing itself," though the species as a *principle* is not a cognized object like the mirror image. Thus, for St. Thomas there are both external principles of cognition (the Eternal Types), and internal principles of cognition (intelligible species), but neither are taken to be principles *as cognized objects* (like, for example, mirror images).

The Introspectibility Thesis suggests that *res extra animam* are cognized "in" the *passions of the soul*, in something like the way in which things are seen in a mirror.[46] In the previous chapter we saw why it is appropriate to call the species, as a formal cause, a principle of the act of understanding. St. Thomas suggests that *res extra animam* are cognized "in" the *passions of the soul*, or more strictly the intelligible species that inform the *passions of the soul*, as "by" a principle of the act of understanding. The passage about the Eternal Types when generalized provides a reason to conclude that *res extra animam* are cognized *in* the intelligible species as *in* a principle of the act of understanding, not in a manner analogous to seeing a reflection of an object in a mirror.

Indeed, this is how St. Thomas deals with the first objection to Question 85, article 2. The objection is: the understood in act is in the one who understands, since the understood in act is the intellect in act. His response is that

> the understood is *in the one* understanding through its similitude . . . and in this manner it is said that the understood in act is the intellect in act, inasmuch as the similitude of the understood thing is the form of the intellect, just as the likeness of a sensible thing is the form of the sense in act. Hence it does not follow that the abstracted intelligible species is that which is actually understood, but that it is a similitude of [what is actually understood].

This is the way one should understand very common statements like "the natures of material things are understood *in* the universal or the universal form;" not as if seeing it in a mirror or mental entity, but as the universal form functions as a principle for the act *by which* the nature of the material thing is known. The "likeness" or "similitude" is "in" the intellect not as something known, but as that "by" which understanding takes place. Thus, the intelligible species are not related to the intellect as "that which it understands" but "that by which it understands."

Res Extra Animam *as Object and Abstraction*

Couldn't the *intelligible species* still be related to the intellect as *that which* it understands, even as it is *that by which* it understands, as is the case with the Divine Intellect? What function such knowing would play in us is not clear, since it is not *by knowing* the intelligible species that the intellect knows *res extra animam*. The difference for human beings is that our intellect is moved to its act by its object, otherwise it is in potency to its act.[47] In the case of the Divine Intellect, it is not *moved* to its act, and is in no way in potency to its act. However, since our intellect is in potency to its act, and as in potency it cannot move itself to its act, something other than it must do so, and this is its *object*. In other words, in human beings the object of understanding must be a being distinct from the act *by which* the object is understood, keeping in mind of course the different ways in which 'existing being' (*ens,*) may be said. Since it is by abstraction from its proper object that the intellect acquires the *intelligible species* that informs its act, it is appropriate at this point to sketch how abstraction is involved in the moving of intellect to its act by its proper object, *res extra animam*.

Earlier, St. Thomas had argued in Question 79, article 2 that the intellect is a passive power, insofar as it receives its actuality, and in article 3 that there must be an agent intellect because

> forms existing in matter are not actually intelligible. . . . Moreover nothing is changed from potency to act except by some actual being. It is necessary, therefore, to posit some power on the part of the intellect, which may make things intelligible in act by abstraction of the species from the conditions of matter. And so it is necessary to posit the agent intellect.

Thus we are not utterly passive—coming to know some *res extra animam* is in part an active engagement and in part a passive reception; the knower actively pursues the knowledge she receives. But, on the other hand, the soul does not understand corporeal things through or by its essence (q. 84, a. 2), nor through innate species (a. 3), nor through species derived from separate Platonic Forms or Avicennian Intelligences (a. 4) nor in the eternal Divine Exemplars (a. 5), but rather the agent intellect must take from sensible things the species to be received by the possible intellect (a. 6).

We should avoid imaginative pictures here that are subject to facile objections; indeed we better try to avoid "pictures" altogether. One such picture might be of the agent intellect extracting the intelligible form from sense experience and putting it some place else (a very mechanistic picture) as if the agent intellect turned to the sense experience, looked for the form, grabbed it out of the experience, and then put it in the possible intellect. St. Thomas writes in response to such a picture:

> Through the power of the agent intellect a certain likeness results in the possible intellect by the agent intellect's turning (*conversione*) to sense experiences (*phantasmata*), which likeness is indeed a representation of those things of which they are the experiences, but [a likeness] only with regard to the nature of the species. And in this way the intelligible species is said to be abstracted from sense experiences (*phantasmata*), not that the numerically identical form was first in the experiences and was made afterwards to be in the possible intellect, in the manner in which a body is taken from one place and put into another.[48]

St. Thomas denies a crudely mechanistic picture of moving forms from place to place. The emphasis is upon the production of the likeness by the power of the agent intellect coming to be related to sense experience in some fashion.

Another picture is of the agent intellect as a light shining upon sense experience illuminating the intelligible species in the experience and projecting it onto the possible intellect, something like how we think of a projector illuminating a slide. But St. Thomas's account does not trade upon this image either. Though he does use the metaphor of "illumination" which he inherits from the Augustinian and Neoplatonist traditions, his description of what this consists in most frequently involves the agent intellect making the possible intellect to be actual with respect to some *res extra animam* by "turning" in some way to sense experience. The most explicit and straightforward account of this occurs in the *Disputed Questions on Truth*. Responding to an objection that our mind cannot receive knowledge from sensible things because they are not actually intelligible, St. Thomas writes that

> in the reception by which the possible intellect receives the species of things from sense experiences (*phantasmatibus*), the sense experiences

are taken as instrumental and secondary agents, while the agent intellect is taken as a principal and first agent.[49]

The "turning to" (*conversio*) consists in the agent intellect using sense experience as a secondary cause employed by its agency. This is an important move for St. Thomas to make since it shifts the discussion away from the metaphorical images of illumination and places it within much larger and general discussions of agency ranging from God's causal concurrence as primary cause in the works of nature as secondary causes, to his concurrence in our free acts. In these discussions he describes the ways in which principal and instrumental causes cooperate.[50] When St. Thomas uses 'illumination' subsequently the force of this causal setting should inform one's reading of it.

St. Thomas goes on in the response to explain the importance of this structure of primary and secondary causality, and the instrumental role of sense experiences in bringing about actual understanding of *res extra animam*.

> And so the effect of the action is left upon the possible intellect according to the condition of each [cause], and not according to the condition of either one by itself. Thus, the possible intellect receives the forms as actually intelligible by the power of the agent intellect, but as likenesses of determinate things from cognition through sense experiences. And so the intelligible forms in act do not exist *as such* either in sense experiencing or in the agent intellect, but only in the possible intellect.[51]

The agent intellect uses as an instrument the sense experience of material beings *extra animam* to render the possible intellect actual, that is, to provide it with the intelligible species that informs its act.[52] St. Thomas's more general discussions of primary and secondary causality make the point that in such a structure the effect will be partially determined by and bear a likeness to all the causes that are *per se* involved, both principal and secondary or instrumental. In this case, the aspect of the effect in the possible intellect due to the agent intellect is the actuality and immateriality of the species, while the aspect of the effect due to the sense experience used instrumentally by the agent intellect is the formal determination of the act, that is the *intelligible species*. In short, the formality comes from the sense experience, while it is rendered actually intelligible by the agent intellect.

This is not a form of the "abstractionism" criticized by Peter Geach in *Mental Acts*. According to Geach *abstractionism* is

> the doctrine that a concept is acquired by a process of singling out in attention some one feature given in direct experience—*abstracting it*—and ignoring the other features simultaneously given—*abstracting from them*.[53]

The key to what Geach is defining is the *selectively attending to* the feature to be abstracted, and the *selectively ignoring of* the features not to be abstracted. Geach's criticism is that one must already possess conceptually the feature X if one is going to selectively attend to it in direct experience, and indeed it seems one must already possess conceptually all the features one intends to selectively ignore in direct experience. At its least objectionable this would appear to be a form of innatism in which abstraction is superfluous, while in its most objectionable it involves a plain contradiction. To his credit, Geach explicitly exempts St. Thomas from his criticism, though I am not sure for the right reasons.[54]

The right reason is because in St. Thomas's account the agent intellect does not selectively attend to some features in sense experience and selectively ignore others. The agent intellect is not a *mind's eye* or *homunculus*. The rejection of the mechanistic image of the transfer of the form from one place to another is the first intimation that suggests that St. Thomas's position is not what Geach is criticizing, since one might equally say that the agent intellect in that view must know what it is looking for, if it is going to perform this transfer. More fundamentally, St. Thomas speaks of "turning to" sense experience simply, not to look at some aspects of it and not others, but to use sense experience as an instrumental cause. Nothing about that account suggests that from the beginning the agent intellect will selectively use some features of sense experience and not others.

St. Thomas's discussion of conceptual functioning suggests that it is a progress from a general and confused act based upon sense experience to a more specific and precise act.

> Everything that proceeds from potentiality to act attains first to an incomplete act, which is between potency and act, then to a perfect act. The perfect act to which the intellect attains is complete *scientia*, through which things are known distinctly and determinately. However, the incomplete act is imperfect *scientia* through which things are known indistinctly and under a certain vagueness.

This discussion occurs in the article that immediately follows the discussion of how the intelligible species are related to the intellect as that by which it understands. According to St. Thomas knowledge proceeds in us from a state of vagueness to clarity, the highest form of clarity being *scientia* in which the principles, causes, and elements of things are known. This is a summary of his commentary on the first book of Aristotle's *Physics*.

In both the *Summa* and the *Commentary on the Physics*, St. Thomas describes how we begin with an indistinct and vague universal grasp of some whole. Progress in knowledge is attained to the extent that the vague universal becomes distinct and less confused and we begin to understand the parts of the whole, as well as its principles, causes, and elements. The example given is "to cognize an animal indistinctly is to cognize an animal *qua animal*, but to cognize it distinctly is to cognize an animal insofar as it is a rational animal or irrational, which is to cognize a man or a lion."[55] In the commentary on the *Physics* St. Thomas makes a visual analogy, "when something far away is seen, first we perceive it to be a body, then an animal, then a man, then ultimately Socrates," and he repeats Aristotle's own example, in which "'a boy first calls every man father and every woman mother, but only afterwards determines,' that is cognizes determinately, 'each one'."[56] The analogy to a child stresses that this is not a process of selective attention, but of developmental growth. We move from understanding *the same thing* generally and without distinction at first to understanding just as generally but now distinctly. If we push this initial generality without distinction back far enough (though not necessarily temporally), we would presumably conceive of things as "some stuff," or more technically, "some being."

St. Thomas asserts that in order for actual understanding to take place the intellect must return to sense experience. The proper object of intellect is material natures existing in things, and so "in order that the intellect understand its proper object it must turn itself to sense experiences, that it may examine the universal nature in the existing particulars."[57] This is the reason most often discussed by commentators. Less often discussed is another reason that St. Thomas gives, that "when someone is struggling to understand something, he provides for himself certain sense experiences in the manner of examples, in which he may examine as it were what he is striving after to understand."[58] This is no longer one's first use of sense experience—some abstraction has already taken place—yet due to its imperfection there is a return to sense experience to perfect that abstraction. Later St. Thomas will describe how this process of refinement also involves the exercise of judgment in the second act of understanding, composing, and dividing as it compares and distinguishes what it knows. In a response to an objection, he makes it quite clear that this second act of intellect also involves the use of sense experience.[59]

I have left out other faculties involved in this interplay of cognition, faculties like memory and imagination. St. Thomas states explicitly that sense experience is only a partial cause of understanding and thus does not limit it, "[S]ensitive knowledge is not the entire cause of intellectual knowledge. And therefore it is not strange that intellectual knowledge should extend further than sensitive knowledge."[60] There is a certain creativity in the intellect's use of sense experience to consider things that it has not experienced.[61] It is not an empiricist account in which knowledge is limited to more attenuated images of the initial impressions of sensation. Sense experience is not a straightjacket upon what we can know, but rather a liberating opportunity to expand and develop it. It provides the tools necessary for developing the understanding that goes beyond it.

St. Thomas's account should not be read as if he is providing a recipe or a set of steps to climb in order to achieve understanding. He is providing an analysis. While it is tempting to read the static character of the analysis back into what is being analyzed, St. Thomas himself constantly warns against attributing characteristics of the mode of knowing to the thing known. One has to appreciate that he is analyzing a living activity of human beings. The anatomist's cat does not show its life in its parts arrayed upon the table; yet its life is only imperfectly understood apart from the static setting out of its organic parts on the dissecting table. St. Thomas is attempting to set out the elements that make up cognition in human beings. But one should not lose sight of the subtle interplay of the faculties he is describing, as they operate back and forth and in concert in a unified act to achieve knowledge both particular and universal.

There are a number of metaphors one might use here. Consider for example the way in which a sculptor's understanding of her instruments deepens and becomes more perfect as she produces statues, and how she may return to improve those statues as her use of her instruments improves. Though there is a sense in which her art is limited by her instruments, there is also a deeper sense in which that limitation makes her art possible. Another analogy is to a child conceived as a human being, who yet develops the capacities to fully participate in mature human life. She is the very same person as she was conceived to be, yet she has grown strong and flourishes. This is after all the metaphor that Aristotle himself chose—a child at first calls all men father.[62]

Looking back to the account of abstraction in its detail, we see a description of sense experience and intellect cooperating to bring about and deepen understanding. It begins with the agent intellect employing sense experience, the complex and distinct features of which may be wholly unknown to the intellect, in order to bring about some minimal understanding; it then proceeds in virtue of that minimal understanding to return to the sense experience, and to employ it even better to deepen that very understanding and make it more graceful. This is not a substitution of concepts, but a reconceiving and reexperiencing of the same things beyond the soul in virtue of the same concept more richly developed. It is a movement as St. Thomas puts it from imperfect understanding to perfect, from the vague universal to the distinct, however long and arduous a task that might be.[63]

Introspective Knowledge: Intelligible Species as That Which the Intellect Understands

The preceding section has been a brief summary of the discussion surrounding the question of how the intelligible species is related to the intellect. For a human intellect, the object of understanding that moves the intellect to its act and the act of understanding that object must be distinct beings, since the intellect is moved from potency to act.

Nevertheless St. Thomas does not give a wholly negative response to the question of whether the intelligible species is related to the intellect as "that which is understood." Intelligible species *are* related to the intellect as "that which it understands." The key is that intelligible species are related to the intellect as that which is understood not primarily, but secondarily. The preceding account describes why material things beyond the soul are what is *primarily* related to the intellect as that which is understood. *Intelligible species,* on the other hand, are only related to the intellect as that which it understands *secondarily,* and only *subsequent* to understanding material things by means of them as formal principles of the act of understanding.

Introspective self-knowledge of the act of understanding is possible. The intellect is capable by *introspection* of coming to understand the means by which it primarily knows external things.

> [T]he intellect reflects back upon itself, according to which it understands itself to understand and the species by which it understands.

And so the species understood is *secondarily* that which is understood. But that which is understood primarily is the [external] thing of which the intelligible species is a similitude.[64]

Earlier he is even more expansive:

> [T]he first object [of the human intellect] is something extrinsic, namely the nature of a material thing. And so that which is primarily cognized by the human intellect is an object of this kind; and secondarily its act, by which its object is cognized, is itself cognized; and through the act the intellect itself is cognized, the perfection of which is this act of understanding.[65]

This parallels very closely the discussion of the same issues in the commentary on the *De anima* at Book III, lessons 8 and 9.

> [W]hat the intellect understands is an essence which is in things, but not the intelligible species, except insofar as the intellect itself reflects upon itself,

and

> the possible intellect, which is as such in potency in the order of intelligible things, neither understands nor is understood except through a species taken into it.

Like anything else, the possible intellect is knowable only insofar as it is in act; "for as the Philosopher shows in the ninth book of the Metaphysics, nothing is intelligible except insofar as it is in act."[66] But of itself the possible intellect is not in act. The intellect becomes knowable to the extent that it becomes actual by receiving within itself the form of its act. But it only becomes actual *as* knowledge of some *res extra animam*, receiving from the latter the species or form of its own act. So the condition enabling the concept to be an informed act that can be known is the very condition for the *thing beyond the soul* to be already known. Consequently the *passion of the soul* cannot be known prior to the *thing beyond the soul* which is known by means of it. It must be known secondarily. Were the *passion of the soul* to be primarily known, and the *thing beyond the soul* only secondarily, the *passion* would cease to be *what it is*; indeed, it would cease to be entirely. But in the kind of mental representationalism under consideration, particularly in its commitment to the Introspectability Thesis, the order is reversed. In that criticism, it is presupposed that mental impressions or *passions of the soul* are known first, and things (*res*) subsequently.

St. Thomas is engaged in a reflective analytic study that in a sense robs concepts of their life,[67] by making them introspective objects of knowledge. Primarily, they are acts of knowing; secondarily, they are objects of other acts of knowing. Consider this analogy. A healthy rotator cuff is necessary for good pitching. Yet most good pitchers learn to throw good pitches without first knowing anything at all about their rotator cuffs. Indeed most of them find any knowledge of their rotator cuff superfluous, until they injure it. Most people can know things, and happily apply their knowledge in language, speculative, or practical activities, without first knowing the means by which such knowledge takes place. The *means* are in a sense transparent to the agent in pursuit of an end. Only by first coming to know something other than ourselves does our capacity for knowing and its exercise come to be knowable to us; our cognitive acts are *introspectible*, but they must first be acts. "In the beginning was the deed."

Moreover, this introspection does not require a new special faculty, apart from the faculty in virtue of which we understand *res extra animam*.

> [T]hat which is primarily cognized by the human intellect is [the nature of material things]; and the act itself by which [the nature of material things] is cognized, is secondarily cognized; and through the act the intellect itself is cognized, of which *to understand* is the perfection. And so the Philosopher says that objects are cognized before acts, and acts before powers.[68]

The very same power or faculty, the human intellect, that knows *res extra animam*, knows its own acts, and knows itself. There is no special introspective faculty, a "mind's eye," standing behind the intellect and its act. In conjunction with the denial of the Third Thing Thesis it is clear that there is no infinite regress of *homunculi* threatening here, though this cannot be fully appreciated until the Internalist Thesis is considered in the next chapter. The concept reflected upon does not "stand apart" from the intellect, since it is the very act of the intellect.

It is true that the intellect understands its acts by further acts: "[H]ence the act by which the intellect understands a stone is other than the act by which the intellect understands itself understanding a stone, and so on."[69] Here again, the original act of understanding, and *that by which it is understood*, the introspective act of understanding are distinct.

Still it is the very same faculty that understands a stone and understands itself understanding a stone.

By our cognitive awareness and attention we are primarily and by nature directed to others, not upon ourselves or our concepts. For St. Thomas an account of human knowing that *begins* with introspection is off target.

> And so it is necessary, that in the cognition of the soul we proceed from those things which are more extrinsic, from which the intelligible species are abstracted, by which the intellect understands itself; so we cognize acts by objects, and powers by acts, and by powers the essence of the soul.[70]

At the risk of introducing a pictorial metaphor, St. Thomas is not interested in getting out of the soul, out to the world. On the contrary, beginning with a human being immersed by his or her acts *in* the world, he is interested in getting *into* the soul. The way in is by considering how the human being, a material being, acts in the world. A human person in his or her cognitive capacity is fundamentally and primarily directed and open to other beings, in particular material beings. We come to knowledge of ourselves through knowledge of others. Only by the actual exercise of this capacity for uniting with another, does the human person by introspection come to self-understanding and self-enlightenment. If for Socrates an unexamined life is not a life worth living, for St. Thomas an unlived life is not a life worth examining.

'Similitude' or 'Likeness' in St. Thomas's Account

In the Introspectibility Thesis, the *passions of the soul* function as likenesses or similitudes. Because we primarily know them and because they are likenesses, it is supposed to follow that we know the things that they are like. The assumption that applies this to the Aristotelian seems to be that if one says that the form in intellect is a likeness of the *res extra animam*, one must by the very fact of using that term hold that the *res* is known by knowing its likeness first. One imagines a mother holding up a faded photograph to her daughter and saying, "Dear, this is a likeness of the father you've never known."

In St. Thomas 'similitude' is not functioning in this way. A *passion of the soul* is only known *qua* likeness by first knowing the *res extra animam* of which it is a likeness. This is reminiscent of what I discussed in the third chapter, while considering Wittgenstein. *Res extra animam* function for St. Thomas as *signs* of passions of the soul; the passions of the soul are known only after the knowledge of things beyond the soul, and in virtue of their likeness to those things beyond the soul.

The form or intelligible species in intellect is not a likeness because of its role in an introspective representational account of knowledge and language. It plays the same general role in any Aristotelian account of how an effect is related to its cause, it is the result or effect in the intellect of a causal engagement with the *object* of intellect, the quiddity of a *res extra animam* that moves the intellect to its act. It is a general principle for St. Thomas that agents act to produce a likeness of themselves in their effects. There is a common but simplistic way of understanding this that renders it manifestly false, namely something like visual or sensual likeness—what "likeness" to a rock does the shattered pitcher bear? But St. Thomas does not understand it in this visual way. The character of the effect is determined by the characteristic of the cause relevant to its agency—the pitcher was shattered because the rock was solid and massive, not because it was gray. Had the rock not been so massive and solid, the pitcher would not have shattered. The formal character of the effect is determined by the formal character of the cause—that is what the likeness or similitude consists in.

So why is a likeness in a cognitive power, in particular in the intellect, a cognition? Not simply because it is a likeness as such, but because of the character of the recipient of the likeness. The effect is received in the recipient after the mode of being of the recipient, not after the mode of being of the cause or agent.[71]

> Change is twofold: one natural, and the other spiritual.[72] Natural, according as the form of the agent of change is received in the subject of change according to natural being, as heat in the heated. Spiritual, on the other hand, according as the form of the agent of change is received in the subject of change according to spiritual being; as the form of color in the pupil, which does not become colored by this. But a spiritual change is required for the operation of the senses, through which the *intention* of the sensible form may be made in the sensible organ. Otherwise, if a natural change alone were to suffice for sensation, every natural body would sense when altered.[73]

Though the passage is strictly speaking about sensation, it extends to intellectual cognition, *mutatis mutandis*.[74] It is the character of the subject of alteration that determines whether cognition will take place in some causal situation.

St. Thomas emphasizes the role of the nature of a cognitive being in any number of discussions within and outside the discussions of human nature as such. Thus discussing the question whether God is a knowing being in Question 14 of the first part of the *Summa* he writes:

> Things that cognize are distinguished from things that do not by this, that things that do not cognize only have their own form, while a thing that does cognize is of such a nature as to also have the form of another thing.

Earlier, in Question 12, article 4, he writes:

> For cognition occurs according as the cognized is in the one cognizing. The cognized, however, is in the one cognizing according to the mode of the knower. Hence the cognition of any cognizer whatsoever is according to the mode of its nature (*secundum modum naturae*).

The having of forms that are "not their own" in cognition is quite distinct from an ordinary material alteration, as the original quotation makes clear when it speaks of a "spiritual" alteration. To have the form of the other in cognition is a formal actualization of the *nature* of the cognizing being. It is not a mere accident, but penetrates to the presupposed nature of the cognizing being.[75]

"Its own form" is the nature that is presupposed to cognition. The form of the cognized thing is "not its own" because the presupposed nature of the cognitive being does not cease to be when it comes to have within it the nature of the cognized thing. It remains the thing it was, while becoming the thing it knows. This is not the case when, for example, having been cold, water becomes hot; the cold ceases to be, when the heat comes to be. In cognition to have the form of the other is, among other things, to have one's own nature perfected, fulfilled, or actualized by the form of the other. It is for one's own nature to be informed by, that is, to become the form of the other.

St. Thomas recognizes in creatures what one might call two "levels" or "moments" in the actuality of a substantial nature, the act of the nature itself, and the act of the powers appropriate to that nature. In creatures the distinction between a substance and its powers rests on the fact that substances are in potency to further act in virtue of their natures.

> For insofar as it is a form, the soul is not an act ordered to a further act, but is the ultimate term of generation. Hence, that [the soul] is in potency to yet another act does not pertain to it according to its essence as a form, but according to its power. And so the soul itself considered as the [underlying] subject of its power is called *first act* ordered to a second act.[76]

Of course here St. Thomas is talking about soul as determining the nature of living things, but the background for it is the general Aristotelian point that the first actuality of a nature determines the natural powers and corresponding potencies of the being having that nature, potencies to second act. The first actuality of the nature of a thing determines what will fulfill it, its natural capacities, its second actuality.

It is this distinction between first act and second act that is the basis for the distinction in St. Thomas between first nature and second nature, the Aristotelian distinction that John McDowell wants to appeal to in *Mind and World*. However, St. Thomas maintains the continuity between first nature and second nature. First nature *becomes* second nature through the exercise of the capacities that first nature has, and the subsequent development of stable dispositions or habits in those faculties. But McDowell wants Aristotelian second nature without a continuity with first nature; these stable dispositions and habits have nothing more than a foothold in first nature, but they are not a development of it.

In St. Thomas when this structure of first and second act is applied to cognitive beings, in particular human beings, the nature is of itself in potency to being completed and perfected by the forms of other things.

> It is not necessary that the likeness of the thing cognized actually be *in the nature* of the one cognizing; but if there is something that is first in potency to cognition, and thereafter in act, it is necessary that the likeness of the thing cognized be *in the nature* of the one cognizing, not actually, but only [*in the nature*] in potency.[77]

Thus the nature of the thing determines what naturally fulfills and completes it. Water is not by its nature incomplete or imperfect even though it may lack heat, and may become hot. By contrast a cognitive being, though complete *in* its nature, is yet incomplete *by* its nature to the extent that it is in potency to the forms of other things, to the extent, that is, that *by* its nature it must formally become other things in order that it may attain perfection. The nature of a thing determines for it what its perfection consists in, what its good is, and imposes upon it an obligation to seek its good or natural perfection. For St. Thomas the *is* of first nature implies an *ought* for second. The good and perfection of beings that are by nature cognitive is to pursue and become *like* other things precisely as *other* things.

The Introspectibility Thesis makes being a "similitude" or "likeness" a necessary function *specific* to the account of knowledge. In St. Thomas's Aristotelian account it plays a *generic* role, indicating that we are talking about how an animal, in particular a rational one, interacts with and responds to its causal environment. Just like any being in its environment it is altered by its environment, and in that sense takes on the likeness or similitude of it. But as the quotation above suggests, not all alterations are cognitions, though some are. What determines whether an alteration is a cognition or not is not its environment, that is, is not that the subject of the cognition becomes like its environment, but rather what kind of being the subject of alteration is. An alteration involving the subject becoming like the agent is a cognition because of the *way* in which the likeness of the cause is received in the subject, and that *way* is determined by the nature of the subject—human nature. Likenesses are received into many things, but into the human animal they are received as cognitions.[78] This is why considering the way in which a human being specifically acts in the environment gives insight into human nature, which is what St. Thomas is interested in pursuing.

Thus, there are two characteristics of *likeness* in the Introspectibility Thesis missing from St. Thomas's Aristotelian account. First, *likeness* is not specific to his account of knowledge the way it is in the Introspectibility Thesis. Second, *likeness* does not provide the *mechanism* by which a known internal representation becomes knowledge of something else. In the Introspectibility Thesis the internal object is not known by likeness, but directly. *Likeness* to something else provides a mechanism that takes the mind from the directly known internal thing to something else indirectly known—thus the internal object becomes a representation. But the reason that the internal thing is directly known is not because it is a representation. Rather, it is directly known because it is directly present to the mind. It is directly known whether it is a representation or not.

It is clear from St. Thomas's account of what is known primarily and what secondarily that likeness just does not play the structural role that it does in the Introspectibility Thesis. The intelligible species is no less directly present to the intellect in St. Thomas's account as the internal object is to the mind in the Introspectibility Thesis. Indeed, it is more so, since as the form of the act of understanding itself, it can't be any more "present" to the intellect. The point is that it is not present to the intellect as an object of understanding, but as the form itself of understanding *res extra animam*.

An Objection: A *Noncognitive* Apprehension of the Intelligible Species

At this point, I want to consider an objection that might be raised to this part of my account of St. Thomas. While there may be no *cognitive* apprehension of the intelligible species prior to the apprehension of *res extra animam*, I have ignored the possibility of a properly psychological *noncognitive* apprehension of the species that precedes and brings about the cognitive apprehension of *res extra animam*. I derive this objection once again from Robert Pasnau's *Cognitive Theory in the Later Middle Ages*. Pasnau writes of the passages we have been considering that "there is probably no part of Aquinas's work that is better known and more widely discussed than his claim that sensible and intelligible species are the *quo* of cognition: not the objects of cognition, but that *by which* we come to know the world. Nevertheless . . . I argue that the standard (indeed, unquestioned) reading of this claim is badly mistaken." The bad mistake is based upon the failure to recognize that St. Thomas does not exclude the non-cognitive but psychological apprehension of species.

In the last chapter we considered an objection arising from Pasnau's claim that the intelligible species is a third thing that stands between the act of understanding and *res extra animam*. We saw how that interpretation of St. Thomas was based upon Pasnau's failure to understand the properly formal character of the species, a failure that arises because of his assimilation of formal causality to efficient. Here that earlier mistaken interpretation aids the Introspectibility Thesis, for if the species is a third thing in between the act of understanding and *res extra animam*, it is a possible candidate for being the object of some sort of mediating apprehension—it lays the foundation for the "act-object theory of" cognition that Pasnau wants to attribute to St. Thomas. In Pasnau's discussion of Question 85, article 2, he attributes to St. Thomas a "noncognitive" apprehension of the species as an object—the species is the object of an act of apprehension at a noncognitive level. Here it is good to recall that in Fodor's revival of mental representationalism the mental representations are not strictly speaking cognized. Rather they are mental entities that are operated upon by the mind's acts, through which operations the mind cognizes the world.

Pasnau grants that according to the *Summa* discussion the intellect does not have a properly cognitive relation to the *species* as a third thing. Rather, it has a quasi-cognitive apprehension of it as a third thing. Thus, he attributes to St. Thomas two modes of *apprehension,* a cognitive mode and a noncognitive mode. So, even though St. Thomas never mentions, much less discusses this noncognitive apprehension, Pasnau can argue that nothing in article 2 actually excludes it. Here the absence of a denial means consent. This allows him to characterize St. Thomas's "Aristotelian theory of cognition" as a "seminaive species theory." For Pasnau

> [a] naive species theory rejects direct realism. It holds, instead, a representationalist theory of perception [and intellection], according to which it is species that we directly perceive, whereas the external world is perceived indirectly.

A sophisticated species theory, on the other hand, allows that

> species may be intermediaries between our cognitive faculties and the external world, but they will be only causal intermediaries. Species will not themselves be the objects of cognition, because they play their role at an entirely subcognitive level. The sophisticated defender of species is a direct realist.

The seminaive species theory he attributes to St. Thomas occupies a middle position, for it

> explicitly rejects representationalism and denies that species are (ordinarily) the objects of cognition, at the same time [that it] takes species to mediate cognition not just causally but psychologically. For Aquinas . . . species themselves are in some sense the *objects* of apprehension.[79]

The important phrase is, of course, "in some sense," since he grants that it is not cognition in the ordinary sense, but rather that it is a form of "apprehension" that counts as psychological.

It might be tempting to say that according to Pasnau I have attributed a sophisticated species theory to St. Thomas, except for my conviction from the previous chapter that Pasnau does not understand the "causal role" played by species when he assimilates it to efficient causation. Notice that he characterizes the species as causal "intermediaries" on the sophisticated theory. But a formal cause is not an intermediary. Thus, the characterization of the "sophisticated species theory" falls with Pasnau's interpretation of species as "causal intermediaries." Indeed, I would deny that St. Thomas has a sophisticated species theory as Pasnau characterizes it. But of course, for the same reason, I would also deny that he has either a naive theory or a seminaive theory, because all three of his classifications of types of species theory share a common misunderstanding of the causal role of the species.

In addition Pasnau never analyzes what St. Thomas means by 'object', and it is difficult to know what Pasnau means by it. He will at times speak of the species as a "mental object," to contrast it with external non-mental objects, which might suggest our contemporary sense of 'thing'. On the other hand, he does think that the species is the specific and proximate agent cause that moves the intellect to its act, which would approach St. Thomas's sense of 'cognitive object'. But on Pasnau's use of 'object' that agency is not sufficient for calling it a "cognitive object," since according to him even the sophisticated theory holds that the species moves the faculty to its act, while at the same time it denies that the species is the cognitive object of the faculty. As Pasnau uses 'cognitive object', it appears to be 'the thing that the faculty considers', not 'the thing that moves the faculty to its act'. Pasnau does not seem to recognize this ambiguity between his use of 'object' and St. Thomas's. In the characterization of the naive theory notice his distinction between the *object of cognition* and the *object of apprehension*. Species "psychologically" mediate cognition, which "psychological" mediation is a mere apprehension, not a cognition. In his sense, the species is not the thing that the intellect considers cognitively, but it is the thing that it considers only *apprehensively*. The key according to Pasnau is that in St. Thomas's discussion of species he does not deny "representationalism as *I* have been conceiving of it."[80]

Pasnau runs up against a severe problem because of this ambiguity between his use of 'object' and St. Thomas's, a problem he does not consider. Since St. Thomas's position is that kinds of powers are distinguished by their acts, which kinds of acts are distinguished by their objects, is the faculty that *apprehends* the species identical with the faculty that *cognizes* the *res extra animam*? If one approaches this question the way St. Thomas would, one would consider their acts. One act is a cognitive apprehension of *res*, while the other is a noncognitive apprehension of *species*. At the level of description these are distinct. However, as we will see in the next chapter, St. Thomas thinks there can be distinct descriptions of one and the same kind of thing. Are they in fact distinct kinds of acts? According to St. Thomas's Aristotelian principle, the appropriate way to determine the answer to that question is to consider their objects. Here a difficulty arises, however, since their objects are *res* and *species* of that *res*, respectively. It would be something like a category mistake to argue either that these are or that they are not formally identical objects, since the *species* does not *have* a form, but *is* a form.

However, if one cannot determine the answer to the question in a principled way from the objects of the acts, one can return to Pasnau's own characterization of the acts to determine whether they are identical. On Pasnau's account they *are* distinct kinds of objects and acts; they must be, otherwise the noncognitive apprehension of the species turns out to be nothing other than the cognitive apprehension of the *res*. But in Pasnau's interpretation cognitive apprehension of an object "entails" belief formation about that object, while noncognitive apprehension of an object does not.[81] Thus they cannot be identical kinds of acts. But if they are in fact distinct kinds of acts, they proceed, according to St. Thomas's Aristotelian principle, from distinct kinds of faculties. Thus, though he does not realize it, Pasnau is in the position of committing St. Thomas to what might be called "shadow faculties," a noncognitive *apprehensive* faculty for every cognitive *apprehensive* faculty a human being possesses—sight and shadow sight, smell and shadow smell, intellect and shadow intellect. It is odd, to say the least, that St. Thomas never discussed these shadow faculties when he asked and answered in Questions 78 and 79 of the first part of the *Summa* whether there are many powers of the soul, how many kinds of powers there are, and how they are distinguished from one another.

We saw that the object of cognition is the thing that moves the power of cognition to its act. If the intelligible species is nothing more than the formal cause of the act, it cannot at the same time be the object of the act. This is just as true for any noncognitive apprehension of species as it is for a cognitive apprehension. This is not an *ad hoc* claim made in the discussion of cognition. It is based on the very general analysis within Aristotelianism of the difference between formal and efficient causation. There is no need to attribute to St. Thomas any non-cognitive but still psychological apprehension of the species, other than a need to make him subject to Ockham's criticisms based upon a mistaken view of their causal role.

Conclusion to the Introspectibility Thesis and St. Thomas

What it is for a concept to be, its essential character *qua* act of intellect, is for it to be a means for knowing something else, primarily *res extra animam*, not for it to be something known. Once something is known by its means, only then can the concept itself be known. This is the fundamental structure of St. Thomas's account and it gives a different sense to the place of "likeness" and "similitude" in St. Thomas's account than the place it has held in the tradition of *mental representationalism* criticized by Wittgenstein and Putnam.

In Descartes, the Empiricists, and the presuppositions of Wittgenstein's and Putnam's criticism, concepts or ideas are what many scholastics call instrumental signs—things known—the knowledge of which indicates or provides knowledge of something else. Wittgenstein's and Putnam's criticisms are based upon taking these instrumental signs as "conventional." To the extent that they then exemplify the Introspectibility Thesis, they become subject to Putnam's criticism. However, because in St. Thomas's account the *concept* signified by an articulated sound is not itself *what* is primarily known, but is simply a *means by which* an extramental thing is known and signified, it is not an instrumental sign. For St. Thomas it might even be appropriate to say that *res extra animam* are instrumental signs of the soul and its acts. Thus St. Thomas avoids the Introspectibility Thesis and the part played by it in Putnam's criticism. McDowell is correct when he writes that "the possibility that goes missing in Putnam's argument could be described as the possibility of mental representing without representations."[82] St. Thomas provides just such an account.

Consider then the "semantic triangle" that is the origin of Putnam's criticism. I showed in the second chapter that St. Thomas does not associate the terms 'primary' and 'secondary' with the semantic triangle, that is words, *passions of the soul*, and *res extra animam*. I showed that he associates them with the first and second acts of intellect respectively. But now, presupposing the results of this chapter, suppose we use primary and secondary where he did not, to characterize the signification of things versus concepts. "We name as we know." What do we primarily know? What are the primary objects of our knowledge? *Res extra animam*. If we name as we know, it stands to reason we primarily name, that is, signify, what we primarily know, and only secondarily signify what we secondarily know. Thus we primarily signify *res extra animam*, and only secondarily signify *passions of the soul*.

Chapter 8

THE INTERNALIST THESIS AND ST. THOMAS'S "EXTERNALISM"

> So it is said in *De anima* III that the soul is in a manner all things because it is destined by birth to know all things
> —St. Thomas

Individuating Concepts

According to the Internalist Thesis there is no intrinsic or necessary relation between a *concept* and *res extra animam*; the *concept* is supposed to be individuated by facts solely internal to the mind. This is Putnam's assumption of Methodological Solipsism, that no psychological [mental] state, properly so called, presupposes the existence of any individual other than the subject to whom that state is ascribed. If a concept is a mental state, it must be something "in itself," upon which relations of significance to the world are built. It must have a description that exhibits what it is in itself, apart from the significance that is invested in it. It is a short step from there to the view that a concept, like anything that has an "in itself" character but is made to stand for something else, requires an interpretation and an interpreter.

There are two features to the response from St. Thomas's account—the formal identity of concept and *res*, and the receptive character of the intellect. Recall that I have argued that concepts are individuated by things external to the intellect. According to St. Thomas,

> the intention of Aristotle is not to assign identity of the concept of the soul through a comparison to articulated sound, as namely of one articulated sound there should be one concept; since articulated sounds are diverse among diverse peoples; but he intends to assign the identity of concepts of the soul through a comparison to things, which similarly he says are the same [for all].[1]

When the intelligible character does not differ between *res* and concept, they are formally identical. Concepts, in turn, are identical when they express identical intelligible characteristics or forms. Form brings a certain identity to things, causing them to be units *of a certain kind*. This is no less true of concepts than it is of *res extra animam*. Nor is this contrary to the well-known thesis in St. Thomas that matter is the principle of individuation. On that thesis, units that do not differ in their formal characteristics are materially distinguished from one another, precisely because they are of the *same* kind.[2] It does not, however, involve the claim that their unity and identity consists simply in, or is exhausted by their material disposition. So form brings a certain unity and identity to the intellect's concepts.

If one leaves out of consideration the particular characteristics a concept may possess in its mode of existence *in anima*, for example, *abstract, predicated universally*, and *in this intellect now*, or *in that intellect then*, a concept is identified with another, or individuated and diversified from another, by the nature it expresses—the *intelligible character*, which is what it is in virtue of the *thing* or *things* for which it is a principle of being. This is the basis of the natural likeness or similitude between *passions of the soul* and *res extra animam*. In this respect, St. Thomas's analysis appears to be a form of *externalism*, as described by Colin McGinn in *Mental Content*. St. Thomas "is individuating mental states (to put it intuitively) by reference to something *other* than mental states."[3] McGinn contrasts *externalism* with *internalism*, the thesis "that mental states are determined by facts relating to the subject considered in isolation from his environment."[4]

Extra animam, of course, individual characteristics accrue to the nature, while *in anima* characteristics of individuality (as an accident of this or that particular intellect) as well as of universality accrue to the nature. The theses from the *De ente* do not constitute a further positive attribution of characteristics beyond those associated with the two modes of being. The theses are essentially a negative characterization: there are distinct individual principles of being and knowing that *do not differ* when *considered absolutely*. This is part of what is meant by the presence of Similitude in the "Semantic Triangle," and why it is natural and not conventional.

However, as we saw in the discussion of the Introspectibility Thesis, our knowledge of the character of our concepts is not immediately or primarily known by us. We must know *res extra animam* before the concepts themselves that are Similitudes of those *res* can be known by intellect. Concepts become known to us only as used; but primarily and initially they are used for knowing *res extra animam*.[5] So knowledge of the identity and difference of our concepts presupposes knowledge of *res extra animam*, in particular the identity and difference of the *res extra animam* of which concepts are Similitudes. Talk about concepts is always already talk about *res extra animam*.

A Difficulty with the Notion of Formal Identity

This claim of *identity of form* or *intelligible character* between *passions of the soul* and *res extra animam* is a difficult thesis. When the general Aristotelian principle is enunciated that effects bear a likeness to their cause, except in cases of generation, there is no suggestion that the effect is *formally identical* to the cause. But here in the discussion of cognition there is. To some of our contemporaries, in particular Putnam, "it makes no sense."

> [B]ut even if we could somehow make sense of the claim that objects and events have intrinsic form, there still remains the question of the relation between that form and the form of whatever it is that thinks about or represents the object. To say that the relation is identity, whether "identity" be taken literally or metaphorically makes no sense.[6]

Often the formal identity of knower and known is introduced into discussion as if it were self-evident and in no need of discussion. Then it functions as a trump card and conversation stopper. Either you see it or you don't. Putnam's exasperation with this seems to me perfectly legitimate.

I do not intend it to function in that way. One way to understand the thesis is to consider what remains to characterize an act of understanding some *X*, if we leave out of consideration the characteristics that pertain to it as an act existing in *this* intellect, at *this* time, at *this* level of generality, and so on. What remains is for us to specify what it is an understanding of. For example, suppose we ask, "How does our understanding of the substantiality of a dog differ from our understanding of the substantiality of a man?" In answering this question, all that remains is to make reference to those features that pertain to a dog *as such*, and those that pertain to a man *as such*; we might answer that the one is an understanding of an animate rational substance, while the other is an understanding of an animate but non-rational substance. If we leave out the intentional phrase "understanding of," these are the very same characteristics we would list if we were answering the apparently different question, "How does the substantiality of a dog differ from that of a man?" We will end up specifying just those features of the natures *extra animam* that pertain to their *absolute consideration*. So *absolutely considered* an *act of understanding* an X does not differ from the X understood, when the latter is *absolutely considered*. Absolutely considered *what it is* for an act of understanding to be of an X, the act's *essence* or *quod quid est esse*, does not differ from *what it is* for the X to be, the X's *quod quid est esse*. Two things are formally identical when the characteristics that pertain to their form do not differ between them.

I am suggesting that we treat this claim of formal identity as an extension to the "modes of being" discussed in the *De ente* (in *res extra animam* and *in anima*) of the way in which we treat the claim that two men or two dogs are formally identical. To do this we leave out of our consideration particular material characteristics that differ among men like *being here* rather than *being there*, *being 180 lbs.* rather than *being 150 lbs.*, and so on. We can do this because, "it is not required for the apprehension of truth that whoever apprehends something apprehends everything that is in it."[7] When we do this we find that for some beings we end up listing the very same characteristics like *being sensitive* or *being rational* or *being reproductive*. When the list of characteristics like these do not differ among beings, we say they are formally identical. When we end up with different lists of characteristics, we say that the beings differ in form.

However, looking at the question of *formal identity* between concept and *res extra animam* this way can lead to its own deceptions. We might be tempted to the following picture. Consider the form *in anima*. How does it compare to the form of the thing? This is the question that Putnam is posing. This is to view formal identity as the bridge spanning the chasm between the mind and the world. In the case of two things *extra animam* the normal procedure would be to consider them together, and note their similarities and differences; we would ask whether their similarities are essential or accidental to their being, and whether their differences are essential or accidental to their being. If we find that they do not differ in their essential characteristics we say they are identical in form, like two men, or two trees. If we find that they differ in their essential characteristics we say that they are not identical in form, like a man and a tree.

However, what is presupposed is that the two things that we discover to be identical in form are independently identified prior to the recognition of their formal identity. The initial identification of one does not depend upon the other. This is surely the case when I say, "These two things are formally identical as elm trees." I have picked them out in some fashion that makes the question interesting whether they are formally identical, and the answer non-trivial, that is, I have picked them out separately in some fashion that does not already display their formal identity. It would not be an interesting question to ask "Are these two elm trees formally identical as elm trees," nor a genuine discovery or advance in learning to answer that question with "Yes, they are." It would be interesting, though, to discover that two things we had initially picked out as elm trees are not, in fact, formally identical—that one for example is a rubber model of a tree from a museum.

Looked at this way, and applied to concepts now instead of trees, the role of the assumption of Methodological Solipsism is clear. The concept would need to be picked out in some independent way, "in itself," so that we could then ask, "How does its form compare to the form of the *res extra animam* represented?" Then we could "discover" that they are identical in form, or not. This picture generates fairly quickly the problems of skepticism and epistemological anti-realism. If the concept is picked out in some way independent of the *res extra animam*, is the *res extra animam* picked out in some way independent of the concept *in anima*? Putnam thinks the Aristotelian is committed to saying no. The *res extra animam* is identified by *attending to* the concept *in anima*. The Aristotelian cannot then turn around and say that we need not be skeptics because we recognize, or discover, or even only hypothesize the formal identity of the *res extra animam* with the *concept*. If this is the correct way to understand the Aristotelian claim, it can only appear *ad hoc* and as wishful, or as Putnam likes to put it, "magical" thinking on his part.

By contrast, after the discussion of the Introspectibility Thesis in the previous chapter, it is clear that St. Thomas's approach to the formal identity of the concept and the *res extra animam* does not proceed according to the picture just described. *Res extra animam* are not identified by attending to the concept. On the contrary, the concept is identified by attending to the *res extra animam*. Anyone who does not attend closely to St. Thomas's account of the *introspectibility* of concepts will not understand his account of this formal identity between concept and *res extra animam*. It is no *discovery* on his part that the form of the concept is identical to the thing known by means of it. Nor is it an *ad hoc* hypothesis generated by wishful thinking hoping to overcome skepticism.

If there is going to be anything like a problem of skepticism generated by St. Thomas's analysis, it is going to be about concepts and the soul. We fail to understand this to the extent that we persist in reading the history of philosophy and the philosophy of mind through Cartesian methodological lenses, however much we may object to Cartesian metaphysics. This Thomistic skepticism would not be a skepticism about the *existence* of concepts, that is, about the *existence* of our acts of intellect. St. Thomas does think we are aware of the fact that we are engaged in certain acts when we think. "Socrates or Plato recognizes that he has an intellectual soul from the fact that he recognizes that he understands."[8] This is knowledge of *the fact* of understanding, but not *the why*, the distinction discussed by St. Thomas in his commentary on the *Posterior Analytics* between knowledge *an sit* (of the fact whether it is) and *quid sit* (what it is). The difficulty is with knowing anything about our concepts beyond the simple fact. The second kind of knowledge is had "in a universal manner insofar as we consider the nature of the human mind from the act of intellect."[9] But St. Thomas thinks there is this difference between the two ways (*an sit* and *quid sit*) of cognizing the acts of the soul, and cognizing the soul through them.

> In order to have the first cognition of the mind it suffices for the mind itself to be present, which is the principle of the act from which it recognizes itself. And so it is said to cognize itself through its presence. But in order to have the second cognition of the mind the presence of it does not suffice, but a diligent and subtle inquiry is required. And hence many are ignorant of the soul's nature, and many more are wrong about it.[10]

In response to an objection, St. Thomas immediately adds, "[T]he mind knows itself through itself, because at length it comes to have a cognition of itself within itself, granted that this is through its act,"[11] reaffirming the earlier point about knowing the soul through its powers which are known through their acts. But these acts are known through their *obiecta, res extra animam*. This is not skepticism as we are used to thinking of it, that is, doubt about the existence and/or knowledge of the other because of how the other is known by us. It is simply a recognition of the difficulty of characterizing concepts beyond their existence, except in terms of the *res extra animam* with which they are formally identical. Perhaps it would be better to call it St. Thomas's *negative psychology*.

If we do not attend to its intelligible characteristics, but identify a concept by the characteristics that pertain to it as an act of intellect, we cannot distinguish one concept from another. Two such concepts will share the character of universality, but we could not distinguish them further into more or less general, or at the same level of generality though distinct tokens of some type. Such a further distinction requires admitting back into consideration their intelligible character. How do we distinguish between the concepts *man* and *animal*, as less and more general respectively, if we prescind from a consideration of their intelligible characteristics? At best we could say I had concept X before concept Y, and so they are distinct. Or having concept X led me to have concept Y, so they are distinct. But these distinctions are relatively uninteresting. Even then, it is not clear to me that one could say that concept X led me to concept Y. Without reference to their intelligible content how do we differentiate concept W from concept X, in such a way that we can say it was concept X and not concept W that preceded and caused concept Y. Indeed, it is not clear to me that I could say that I had one before the other. How do we answer the question of which concept came before the other, X or Y, if we prescind from the intelligible characteristics or content?

The reality of a concept is to be nothing other than the means of knowing something else. It has no "in itself" characteristics apart from, on the one hand, the characteristics it possesses as an act of intellect, which are not sufficient to differentiate it from any other concept, and on the other hand, the intelligible characteristics that determine it *to be* knowledge of some *res extra animam*. If I told you that there is something in the next room that is a rational animal, I would at the same time be providing you with the only cognitive access to it that you have, and by that very fact be telling you that it is formally identical to yourself. It is no discovery or hypothesis on your part when you conclude that whatever else it might be, it is formally identical to yourself. The cognitive access I provide you with allows you to ask further questions, like "Well, since there is poison gas in that room, shouldn't we get him or her out?" Of course in this example, you are free to go into the next room and confirm the truth of what I said, just as you might confirm that what we had thought was a tree like this one is but a rubber model, or what we had thought was a man like ourselves is but a cleverly constructed robot. We have many other ways of achieving cognitive access to these things.

But the discussion of the Introspectibility Thesis showed that this is precisely where concepts differ from *res extra animam*. What we know formally of our concepts is determined by what we know of their *obiecta*, *res extra animam*. We cannot then turn around and ask, how do our concepts compare with their *obiecta*. We do not compare a concept with a *res extra animam* and discover or hypothesize a formal identity between the two. On the contrary, we assign "the identity of concepts through a comparison to [extra-mental] things." Formal identity cannot function as an *explanation* for how concepts "hook on to the world," since it is *presupposed* in determining the identity of those concepts. It is the presupposition for making concepts cognitively accessible to us in the first place. To then use it as an escape route, out of a trap one has gotten oneself into by adopting other conflicting presuppositions, begs the question if one feels compelled to answer it.[12]

St. Thomas, following Aristotle, holds that the first act of intellect cannot be evaluated as true or false. As St. Thomas explains it in the *Disputed Questions on Truth*, the first act is too much like the *res extra animam* to be called "adequate" to it. This is not a *discovery* about the first act and truth, but a precondition for talking about the intellect and how it may be truly related to *res* in the second act.

> The intelligible character of truth consists in an adequation of a thing (*res*) and intellect; but something identical is not adequate to itself, rather equality requires diversity; hence the intelligible character of truth is first found in the intellect when the intellect first begins to have something proper to itself which the thing beyond the soul does not have, but something corresponding to [the thing beyond the soul], between which an adequation can be found. But the intellect forming quiddities, only has a likeness of a thing existing beyond the soul . . . when, however, it begins to judge of the apprehended thing, then the judgment itself is something proper to it, which is not found in the thing beyond the soul.[13]

What the intellect has proper to it, not found in the thing beyond the soul, is the subject predicate structure of the judgment, a structure made possible only by conceiving some *res extra animam* in diverse ways in the first act of intellect. St. Thomas will go on to explain here and elsewhere that the truth of the judgment consists in the way in which the unity of subject and predicate in the judgment is adequate to their unity in the thing beyond the soul; diversity because of subject-predicate structure, truth because of adequacy to unity in the thing. Thus in the case of *man*, *rational*, and *animal*, *man is a rational animal* is true because *man*, *rational*, and *animal* are in fact one in *things beyond the soul*, and the unity of the subject-predicate structure of the judgment adequately captures that unity in the *res extra animam*.[14] On the other hand, *man is not a rational animal* is not true because its unity is not adequate to the unity in the *things beyond the soul*.[15]

Recall that the passions of the soul are not the same for all because of the evident fact of diverse contradictory opinions. St. Thomas resolves the issue by an appeal to the difference between first act and second act. The first act is formally the same in all, again because this is a presupposition for talking about the intellect. The second act may differ because of its subject predicate structure. Indeed where diversity of opinion is not based upon equivocation, and thus not merely apparent, but real, it must be presupposed that the elements in isolation, that is, subject in isolation and predicate in isolation, are formally identical between the disputants. They differ in the judgments made that involve them. Real differences of opinion presuppose a certain level of prior agreement.

Thus, for St. Thomas, because of what concepts are, the formal identity of concept and thing is a precondition for our talking about concepts, and in that sense is presupposed. Identifying concepts in this way, specifying their essential characteristics by reference to what is known by means of them, gives us cognitive access to them, and enables us to proceed to ask the interesting questions about human knowledge that the Aristotelian is interested in. St. Thomas finds us acting in the world, and wants to give an account of us and our souls in the world. What is an interesting question for St. Thomas is how the mode of being of the knower differs from the mode of being of the thing known, on the presupposition that his acts of knowing are formally identical to the things known—how the form of *dog* can be present *in anima* without the concept so informed having to breathe and take nourishment, as a dog *extra animam* so informed must do? How can *this* material substance, the knowing man, be formally identical with *this other* material substance, the known dog, as well as *this still other* material substance, the known cat, without the cat being formally identical to the dog? How can a man, in some fashion, "be all things?"[16]

These are the interesting questions, as St. Thomas pursues greater knowledge of human nature. But they cease to be interesting if we start off with the thesis that the concept is not formally identical with *res extra animam*; they cease to be interesting because according to St. Thomas we no longer have any cognitive access to what we are talking about, and the background against which these questions arise.

Still More Difficulties with Formal Identity

I have been stressing the importance of formal identity for our cognitive access *to* concepts. But formal identity is not enough to establish that concepts are not solely individuated by facts intrinsic to the mind. The oak in my backyard is formally identical to the oak in my mother's backyard, but neither tree depends upon the other for the conditions of its formal identity. The tree in my backyard can be individuated by facts solely internal to it, even though it is formally identical to other trees. Formal identity doesn't get us dependence. This is yet another indication that "likeness," even "formal identity," is not playing the "obvious" role in the Thomistic-Aristotelian account that some of its advocates may envision, and that Putnam finds makes no sense. It may well be true that I can only describe to my son William the tree in my mother's backyard that he has not seen by describing the tree in our backyard that he has seen; it may be true that what allows me to do this is the formal identity of the two trees. But that only makes the form of our *speech* about her tree depend upon the form of our tree. It does not make the form of her *tree* depend upon our tree. To this point I seem only to be making a claim about the individuation of our knowledge and speech about my mother's tree, not a claim about the individuation of that tree itself.

Consider the divine intellect. It is certainly true that the form of our *talk* about the divine intellect and divine ideas is individuated and depends upon the created beings that are known by God. This is because we name as we know. It is only through our own knowledge of the created objects of God's knowledge that we come to know anything about the intellect of God. Further, "in some sense"[17] the divine ideas are formally identical with their created objects. Does it follow therefore that the act of the divine intellect knowing its created objects depends upon those objects for its individuation? No. Indeed it is just the opposite—the knowledge that the divine intellect has of its created objects *is* individuated by "facts" solely internal to that intellect. Paradoxically the created objects of the divine intellect depend upon its knowledge of them for the conditions of their individuation—they are not individuated solely by "facts" internal to themselves.[18] So in both the most mundane and most transcendent contexts, it is clear that "formal identity" is not doing the work.

Recognizing what is meant by formal identity here and its role in making concepts cognitively accessible is only the first step to understanding St. Thomas. It is not sufficient for overcoming the Internalist Thesis. It still seems possible that this relation of formal identity could be a contingent relation into which a self-standing mental representation enters. The mental representation might have its own "in itself" identity apart from this relation, an "in itself" identity which is cognitively inaccessible to us. The formal identity we have been discussing might provide cognitive access to the representation taken as actual knowledge in a context, while "in itself" it has a certain identity inaccessible to us. Recall that Fodor said we could not "express" narrow content, but only sneak up on it, referring to it obliquely by setting wide content within quotation, and other such devices. But that fact about our speech about narrow contents did not imply that the narrow contents are not individuated by facts solely internal to the subject in which they exist; indeed that is precisely the point about narrow contents—they are so individuated and can thus be identical across diverse nomological contexts.

Passio and Reception

What is crucial for this discussion is the recognition of the receptivity of the form in St. Thomas's discussion of Aristotle. St. Thomas stresses that the concept is a passion of the soul, with regard to acquiring its formal characteristics. This is evident from the *Peri hermeneias*, where the things signified$_1$ by words are called "passiones animae." St. Thomas of course recognizes that this is an "extension" of the term *passio* from its ordinary Aristotelian use.[19] But it is clear from the general account of how the intellect comes to a universal grasp of *res extra animam* that its formal structure is received from *res extra animam*.

But 'received' invites another deceptive picture, namely, of the form being *exchanged*. However, as I pointed out in the previous chapter, one has to keep in mind that what it means in this context is that the intellect which potentially knows *X*, actually knows *X* because of its encounter with *X*.[20] In this context, it would be appropriate to speak of the intellect's active response. The intellect does not passively receive something from its object as a transfer, leaving the object thereby diminished. Nor does it actively destroy its object, in the process of transforming itself in light of its object. It transforms its own potentialities to know into actual knowledge by using its encounter with its objects as an instrument. It certainly does not receive the matter of its object.

There is no single *being* which is at once both the form of the *res extra animam* and the form of the concept *in anima*. One has to avoid a similarly misleading picture when one considers how "receive" is used in this context. There is not something that is extracted from the *res extra animam* which is then placed in the intellect to construct the concept. Neither, *a fortiori*, is there a single *being* which *was* the form of the *res extra animam* and now is the form of the concept *in anima*. The form of the concept is "received" from the thing in this sense, namely, that the intellect actively responds to some *res extra animam* in a fashion appropriate to the *res extra animam* that is encountered. It responds within itself, and brings itself from potency to act, in such a way that the character of the response is determined by the formal character of the *res extra animam*. The formal character of the act of understanding is "received" from the *res* understood, because if it had encountered some other *res*, differing formally, it would have responded in a formally different way.

Embedded as presuppositions in this account are the theses that I know some *res extra animam*, and that had I not encountered in some fashion the *res* that I know, I would not know them; in other words, my knowledge of them is not innate. My knowledge of them is not "really something else," as in Fodor, that happens to be knowledge of them because of the causal nexus that happens to obtain, and where the very same innate structure would be knowledge of something else in a different causal nexus. It comes to be, as such, from a mere potency to be. Because the intellect potentially knows all things, it actively responds to its encounter with *X*. It responds in an *X*-like fashion, rather than a *Y*-like fashion, which it would have done had it encountered *Y*. Something comes to be as a response to an encountered object, in a way determined by the object that "moves it to its act," that moves the knower to respond.

So the formal characteristics of the act of understanding depend upon the formal characteristics of the object understood. But not vice versa. There is in fact an asymmetry at play in the case of the formal identity involved in knowledge, which is not present in the case of my backyard tree and my mother's backyard tree. It is a formal identity which results from an active response of the one to the other. Again one sees that what is crucial is the nature of the subject of the action. The response is a *likeness* because an effect is like its cause. The *likeness* constitutes *knowledge* because that is the way in which the human animal responds to its encounter with its objects. The response is *formally identical*, because of the totality of *what it is* to be that response. The formal identity and the characteristics of our talk about our acts of knowing are forced upon us by what our acts of knowing really *are*. But to speak of "forced" here is wrong. It is granted to us to speak of our acts of knowing in the way in which we do, because of what our acts of knowing *are*.

Take away the trees, and our intellect ceases to have the form in intellect of trees. This is the importance of the claim that the form *in anima* is a *passio animae*. Whatever structure the intellect takes on in proceeding from potency to the act of understanding is a result of the particular way in which the human person interacts causally with its environment. It is the appropriate response of the human person to her environment, by means of her intellect. Something is known only insofar as it is in act. An act of understanding is *in act* precisely *as an act of understanding some X*, as an act of the *specifically* human response to that *X*. It becomes an object for our reflective understanding just insofar as it is actually an understanding of *X*. If it is not an understanding of *X*, it is nothing, having no formal character and therefore unknowable. And here is the asymmetry. Take away the thing known (that is, suppose it was never encountered by the knower, not that it once was and is no longer) and you take away the response—the form of understanding. The form of understanding, the intelligible species, is the formal principle of the active response of the human person to the *res* that it engages in its experience.

This specific response is not divorced from the other ways that a human person engages the world, other ways that he or she may share with "lower" animals, and even inanimate things. Indeed according to Aristotle and St. Thomas it presupposes the other ways of responding, inasmuch as it is united with them in the unity of the substantial form. It informs them, and makes these other ways of responding the responses of a *rational animal*. On the hypothesis that there are other substances that engage in understanding, even perhaps material instances, it can only be called analogous to our understanding, unless all the faculties presupposed to understanding are integrated into the unity of the substantial form in just the same way as in us. But then, against the hypothesis, they just are us!

Even More Difficulties for St. Thomas's *Externalism*

There are, however, further difficulties under the surface here, for as it stands, this account appears too simple. There could be complex concepts constructed out of simpler elements, where we do not have any knowledge of *res extra animam* prior to the employment of those concepts.

Consider that concepts involving fictional beings, unicorns for example, seem to be employed where no knowledge is even possible of a *res extra animam* for which the concept may be a Similitude. This possibility is presupposed by Aristotle's account of scientific procedure, where one starts with a *nominal* definition prior to showing the existence of the subject of a science; in such instances one may have knowledge of the existence of a subject only *per accidens,* and not *per se.*[21] Today we can think of the search for various sub-atomic particles. In practice this may be only a formal distinction of stages in the process, since there may be no real doubt about the existence of the subject, but it seems possible that there are or will be instances in which the existence must be shown. It might also be necessary in macroscopic sciences to show, starting from the ordinary use of a term, that there is a subject with sufficient unity to constitute an appropriate subject of investigation; the common use of a term might be so broadly cast that there is no sufficiently unified subject to be studied;[22] restricting the use of the term will be necessary to demonstrate a subject of sufficient unity. Spiders are called insects, as 'insects' is commonly used. Yet biologists find it necessary to exclude spiders from the study of insects, because upon examination they are sufficiently different from the vast majority of other things that fall under the common use. By restricting the use of the term in this way, it appears that these biologists are establishing the existence of an appropriate unity for their science of insects.

The facts of our cognitive life show that a complex concept may be employed prior to one's knowledge of, in fact in the very process of showing the existence of, *res extra animam* of which the concept is a Similitude. But if we cannot know the identity and difference of our concepts prior to a knowledge of the *res extra animam* of which they are Similitudes, it appears that we cannot know in various areas of discourse whether or not we are equivocating, whether or not we are thinking or talking about the same thing. Of course this bears directly upon the question of definition in the sciences. It seems that we must know our concepts before using them.

St. Thomas does not discuss this specific problem, though he recognizes a certain creativity in intellectual knowledge that goes beyond a mere mimicking of sense knowledge. I think his answer to Boethius's difficulty about truth and opinion shows the way to work through the difficulties raised here. In lesson 2, passage 20 of the *Commentary on the De interpretatione,* he considers the obvious points that combinations and divisions are conceived by the intellect, that men hold many different opinions on the same thing, and that they hold many false opinions. Based upon this evidence, the objection is raised that not everything conceived by the intellect is the same in all men. St. Thomas responds to these difficulties by pointing out that Aristotle has the *simple* conceptions of the intellect in mind when he is assigning the identity of concepts among all men,

> but because the false can be in the intellect, according as it composes and divides, but not according as it understands the *quod quid est*, that is the essence of a thing, as is said in III *De anima*, this [thesis] ought to be referred to the simple conceptions of the intellect (which incomplex sounds of the voice signify), which are the same for all.[23]

St. Thomas's point is that Aristotle did not mean to claim that every person has all and only the concepts had by every other person.[24] Rather, it is the claim that whenever two persons understand the same intelligible character, their understanding does not differ among them. While these difficulties that we are considering now show that complex concepts may be *used* prior to a knowledge of *res* to which the concepts bear the relation of Similitude, and even when no knowledge of an appropriate *res extra animam* is possible, they do not show or require that such complex concepts can themselves be *known* prior to their use. If I think about a fictional being, like a unicorn, I am not *thinking about* a complex concept involving the concepts of horse, head, and horn. I am employing the concepts of horse, head, and horn *to think about* something that is a horse with a horn on its head. If I think about gold, prior to encountering it, I am not thinking about my concept of gold, I am thinking about a yellow metal, using the concepts of *yellow* and *metal* to think about it. We saw in the discussion of the Third Thing Thesis that there is a minimal sense of *being* in which we can think of these things as *beings*, namely ens_p, without thinking of them as either *res extra animam* or *res in anima*. It is clear from that discussion that if some ens_p is not a *res extra anima*, it does not follow that it is a *res in anima*.

What then is the importance of the text from lesson 2, passage 20? Human activity is surely involved in the construction of complex concepts, particularly those involving fictional beings and nominal definitions, just as much as in the formation of opinions and false beliefs. But since we are abstracting from conditions that pertain to the knower's act, in analyzing the use of such complex concepts and their elements presumably we will arrive at simpler and more fundamental elements, just as when we analyze complex judgments that make an assertion. At some point, the analysis will arrive at simpler elements, knowledge of which cannot be had prior to a knowledge of the *res extra animam* of which these simpler elements are Similitudes.

So the identity of a complex concept will be determined by the respective identities of its more fundamental and simple elements. Consequently, the identity of a complex concept will be remotely determined by the *res extra animam* of which its simpler elements are Similitudes. Presumably we can determine that we are not equivocating when we use complex concepts by determining what simpler elements are involved in the use of the more complex—not by an inspection of the complex concepts apart from their use, but by an analysis into the simpler elements involved in the use of the complex. So, it remains true that our understanding of the conditions of identity and difference of all of our concepts presupposes our knowing the *res* of which our concepts are Similitudes, or at least the *res* of which our concepts' more fundamental elements are Similitudes.[25] Concepts in their simplest forms are acts of simple apprehension, acts that apprehend principles of *being* in the things known by means of those concepts, and which cannot be false.[26]

This response accords well with the *developmental view* of concept use in St. Thomas sketched in the previous chapter. I think this view is in some ways foreign to the static view that animates many modern and current discussions. In that static view, concepts do not develop; they are what they are; any development in them would involve the ceasing to be of the old, and something new coming to be which is not a real development in the same thing. At its most extreme, when this static character is coupled with *conceptual holism,* that is, the thesis that the content of a concept is determined by its place within the whole conceptual scheme within which it fits, a development of conceptual function appears as a substitution of an entire conceptual scheme for another. The new might be related to the old in some way, but it is a new and different thing. Since conceptual change involves a change at the level of conceptual schemes, it is difficult to think of how the old and the new might be related. Of course, I am not suggesting by contrast an atomistic account of conceptual content.

In St. Thomas's developmental view of the transition from the "vague universal" to the distinct, the content of the universal concept is determined neither holistically nor atomistically, once and for all. Because for St. Thomas the universal concept is individuated by *res extra animam,* the unity that obtains through its development, the unity that constitutes it as the same concept once vague, now distinct, is determined by the ways in which the unity of the *res extra animam* discloses itself through its encounters with the active knower in experience through time.

Concepts as Acts of Simple Apprehension

Peter Geach, in *Mental Acts*,[27] wants to deny that there "are acts of 'simple apprehension', in which concepts are somehow exercised *singly* without being applied to anything." His discussion comes very close to simply identifying concepts with capacities for linguistic expression. He avoids a complete identification by recognizing that it is a sufficient but not a necessary condition for "having the concept of *so-and-so* that [a person] should have mastered the intelligent use . . . of a word for *so-and-so* in some language." However, the example he uses where the necessity fails involves a pathology, namely when someone with aphasia can still play chess. Nonetheless, his overwhelming emphasis on concepts as capacities for linguistic expression remains. "[T]he central and typical applications of the term 'having a concept' are those in which a man is master of a bit of linguistic usage," and "I shall apply the old term 'concepts' to these special capacities [to express a judgment in words], for intelligently using the several words and phrases that make up the sentence."[28]

In this context he is concerned that such acts of simple apprehension would not be "applied to anything" in the form of a linguistic judgment, as if there could be complete linguistic acts that involved simply uttering single words. However, it is not clear that he is in fact denying what St. Thomas is asserting, because of his near, if not quite complete, identification of concepts with capacities for linguistic expression. In fact, commenting on Aristotle's claim that verbs signify$_1$ something because in uttering them "he who speaks establishes understanding, and he who hears rests,"[29] St. Thomas raises an objection very much like Geach's.

> [B]ut this seems to be false, because a completed sentence (*oratio*) alone causes the intellect to rest, but not if either a noun or a verb is said of itself: for if I should say 'a man', the soul of the listener is filled with suspense concerning what I wish to say of him; and if I should say 'he runs',[30] [the listener's] soul is filled with suspense concerning whom I say it of.[31]

Responding to this objection, St. Thomas draws our attention once again to the two operations of intellect.[32] It is the first "which is the conception of something, and according to this [operation] the soul of the listener is set at rest," when it considers a noun or verb in itself, since "it was in suspense before the noun or verb was brought forth and its bringing forth terminated." On the other hand, "with respect to the second operation, which is the composition and division of intellect," the soul does not rest upon hearing a noun itself or a verb itself, since this operation is not completed until something is said *about* something else. By means of the first operation, signified$_1$ by a noun or a verb, we understand something. By means of the second act, signified$_1$ by a complete sentence, we understand something *about* something.

This distinction between first act and second act need not be interpreted as involving a temporal distinction and genesis, as in a play the second act follows the first. In this respect, Geach may well be correct; the first act may be reached by analysis from the second.[33] However, as St. Thomas interprets him, Aristotle is simply drawing our attention to the fact that what is signified$_1$ by nouns and verbs is simple or incomplex in comparison to what is signified$_1$ by complete sentences.[34] All that is necessary to recognize *acts of simple apprehension* is that there be a distinction between understanding something, and understanding something *about* something.

For St. Thomas it is this distinction that makes error possible; error arises in the second act, not the first. Something is false if it says of what is that it is not, or of what is not that it is. If nothing is understood in the subject and predicate terms considered separately, how is it possible to say of *something* that is, that *it* is not? Error and babble are not the same thing. For St. Thomas, as for Aristotle, we must in some way know what we are talking about and what we want to say about it, if we are even going to have the chance of messing it up.

This distinction also undermines Dummett's claim about the Aristotelian tradition.

> A continuous tradition, from Aristotle to Locke and beyond, had assigned to individual words the power of expressing 'ideas', and to combinations of words that of expressing complex 'ideas'; and this style of talk had blurred, or at least failed to account for, the crucial distinction between those combinations of words which constitute a sentence and those which form mere phrases which could be part of a sentence.[35]

St. Thomas is well aware of the "crucial distinction," and the need to give a different account of what is signified by each.

St. Thomas's *Externalism* in Summary

Throughout this discussion, I have been consciously drawing attention to St. Thomas's position that concepts in their simplest forms are acts of simple apprehension, acts that apprehend principles of *being* in the things known by means of those concepts. I mentioned earlier that St. Thomas's analysis of Aristotle appears to be a form of *externalism* as described in McGinn's *Mental Content*, though McGinn does not explicitly consider him. McGinn goes on to distinguish types of *externalism*, namely *weak* and *strong*. The *weak externalist* holds the thesis "that a given mental state requires the *existence* of some item belonging to the nonmental world, and that its identity turns on that item."[36] This suggests that there is some single being, or perhaps type of being *extra animam* that determines the identity of concepts. The *strong externalist* holds the thesis that "a given mental state requires the existence *in the environment of the subject* of some item belonging to the nonmental world, and that its identity turns on that item."[37] Not only must there be some one item in the "nonmental world" determining the identity of the concept, but it must be an item "in the environment of the subject." The "environment of the subject" might be distinguished by any number of criteria, including the one McGinn cites, namely "causal connection" to the mind involved.

St. Thomas's willingness to countenance nominal definitions, where no being *extra animam* has yet been found, or where the being involved is at best known *per accidens*, and concepts involving fictional beings, suggests some difficulties in situating St. Thomas vis-à-vis these detailed classifications of McGinn's. In the case of concepts involving fictional entities like unicorns, there is no single being or type of being *extra animam* that determines the identity of the concept—its identity is determined by horses and horns; in the case of a nominal definition, we might find out as a matter of scientific progress that there is no *single* being or type of being in the world that determines the identity of the concept employed in the definition, for example, insect or fish where there are too many, or phlogiston and caloric where there are none at all. For these reasons, St. Thomas does not seem to be a *weak externalist* in McGinn's sense. Since *strong externalism* includes within it the conditions of *weak externalism*, adding that the concept must be individuated by a single being or type of being *in the environment* of the one possessing the concept, the problems of nominal definitions and fictional beings apply *a fortiori* to it. So *strong externalism* also seems inadequate, at least in the instances of concepts involving fictional beings and nominal definitions.

However, as this analysis has indicated, such concepts are constructed out of simpler concepts, which at some level are individuated by *res extra animam*. In particular, St. Thomas's causal analysis of the use of *passio animae* indicates further that these simpler concepts are individuated by *res* in the environment of the one possessing the concepts. This does not preclude the sort of "division of linguistic labor" that Putnam emphasizes, in which experts may be the ones possessing the relevant concepts individuated by *res* in the environment of the experts.[38] The experts may then communicate to non-experts the appropriate scientific use of words signifying$_1$ those concepts. But presumably the experts can only communicate that use in terms of concepts already possessed by the non-experts. These latter concepts would presumably be analyzable into simpler elements, and so on.

In any case, the simpler acts of apprehension not only have their conditions of identity determined by *res extra animam*, but by *res extra animam* in the "environment of the subject." So with respect to these simpler concepts, St. Thomas seems to be a *strong externalist* in McGinn's taxonomy. And with respect to the complex concepts involved in fiction, and nominal definitions, we might call his position *fundamental strong externalism*, in order to emphasize that even in these latter cases the fundamental elements of these concepts are *strongly external* with respect to their individuation. McGinn himself affirms *strong externalism* with respect to natural kind concepts like *cat*, *water*, and *electron*, and *weak externalism* with respect to a number of other concepts, for example, artifact concepts like *table* that are defined functionally. However, he is also interested in additional "mental states," beyond what I have been calling "concepts." He moves between *externalism* with respect to properties, to *internalism* with respect to what might be called "first person privileged mental states," like emotions, pains, pleasures. For the most part he proceeds on a case by case basis.

When one considers the fundamental and strongly external character of the identity of concepts in light of the *De ente et essentia*, the *Summa*, and the *De anima* commentary, as well as all that has been said in this section, it is sufficient to remark here that the recognition of formally identical concepts in distinct intellects or in the same intellect at different moments of exercise is no more controversial than recognizing identical substantial forms in Plato and Aristotle, even though as substantial forms of numerically non-identical beings, they are numerically non-identical.[39] Which is not to deny that for some the recognition of identical substantial forms in Plato and Aristotle is a very controversial matter indeed.

Unity of Form or Essence

In this part of the chapter I want to consider a very recent objection raised by Putnam to the Aristotelian account of words, thoughts, and things. This most recent objection is directed to the metaphysical underpinning of the account, specifically, its dependence upon the uniqueness of forms in *res extra animam*.

We have seen that the forms that bring identity to the intellect's act of understanding do not arise from the intellect itself but from the *res extra animam* understood, in which these forms are principles of being. These conditions of identity for concepts, whether simple or complex, are determined by the Similitude of the concepts if simple enough, and by the Similitude of their elements, if complex enough. For St. Thomas this Similitude *in anima* is cognition or understanding precisely because it is constituted as such by the determinate and specific *response* of a rational animal to its environment. Thus, this Similitude functions as his *ticket in* for describing specifically human nature. In the language of St. Thomas's *Disputed Questions on Truth*, the forms of *res extra animam* are the *measures* against which the intellect's complex activity of forming opinions, beliefs, judgments, and reasoning is judged to be adequate or not, that is, true or not, not simply because events in the intellect happen to be correlated with events in the world in a nomologically consistent and asymmetric way, nor because the form of the intellect's activity happens to correspond to those *res extra animam*, but rather because those *res extra animam* provide the form of the intellect's activity and render the intellect itself knowable. Criticizing current causal accounts of mental representation, Putnam writes that external causes, as presently conceived by the philosophical community, do not have enough "form" to be "self-identifying."[40] This is not the case in St. Thomas's Aristotelian account, though "self-identifying" is probably not the right phrase to characterize the situation to the extent that it seems to ignore the active response of the knower to the known.

Putnam now seems to recognize the analysis of form as a virtue of the Aristotelian account. He grants that the Aristotelian may have a robust enough *form* to overcome his earlier objections, because he realizes the Aristotelian form is not the "logical form" of the *Tractatus*, which he now thinks may have been the real target of his earlier criticism. He does not, however, say what the positive characteristics of the Aristotelian form are that allow it to overcome the objections, just that it is not the Tractarian form. However, Putnam still sees the "Neo-Aristotelian" and the

Tractarian Wittgenstein as trying to overcome the skeptical problem, the problem of how language or mind "hooks on to" the world. Now he simply has a new objection to the robustness of the Aristotelian form. Putnam denies or at least doubts that "substances have a unique essence."[41] Since this objection is supposed to apply *mutatis mutandis* to forms of all the other categories, this is no longer an objection to the "philosophical psychology" of the account, as the previous objections were, but rather a properly metaphysical objection. Despite my argument that St. Thomas is not trying to overcome the skeptical problem, it is possible to consider the objection itself. Broadly, it might be put this way—there are no unique forms in *res extra animam* for the concepts that are supposed to be individuated by them. The emphasis is on the "unique." Putnam is not denying that there are Aristotelian forms in *res extra animam*; for the sake of argument he gives to the Aristotelian his forms. What he denies is that for any concept there is a unique form in *res extra animam* that determines its identity.

As evidence for his denial of unique essences, he cites the fact that different sciences count different characteristics as "essential" to their subjects. In particular, he cites how evolutionary biology would count the history of the development of a population of dogs as essential to being a dog, while genetic biologists would count genetic structure, and discount history. Because they count different characteristics as essential, they distinguish different essences. But because they distinguish different essences, it appears that the two different scientists must be talking about different things; thus, there is no unique essence to determine the identity of the concept *dog*, and subsequently to determine the identity of the term 'dog'. Putnam simply cites the plurality of sciences, but I think it fair to construe his general objection informally as follows:

1) Sciences give essential descriptions of X's (e.g., dogs).

2) The essential descriptions of X's given by the different sciences are manifestly diverse.

3) Therefore, the forms that constitute the identity of the descriptions are diverse.

4) But the forms that constitute the identity of the descriptions are supposed to be identical to essential forms of *res extra animam*, according to the Thomistic-Aristotelian.

5) But then from 3) and 4) the essential forms *in res extra animam* must be diverse since identical to diverse forms *in anima*.

6) Therefore, X's, insofar as they fall under different sciences do not have unique essences.

7) Therefore, the uniqueness of the concept X employed in the different scientific descriptions of X's is not determined by a unique essence of X.

8) Therefore the Thomistic-Aristotelian view is false, since the Thomistic-Aristotelian assumes the contradictory opposite of 7).

His objection is not unrelated to his problems with different conceptual schemes and what he calls *Metaphysical Realism*. There are no essences "out there" awaiting discovery; the conceptual scheme of a particular science determines for itself what counts as an essence for it. To be fair, Putnam thinks that once the science has made unto itself an essence in view of some interest of the scientist, everything else falls into place; it's not just anything goes.[42] Still, the distinct conceptual schemes of the distinct sciences determine distinct essences.

There are a number of possible responses to his argument. To begin, consider what the objection presupposes, namely, that even as the *essence of dogs* changes from science to science, still the concept *dog* remains uniform or unique across those sciences. That presupposition is embodied in 7) above, in the phrase "different scientific descriptions of X's." But given Putnam's objections to *Metaphysical Realism*, it seems mildly out of place for him to assume that the concept X, as well as the set of objects that fall under it, are identical across the multitude of sciences; that could only be the case if those sciences take place within a larger conceptual scheme that holds the concept X fixed, while allowing for the variation of the *essence of X* within the sub-conceptual schemes of the sciences. But what is that larger conceptual scheme within which the sciences find a home? Putnam presupposes it, but he does not tell one what it is. And why couldn't it similarly hold fixed the *essence of X* across the sciences? But more on that later.

One possible response to Putnam's new objection would be for the Thomistic-Aristotelian to argue that the multiplicity of essential descriptions does not indicate a multiplicity of essences, but in fact the descriptions describe one essence. Or the Aristotelian might respond that even if they do not all describe one essence, what they describe are related to one essence, and so they are called essential descriptions analogously because of their relation to an essential description properly so called. I will argue that St. Thomas's response would be a combination of the latter two.

How Many Kinds of Dogs?

Let us examine more closely the presupposition involving the uniqueness of the concept *dog* by contrast with the plurality of *essences* of dog; but in order to make a point below, I want to change the example from Putnam's dogs to bats. The geneticist Jill talks about bats and says the essence of bats is 'such and such genetic structure'; call it *essence$_1$*. The evolutionary biologist Jack also talks about bats and says the essence of bats is 'such and such history of evolutionary descent'; call it *essence$_2$*. The claim is that *essence$_1$* is not identical with *essence$_2$*. Thus the essence cannot be determining the identity of the concept *bat* which is used by both Jack and Jill when they use 'bats' to talk about bats.

Notice, however, what Putnam presupposes or takes for granted here. Jack and Jill are providing accounts of the essences(s) of *the same kind of thing*, bats. Consider the following statements:

> Jill the geneticist says, "Bats have an essence consisting of such and such genetic structure."
>
> Jack the evolutionary biologist says, "Whales have an essence consisting of such and such history of evolutionary descent."
>
> Jack the evolutionary biologist says, "Bats have an essence consisting of such and such history of evolutionary descent."

An uncontroversial claim is that Jack's statements are substantively different. In the context of this discussion we would not find Putnam's claim about the plurality of essences interesting if it involved Jill's claim about bats and Jack's claim about whales instead of their claims about bats. No Aristotelian is bothered by a plurality of essences for *different kinds of things*, bats and whales. If Jack and Jill aren't talking about the *same kind of thing* when they give their respective accounts of the essence(s) of bats, Putnam's objection is based upon a straightforward fallacy of equivocation on 'bats'. In that case, even those not convinced by the Aristotelian's larger claims should recognize that Putnam might as well have said that

Jill the geneticist says that bats have an essence consisting of such and such genetic structure, while Billy the manager says that they are made of wood and have such and such a geometric structure. The Aristotelian expects to find a plurality of essences among a plurality of *different kinds of things*. The trouble, if there is one, is with a plurality of essences for *one kind of thing*.

So, while denying a unique essence of dogs Putnam must assume a unique concept *dog* within and across the sciences, as well as a corresponding unique set of dogs that fall under it, the members of the unique kind.[43] But given that he begins with the diversity of descriptions, and given the phenomenon of equivocation in language, which he recognizes and exploits in his objections to *Metaphysical Realism*, why does he begin by privileging the one, namely, the kind, over the other, namely, the essence? It would be more plausible for one who begins with the descriptions to deny that the plurality of sciences employ the same concept *dogs* and speak about the same kind of thing, dogs. But what Putnam is trying to do is drive a wedge between the kind and its essence. So he holds one fixed, the kind, while multiplying the other, the essence(s).

How Many Natures or Essences of Dogs?

While Putnam presupposes this *same kind of thing*, and that there is no equivocation between the sciences about *it*, even then it is not clear that he can get the objection going. This *same kind of thing* is supposed to have a plurality of essences or natures. He writes, "not just anything can be called the nature of dogs. On the other hand, it does seem that more than *one thing* can be called knowing the nature of dogs."[44] Why should a fact about what can or cannot be "called knowing the nature of dogs" be taken as evidence for what is or is not the nature of dogs? One is a fact about us, the other, if it is anything at all, is a fact about dogs.

I don't know whether Putnam ultimately agrees with Locke that "we make unto ourselves the species of things," but however he wants to set up the possible plurality of essences from the plurality of sciences, such a commitment is not forced upon us by any particular science. Presumably it is not a well-confirmed hypothesis of any particular natural or social science under which some being falls that there may be a multiplicity of natures for that being corresponding to the multiplicity of sciences other than that particular one. What exactly would the experiment be that might falsify or confirm that hypothesis? How does such a hypothesis exhibit within a particular science the theoretical virtues of simplicity, predictive power, and fruitfulness for further research programs? It is clear that such a thesis is an extra-scientific *philosophical position* that one takes and that requires argument, if it is going to be used against some other position; it is not simply *read off of* the descriptions and success of contemporary sciences. In short, it requires philosophical argument to show that a multiplicity of essences is a necessary condition for the possibility of some kind of being falling under a multiplicity of sciences.

Putnam gives the counterfactual example of a dog synthesized from chemicals off the shelf, resulting in something with exactly the same DNA as his dog Shlomit.

> [T]hen, from a molecular biologist's point of view the resulting 'synthetic dog' will count as a dog. . . . From the point of view of an evolutionary biologist, the situation is different. I suspect, in fact, that evolutionary biologists would not regard a "synthetic" dog as a dog at all. From their point of view, such a thing would simply be an artifact of no interest.[45]

Notice that the evolutionary biologist will not even include the synthetic dog in the set of objects that constitutes the species or natural kind; whether or not Putnam is right about the judgment of the evolutionary biologist concerning the synthetic dog, the individual is excluded from the *kind* because of what the scientist takes the *essence* of the *kind* to involve. This exclusion points out the close connection here between the *essence* and the *kind of thing* for the practicing scientist, the connection through which Putnam wants to drive a wedge, and yet upon which he must rely in order to get the objection going.

This example of Putnam's makes plain that no individual science is doing the "counting" of the plurality of the things that count as "knowing the natures of dogs," since no science considers anything a nature or essence except what it describes within itself. And within itself, in virtue of that nature or essence it counts objects as falling under it. So what is the "point of view" from which the plurality of what counts as "knowing the nature of dogs" is recognized? *A fortiori,* and once again, what is the point of view from which the conceptual unity of 'dog' is taken, if not from any one science?

At best his objection seems to be a restatement of the problem that Putnam has with *Metaphysical Realism*. Each distinct science as a conceptual scheme is determining for itself what counts an essence for it; and his example of the synthetic dog even suggests that each science as a conceptual scheme is determining for itself what counts as a dog. The lack of unity among the sciences speaks to the lack of unity in what counts as essences and as dogs for them.

Recall St. Thomas's discussions of how error and truth and falsity enter into differences of opinion and disagreements. These phenomena in our discourse, if they are to be real, require some measure of presupposed agreement about the subject.

What St. Thomas takes to be a vice in argument, equivocation, Putnam sees as a solution, not with regard to 'dog', which he presupposes is used univocally, but with regard to 'essence' or 'nature'. For him the dispute is resolved by pointing out to the disputants that they are both in possession of *a* nature or essence of dogs, but that they are wrong in thinking there is only *one* such essence. They can come to agreement if they recognize that they are studying *the same kind of thing*, dogs, though *what* they are studying differs essentially between them. But if that dispute is solved in this way, then the question remains why the geneticist should try to take account of, or relate her work to the work of the evolutionary biologist, and vice versa, since they are studying things that are essentially different. There is no reason to think that knowledge of the one essence or nature is related to knowledge of the other, except *per accidens*. Putnam cannot fall back upon the unity of the kind, since by supposition the kind has no essential unity. And his own example of the synthetic dog suggests there may not even be any unity of the kind.

But are the terms 'essence' and 'nature' subject to this radical equivocation in the practice of the sciences? Both scientists in Putnam's scenario purport to be studying dogs, not whales. Certainly the essence of whales as studied by a *geneticist* will be an essence distinct from the essence of dogs as studied by an *evolutionary biologist*. Should the *evolutionary biologist* pay no more attention to the research of the *geneticist* studying dogs than he does to the research of the *geneticist* studying whales? Isn't it the case that progress in one area of scientific endeavor often proceeds because of the attention the scientists involved pay to the results obtained in closely related sciences *studying the same kind of thing*? Is it really a solution to be desired for the geneticist and the evolutionary biologist to go their separate ways counting different things the natures or essences of dogs? Isn't it one of the theoretical virtues we expect of our scientific theories that they will resonate with other sciences studying the same kind of thing? This fact about scientific ways of knowing becomes difficult to reconcile with a plurality of utterly unrelated natures or essences that are known about the *same kind of thing*.

Retrieving the force of Putnam's objection is only bought at the price of recognizing some sort of unity that brings together the essential subject matter of the sciences. Otherwise why are the scientists talking with one another at all? What is that unity? If it is not a unity in the *res* studied, but only the unity of the *concept* divorced from the *res*, then that just starts to look like some form of Idealism that no Aristotelian, and I suspect few practicing scientists, have any interest in. Here one is reminded of Quine's memorable statement to the effect that meaning is what essence becomes when it is divorced from being and wedded to the word.

Talking about dogs in diverse ways is different from talking about dogs and whales in diverse ways. Rather than drawing the philosophical conclusion from the multiplicity of sciences that there are or may be a multiplicity of unrelated essences for some *one kind of thing*, it seems just as plausible, if not more so, to conclude from the unity amidst diversity that the sciences display, that there is some unity to that *one kind of thing* that relates the plurality of essential descriptions to one another.

Putnam's objection tries to drive a wedge between the *kind of thing* talked about and *what its members essentially are, what it is for [them] to be* (*quod quid est esse*). Once this gap between the kind of thing and its essence(s) is introduced, the question arises as to what the unity of the kind consists in, such that the many scientists can say of *it* that *it* has a plurality of essences. Putnam does not, however, take into account the presupposed unity behind his objection, and it is here that the fundamental difficulty with his objection comes into view. Proceeding from a multiplicity of *ways of knowing* to a multiplicity of *ways of being*, that is, essences, Putnam himself seems to be introducing a different kind of third thing, his wedge, the set of third things that stands between the underlying unity of a kind and our scientific ways of considering it. The mediating concepts of the Third Thing Thesis and the Introspectibility Thesis show up again, but now under the guise of essences projected by the sciences out into the world. And the Internalist Thesis is resurrected in the multiplicity and diversity of these *third things*, these projected essences, to the extent that they are not determined by or related to the underlying unity of the kind, but are related to the multiplicity and diversity of human modes of understanding. "They just stand there."

It just looks like an unquestioned "metaphysical" assumption on Putnam's part that because separate scientific investigations consider different characteristics essential to their subject, that therefore there must be, or at least may be a multiplicity of essences for any one kind of thing. But Putnam, before anyone else, ought to recognize this as a fallacy. After all, he is the one who reminded us in "The Meaning of 'Meaning'," as he developed the Twin Earth paradox, that a difference of *intension* does not imply a difference of *extension*. The extension here is not the underlying presupposed unity of the kind, but the natures or essences. The plurality of essences or natures, standing in-between the underlying unity of the kind and the plurality of sciences, is supposed to follow from there being "more than one *thing* that can be called knowing the nature of [X]." St. Thomas was well aware of this fallacy, for he often reminds us not to commit it, that is, not to confuse the *ways* of knowing with the things known. A plurality of ways of considering Y does not imply a plurality of Ys considered. But here Putnam has good company, since according to St. Thomas this was the characteristic mistake of the Platonists. Thus, one might call it the Platonist Fallacy. Even if it were possible for there to be a multiplicity of natures or essences for some *one kind of thing*, that possibility does not follow, as Putnam seems to believe, from the multiplicity of ways of knowing that thing.

One Essence?

Putnam might respond that *he* does not intend to suggest that there are a multiplicity of essences from the multiplicity of sciences, but that it is the Thomistic-Aristotelian who is committed to this when he argues that the concepts used in language are formally identical with *res extra animam*. Look at the transition above between 2) and 3) in the informal outline I gave of Putnam's objection. From the diversity of distinct descriptions, one infers a diversity in the forms involved in the descriptions that allows one to argue to the plurality of essences in *res extra animam* given the Aristotelian assumption of identity in 4). So it is the Thomistic-Aristotelian who is forced into this fallacy by his larger assumptions.

However Putnam cannot deny that he himself argues from the plurality of descriptions to the plurality of things described. His discussion shows that *he* thinks it is true that what counts as the *nature* of X is determined by the science in question, and the plurality of such descriptions indicates a plurality of things described. To stress this point he compares the "interest" dependence of natures to the discussion of the interest dependence of causes that followed upon the work of Hart and Honoré in the 1960s. The plurality of essences position is supposed to come from a coalescing philosophical consensus from *outside* of Aristotelianism as a fundamental objection to it. "The greatest difficulty facing someone who wishes to hold an Arisotelian view is that the central intuition behind that view, that is, the intuition that a natural kind has a *single* determinate form (or "nature" or "essence") has become problematical."[46] Thus, it is not an internal objection to the Thomistic-Aristotelian, but an external one. And, as presented by Putnam, it commits what I have labeled the Platonist Fallacy.

But look carefully at the transition from 2) to 3) in light of the discussion of introspectibility in the previous chapter, then also the transition from 3) and 4) to 5). The objection moves *from* the description *to* its conceptual form *to* the world. A conclusion is drawn about the form *in res extra animam* by starting from the form in the concept, the direction that leads to a misreading of St. Thomas. This transition presupposes that one *reads off* the identity of the forms involved in conceptual functioning simply by looking hard at the "scientific" descriptions. Even if concepts are projected *out* of the mind and *into* public language use after the *Linguistic Turn*, this transition between 2) and 3) in Putnam's latest objection embodies his own commitment to the Introspectibility Thesis, no less than it was presupposed in the target of his earlier objections when concepts were thought to be "in the head."

Putnam's commendation of the *Linguistic Turn* is that

> [it] has the advantage of being a public study, and more in the spirit of modern social science. If the way to find out what the concept Cause is . . . is to introspect one's own images, etc., then we would hardly expect to get very reliable reports, let alone agreement. If analyzing the concept Cause is rather a matter of studying the way in which we use the word . . . then the hope for reliable reports . . . is much greater.[47]

Now, against the background of his multiplicity of essences objection, this passage has a different resonance to it than appeared at first sight. Rather than introspecting our "images, etc.," we introspect the linguistic concepts of our scientific language, and by such introspection we come to know the world as mediated by that scientific language as it determines for us the natures or essences of things. Even though concepts are "in the mouth," they are just as introspectible and mediating as they ever were; giving them the aura of scientific descriptions doesn't make them less so.

However, given the discussion of the last three chapters, it should not be a surprise that St. Thomas denies the transition from 2) to 3). Indeed, a careful look at his work shows that he denies it in a discussion remarkably similar to the considerations from science that Putnam employs—the medieval problem of the multiplicity of substantial forms in a human being. It is a useful discussion because it makes quite clear what he takes to be the fundamental character of forms as essential metaphysical principles. The question took the form, "whether in addition to the intellectual soul there are in a man other souls differing essentially."[48] The medieval disputants were certainly well aware of the different scientific investigations under which human beings may fall. The problem centered on at least three different ways in which a human being might be described "scientifically," namely, as sharing vital activities in common with plants and animals (for example, nutrition, growth, reproduction), as sharing vital activities in common with animals but not plants (for example, sensation), and as having a vital activity apart from plants and animals (reason). To each of these classifications, some principle of essential unity is assigned, namely a vegetative kind of soul, a sensitive kind of soul, and a rational kind of soul. In the discussion, all of the arguments for a plurality of substantial forms come from Aristotelian considerations about how these types of substantial form differ from one another as described by the different sciences.

The classifications are manifestly different. If one takes these diverse classifications to be providing essential descriptions, it appears that one must hold that in a plant there is one nature or essence determined by the vegetative soul, while in an animal there are two natures or essences determined by the vegetative soul and the sensitive soul, and in a man there are three natures or essences determined by the vegetative soul, the sensitive soul, and the rational soul. It seems to be the case "that more than *one thing* can be called knowing the nature of" human beings, and the objection goes that therefore there is more than one thing that is the nature in human beings.

Thus, in their own way, one can see the medieval dispute posing the sort of problem Putnam would now pose from the difference between a genetic explanation and an evolutionary one, among others. But St. Thomas responds:

> [I]t ought to be said that it is not necessary to assume a diversity in natural things from the diversity of intelligible characteristics or logical intentions which follow upon our manner of understanding, since the intelligible character of one and the same thing may be apprehended in diverse ways.[49]

In other words, a difference of *intension* does not determine a difference of *extension!* Here St. Thomas recognizes and clearly avoids the Platonist Fallacy. He focuses upon the source of plurality as arising from our ways of considering "one and the same thing." Applying this position of St. Thomas to the transition from 2), 3) and 4) to 5) in Putnam's objection, it is clear that St. Thomas denies the introspective character of the transition.

This denial on St. Thomas's part serves as a confirmation of the general account I have been giving throughout this work. Consider again his discussion in the commentary on the *De interpretatione:*

> [T]he intention of Aristotle is not to assign identity of the concept of the soul through a comparison to articulated sound, as namely of one articulated sound there should be one concept; . . . but he intends to assign the identity of concepts of the soul through a comparison to things, which similarly he says are the same [for all].

Still, nothing in St. Thomas's response to the plurality of substantial forms argument indicates that there cannot be a plurality of essences for any one kind of thing. He is simply remarking that such a plurality does not follow from the plurality of our modes of description as Putnam would have it; he is remarking that the argument to that conclusion is unsound.

St. Thomas does not simply rest content, however, with denying that the multiplicity of ways of knowing one and the same thing provides an argument to a multiplicity of natures or essences for that thing, he gives three arguments why. The first is the most pertinent to Putnam's objection, and focuses upon the fundamental character of substantial form as essential principle, as providing the "what it is for the thing to be" (*quod quid est esse*).

For St. Thomas a nature or essence is an intrinsic formal principle of being, unity, and act. It provides, or better, *is* the real unity in *being* of any set of essential characteristics had by some *thing*. For any one kind of thing, consider the set of all characteristics taken to be essential to that kind of thing by all the sciences; one might ask whether there are as many essences of that kind of thing as there are subsets of that set, excluding the null set—nutritive essence, reproductive essence, nutritive-reproductive essence, nutritive-reproductive-sensitive, and so on. St. Thomas's answer is no. Form as a metaphysical principle is an intrinsic principle of actual being and unity, the determination or limitation to this one kind of being (*ens*). It is the reason why all the essential characteristics come together as *one* in this kind of being.

> Something is absolutely one being only through one form, through which it has being (*esse*). For a thing (*res*) is a being (*ens*) and is one being (*unum*) from the same principle. And so those things that are described from a diversity of forms are not absolutely one, as for example a white man. If therefore a man were to have a form in virtue of which he is living . . . and another form in virtue of which he is an animal . . . and still another in virtue of which he is a man . . . it would follow that the man would not be absolutely one, just as Aristotle argues against Plato in book VIII of the *Metaphysics*.[50]

St. Thomas focuses upon what the fundamental metaphysical character of the soul as a substantial principle is—to determine being and unity as "what it [is] for [this] to be" (*quod quid erat esse*). For *this* to be, is for it to be human. "In things composed from matter and form, something is one through the form, and derives both its unity and species (natural kind) from it."[51] 'Species' here does not have the sense it has had throughout this work, namely *form of the act of understanding*. Here it has its technical logical sense of a *class or kind within a larger class or genus*. A thing is a member of a *kind of thing* because of its form, but more fundamentally that form is the unity of its *being*, the unity of the characteristic acts of a being of that one kind.

St. Thomas asserts that if something is described through a diversity of forms, it is not absolutely one. But notice the difference between saying that and saying that any diversity of descriptions involves a fundamental diversity of forms; the latter St. Thomas flatly denies. Diverse forms in a description are judged diverse because of the judgment of the diversity of those forms in *res extra animam*, not *vice versa*. Consider St. Thomas's examples 'a man' and 'a white man'. According to his initial response, we cannot simply read off of 'a white man' that it involves a plurality of forms *in res extra animam*. On the contrary, we judge that the description 'a white man' involves a diversity of forms because we already presuppose that *white* is a diverse form from *man* in *res extra animam*.[52] Apart from our knowledge of the diversity of *white* and *man* in *res extra animam*, we cannot judge the description 'white man' to involve any more forms than does 'man' alone, or for that matter 'white' alone. The appropriate question to ask Putnam, then, is what knowledge of the essence of dogs *in res extra animam* he is relying upon to judge that the different sciences are describing a multiplicity of essences? In the transition from 2), 3), and 4) to 5), Putnam's objection rests precisely upon what St. Thomas denies with his statement that "it is not necessary to assume a diversity in natural things from the diversity of intelligible characteristics or logical intentions which follow upon our manner of understanding."

St. Thomas's argument against a plurality of substantial forms is straightforward. Form determines something to be and to be absolutely one. If something has a number of forms it is not absolutely one. A man is absolutely one. By *modus tollens*, a man does not have a number of forms. It is important to understand this conclusion correctly. St. Thomas is not denying that the being who is a man has a number of accidental forms. But *being a man* is not constituted by those other accidental forms, *white* for example. *Being a man*, or any substance for that matter, consists of a certain absolute or simple *unity of being* that is a simple unity of the actual characteristics of being that *kind of thing*, despite those accidental forms; it is a simple unity of form not a complex unity of diverse forms. That is why he brings up by contrast the example of *a white man*. Here there are two or more diverse beings forming a complex unity, a *per accidens* unity, *a man* and *an instance of the color white* as an accident inhering in the man. These do involve, as he says, diverse forms, determining a complex unity between a man and his accidents. The man, in *being a white man*, can be said to be related to his color as to something *other than what he is*. The point of saying that a man does not have a number of substantial forms is to say that in *being a man*, rather than in *being a white man*, he *cannot* be said to be related to something other than *what he is*, his nature or essence. He *is* his essence or nature.⁵³

If the activities characteristic of life (nutrition, growth, and reproduction), and characteristic of animals (sensation), and characteristic of human beings (reason) are determined by diverse and multiple forms, a man would be constituted from diverse and multiple beings, like the *white man*. Notice that for St. Thomas there is no difficulty saying that a *white man* is constituted from diverse and multiple beings since he recognizes the analogous uses of 'being', and how they are related. For St. Thomas, to be *a man* and to be *a white man* are not to be fundamentally the same thing. But, to be *an eating man*, *a growing man*, *a reproducing man*, *a sensing man*, and *a rational man* are fundamentally to be the same thing; they are to be *a man*. This is true even as these activities differ among themselves, are exercised at different times, and sometimes in conflict. Though not the same activity, they are fundamentally activities of the same being, because they are the acts characteristic of, and determined by the fundamental unity that is *what it is for that man to be* the kind of thing he is, his essence.

This is a perfectly general argument and can be applied to any substance, and *mutatis mutandis* to other categories. In St. Thomas's terms, for Putnam to assert a multiplicity of natures or essences in beings of any one kind of thing is for him to deny the substantial unity of those beings. His dog Shlomit is not one substantial being, but a number of substantial beings.

Here the difference and conflict between what Putnam and what St. Thomas mean by 'essence' or 'nature' becomes clear. For Putnam, *essence,* if it has any place at all, is a fundamental part of a classificatory scheme, a special set of abstract properties under which the values of the bound variables of a conceptual scheme formally considered (a science, for example) may fall; in that sense, that is, as abstract sets of properties, they are extrinsic to the concrete objects that fall under them, whatever else might be characteristic of their abstract ontological status. Such abstract properties are special by contrast to other abstract properties that are not special to the extent that the latter do not fall within the purview of particular sciences. Because the different sciences determine different abstract sets of special properties, the different sciences determine for themselves different essences or natures. In any particular science, the abstract properties specified by some other distinct science count as accidental to the set of abstract properties specified as essential by the first science.

For St. Thomas, by contrast, *essence* is an intrinsic principle of a being, its actuality limiting but also enabling the being to be as it is, that is, to do the characteristic activities constitutive of *what it is, to be* the kind of being it is. St. Thomas's example of the white man is well chosen. If there were a number of substantial forms for a man, then they might as well be accidentally related to one another. Being rational in a human being would be no more related to eating and reproduction than it is to being white, which is to say not at all. Then why should we expect in a normative way one's *eating* and *reproducing* to be *informed by reason,* that is, to be rationally ordered toward the good of being human, since we do not expect in a normative way one's being white to be ordered in a rational way toward the good of being human? Reason is expressed and manifested in these activities; it does not stand apart from them.

Finally, I think it important to point out that St. Thomas need not hold that every contemporary science is directly describing the unique essence of some kind of thing. I have simply argued two points. First that he would deny that the plurality of sciences implies that they are describing distinct essences of the same kind of thing. Second that the unity of human life is for St. Thomas sufficient evidence to deny a plurality of essential forms in human beings; and that this argument is perfectly general as his reason for denying a plurality of essences in any particular kind of thing. That said, it is also open to him to argue that a science like evolutionary biology is providing an *essential* description in an *analogous* sense, namely, that it gives a description of how the essence *properly speaking* of that kind of being came to be through a historical process of descent. The recognition of analogy in the use of the terms 'essence' and 'essential' may well provide the means for ordering the sciences one to another, in terms of the related ways in which they investigate the essence or nature of a kind of thing.

Humility and the Essences of *Res Extra Animam*

St. Thomas's rejection of the kind of transition from 2), 3), and 4) to 5) in Putnam's objection presupposes the possibility of considering "one and the same thing" in a multitude of ways. This possibility, set in its metaphysical context, in turn presupposes that we do not have an immediate comprehensive grasp of the entirety or fullness of the nature or essence, a presupposition for which the evidence in St. Thomas is overwhelming. Writers are fond of quoting his famous statement that "our cognition is so weak that no philosopher was ever able to perfectly find out the nature of a single fly."[54] Often this is taken as a denial or at least a doubt on St. Thomas's part that we know anything at all of the essence of a fly, much less the essence of anything else. A rather large claim to be built upon the back of the poor fly. But it goes unremarked that he writes this in the context of a sermon on the Apostle's Creed while discussing the credibility of belief in what one does not see for oneself. There it would seem he has a stake in emphasizing the weakness of the human intellect in itself and apart from social contexts, particularly the social context of religious faith. But even so he simply says the philosophers have not "perfectly" found out the nature of a fly. For St. Thomas 'imperfect' does not necessarily mean 'faulty', 'erroneous', 'failing', or even 'inadequate', though it may. Its core meaning is simply 'not complete'. His claim is that we lack a comprehensive and final grasp of the entirety of the essence of something,[55] as if the intellect in its first moment penetrates to the very core of the essence with laser-like force.

Recognition of St. Thomas's humility before the essences of things, even a fly, is inherent in his account of the transition from the vague universal to the clear. It is also inherent in his discussion of the necessity for the human intellect to know by "composition and division."

> [T]he human intellect in the first apprehension does not seize on the spot (*statim*) a perfect cognition of a thing, but in the first place apprehends something of it, namely the quiddity of the thing itself,

which is the first and proper object of the intellect, and then it understands the properties, accidents, and encompassing essential relations of the thing. And according to this it is necessary that it compose or divide one apprehension with another, and from one composition or division proceed to another, which is to reason.[56]

Here he explicitly contrasts this human mode of knowing with the angelic and divine which "in the cognition of the quiddity of a thing cognize of that thing whatever we are able to cognize when we compose and divide."[57] The distinction between first act and second act of intellect provides the underlying structure of the passage. For St. Thomas the truth of a "composition" in its simplest form occurs when the intellect judges that what it has understood diversely is one in reality.

> For predication is something that is completed through the action of the intellect composing and dividing, nevertheless having as a foundation in a thing the unity itself of those things which are said of one another.[58]

It is tempting to think that an Aristotelian like St. Thomas, who places so much emphasis upon essence or form as providing the individuating principles for concepts, also thinks we have some direct quasi-angelic or divine intuitive grasp of the essences of "things in themselves." This can breed a temptation to see him as overly rationalistic and to stress his "intellectualism." This discussion should lay that temptation to rest, in particular when one recalls the transition from the vague universal to the distinct, the embodied character of nearly every aspect of the process of coming to know *res extra animam*, and St. Thomas's explicit contrast of the human intellect with the angelic and divine intellects precisely on the point of "not getting it all at once." It also confirms in many different ways the points I have been making throughout about his appropriation of Aristotle, most particularly that the *identity* of knower and known is not playing the *skeptical-problem-solution* role that many of its advocates and critics alike attribute to it.

As St. Thomas understands Aristotle's *Posterior Analytics*, to have a perfect scientific understanding of some thing is to know all of its elements, principles, and causes. Our understanding of the nature or essence of a thing is partial and always on the way towards complete or perfect comprehension, because of the diverse ways in which the richness of that nature discloses itself by the actualities of its being that proceed from it to the response of our embodied cognitive capacities. Joseph Pieper captures this theme in St. Thomas particularly well by emphasizing that things "are so utterly knowable that we can never come to the end of our endeavors to know them. It is precisely their knowability that is inexhaustible."[59] The Thomist desire for knowledge is tempered by humility before its object.

Conclusion

In his recent Dewey Lectures, "Sense, Nonsense, and the Senses: An Inquiry into the Powers of the Human Mind," Putnam suggests that he cannot follow the Aristotelian. He suggests that he could have titled his lectures "Aristotelian Realism without Aristotelian Metaphysics," and at the same time titled them

> 'Deweyan Realism' for Dewey, as I read him, was concerned to show that we can retain something of the spirit of Aristotle's defense of the common-sense world, against the excesses of both the metaphysicians and the sophists, without thereby committing ourselves to any variant of the metaphysical essentialism that Aristotle propounded.[60]

Can Aristotle's "defense of the common-sense world" be divorced from its metaphysical setting? An Aristotelian account of knowledge divorced from its larger philosophical context would seem to be "Aristotelian" in name only. Why not just call it a Putnamian defense of the commonsense world? Perhaps the difference is that for Putnam metaphysics has its starting point in highly abstract reflections upon the latest results of contemporary sciences, which are taken to correct, reduce, or eliminate the uninformed beliefs of "common sense" or "folk theories" about the nature of reality and the real essences of things. Against that understanding of science, Putnam would jump in to defend common sense against metaphysics. It is a mistake, however, to assimilate Thomistic-Aristotelian metaphysics to the picture Putnam has, in which "common sense" is under attack and must be defended. Where Putnam wishes to separate Aristotle's "defense of the common-sense world" from its largely metaphysical context, the virtue of St. Thomas's account is that he seeks to explore ever more deeply that very context.

At the end of the day, it represents progress to recognize where the real argument lies. Some may not wish to hold on to the metaphysical background involved in St. Thomas's Aristotelian account, but such a rejection will have to be on grounds other than that of *internalism*.

Chapter 9

CONCLUSION: TOWARD A MORE PERFECT FORM OF EXISTENCE

> He is truthful insofar as what he says of himself is true, not only by his words, but also by his deeds, that is, insofar as both his words and his exterior acts are in agreement with his character.
>
> —*St. Thomas Aquinas*

I have claimed that a substantive philosophical and historical claim stands behind the methodological program of the *Linguistic Turn,* that traditional philosophy had regularly misconstrued how it is that language is meaningful as it posited "in the head" a realm of objects called meanings. Even as many philosophers have given up the project of solving the problems of philosophy by a greater attention to ordinary use or a reformation of confused ordinary use to an ideal, they retain this view of "traditional philosophy." That view remains a barrier for understanding what historical figures in philosophy might have to contribute after the *Linguistic Turn.* In the preceding chapters we have seen why St. Thomas's Aristotelian discussion of how language is related to the world via the mind is not subject to some of the most powerful objections directed at the Aristotelian tradition, and thus why it does not easily fit into that view of "traditional philosophy."

Now, in conclusion, I want to return to the passage of St. Thomas in Book I, lesson 2 of his commentary on the *De interpretatione*. St. Thomas implicitly warns of the difficulties that attend treating concepts and words under the same *ratio* (intelligible character).

> [H]ere it ought to be observed that he [Aristotle] said written words are notes, that is signs of articulated sounds, and similarly vocal utterances [are signs] of passions of the soul; but passions of the soul he says are *likenesses* of things. . . . But letters are such signs of vocal utterances, and vocal utterances [are such signs] of passions of the soul, that no *ratio* (intelligible character) of *likeness* is observed in them, but only the *ratio* (intelligible character) of institution, just as in many other signs, like the trumpet as a sign of war. However, in passions of the soul it is necessary to observe the *ratio* of likeness to the represented things, because they naturally designate them, not from institution.[1]

Likeness is a natural relationship between concept and *res*. What is natural is the way in which the human subject specifically responds to her encounter with her environment. St. Thomas is commenting on the fact that Aristotle does not use the same term to characterize the relation between passions of the soul and things beyond the soul that he uses to characterize the relations of written words to spoken, and spoken words to passions of the soul.

In the context of the *Commentary,* St. Thomas's remarks have no special relevance that would single them out for special consideration over and above the rest of the passages. He notes a difference in the text, explains it briefly, then moves on. But seven hundred years later, we cannot help but read it in a different light. All we have to do is consider once again Putnam:

> What makes it plausible that the mind . . . thinks . . . using representations is that all the thinking we know about uses representations. But none of the methods of representation that we know about—speech, writing, painting, carving in stone, etc.—has the magical property that there *cannot be* different representations with the same meaning. None of the methods of representation that we know about has the property that the representations *intrinsically* refer to whatever it is that they are used to refer to . . .[2]

and

Conclusion: Toward a More Perfect Form of Existence

> [e]ven a large and complex system of representations, both verbal and visual, still does not have an *intrinsic*, built-in . . . connection with what it represents. . . . And this is true whether the system of representations . . . is physically realized—the words are written or spoken, and the pictures are physical pictures—or only realized in the mind. Thought words and mental pictures do not *intrinsically* represent what they are about.[3]

For Putnam, all representations, including Aristotelian concepts, are "institutionally" realized and require an interpretation. John McDowell recognizes that for Putnam all representations, including mental representations, are *symbols*. "A symbol is an item whose intrinsic nature is characterizable independently of its representational properties."[4]

Putnam's view rests on his assumption of methodological solipsism that no psychological [mental] state, properly so called, presupposes the existence of any individual other than the subject to whom that state is ascribed. If a mental state bears a relation to something else, then that relation must be extrinsic to the intrinsic identity of the mental state. It must be symbolic. McDowell points out that there is an air of residual "scientism" about this methodological solipsism, despite Putnam's desire to counter the latent scientism of fashionable philosophical accounts of the mind. McDowell writes:

> There is an equivalence implied . . . between "psychologically real" and "scientifically describable," which cries out to be questioned: it looks like simply an expression of scientism about what it might be for something to be psychologically real. (We do not need to surrender the term "psychological" to *scientific* psychology.) But as far as I can see Putnam leaves the equivalence unchallenged.[5]

One reason Putnam leaves it "unchallenged" may well be the overwhelming influence of modern philosophy upon his vision. He cannot consider any account of how language "hooks onto the world" via thought in any way other than the fundamental structure set out by the Empiricists, and criticized by Wittgenstein. We have seen what effect this vision has on Putnam's characterization of the "Aristotelian semantic triangle." But we have also seen that on this account, the Aristotelian concept ceases to bear the specific characteristics that St. Thomas attributes to it, in particular, the features in virtue of which it is a *mental thing,* in virtue of which it is known, and in virtue of which it has an intrinsic dependence upon *res extra animam* for the conditions of its identity.

'Sign' is often applied indiscriminately to words and concepts or mental impressions even by defenders of the Aristotelian position,[6] suggesting at least implicitly a fundamental similarity of the relations between words and concepts and concepts and *res extra animam*, as if species of the same genus. Locke and Berkeley refer to both words and ideas as signs. Certainly Wittgenstein's criticism treated mental impressions and pictures indifferently. One can see the plausibility of Dummett's claims and Putnam's criticism. However, St. Thomas believes that it is a fundamental mistake to treat words and concepts as on a par with one another as signs. Indeed, I believe it is inappropriate to treat the concept as a sign at all. For St. Thomas, the consideration of institutionalized representations like writing and speaking, painting and sculpting, and so on, is intrinsically different from the consideration of concepts, precisely because of the fault line between how a concept is a "likeness" and how words, paintings, sculptures, and so on, are "institutionalized signs."

That the Thoughts of One Man Should Be Made Known to Another

It is by no means an uncommon objection to hear that in the Aristotelian or Thomistic traditions there is little recognition of the social character of knowledge and its acquisition, that it is excessively individualistic and does not take account of the ways in which social realities, particularly language, condition learning and knowledge. This charge comes even from some of its friends who seek to show St. Thomas's continuing relevance to our contemporary discussions. Although I have benefited in many ways from reading John Haldane's work, he suggests this very objection when he discusses St. Thomas's account of concept acquisition: "Aquinas himself speaks of the active and passive intellects as powers of one and the same thinker, which raises a question as to whether he is over-individualistic in his conception of the mind."[7]

My work begins with a discussion of language and almost immediately dives into a discussion of intellect and knowledge, which according to St. Thomas are, respectively, a power and a habit of an individual. This could be seen to echo the traditional subordination of the *social* and *public* nature of language to individualistic thought expressed in the opening passages of the *De interpretatione*, as if language is of little importance, and "thinking" or understanding is a private, inner, self-standing act distinct from language. The autonomous individual learns what she does on her own, and only thereafter communicates it to another through language.

This passage in Aristotle can no doubt be read in such a way. Still, though I agree that Aristotle and St. Thomas did not devote the bulk of their philosophical energies to discussing the problems of language and its social conditions, I think the objection is in danger of seriously miscasting the importance of language to them.

One thing the present work should have made clear is the intimate relationship between rational animals and the world, not just *qua* animal, and not just *qua* rational, but specifically *qua* rational animal. As *rational animals* our specific mode of animal life is incomplete or indeterminate and in potency without our encounter with and our being informed by the world in which we live, and move, and have our being. I think Aristotle and St. Thomas would heartily agree with McDowell in his saying that, "where mental life takes place need not be pinpointed any more precisely than by saying that it takes place where our lives take place."[8] McDowell adds:

> My aim is not to postulate mysterious powers of mind; rather, my aim is to restore us to a conception of thinking as the exercise of powers possessed, not mysteriously by some part of a thinking being, a part whose internal arrangements are characterizable independently of how the thinking being is placed in its environment, but unmysteriously by a thinking being itself, an animal that lives its life in cognitive and practical relations to the world.[9]

Yet for all his efforts to eliminate the dualism of the internal and the external in our philosophical reflections upon knowledge, McDowell still does not seem to escape the echoes of that dualism. He writes of a thinking being, but still pictures this thinking being over against an object-world to which it is *related*. The thinking being is not displayed as dwelling within *the* world, and entering into relations with other parts of that world. We are still left with the dualism of a first world that seems complete, represented by McDowell's concession elsewhere to the law-likeness and determinism of *first nature*, and a second world of the thinking animal that is incomplete, represented by his upholding of the freedom and spontaneity of *second nature* within the space of reasons and spontaneity. We have a "foothold" in the first because of our animal nature, but because of the autonomy of our rational nature that "operates freely in its own sphere," "we need not connect [our] natural history [second nature] to nature as the realm of law [first nature] any more tightly than by simply affirming our right to the notion of second nature."[10] We are asked by McDowell to see these two worlds and natures as slightly more intimately related by this "foothold" than the external-internal obsessions of modern philosophy have hitherto allowed, but the dualism remains. The human spirit over the bent world broods.

For Aristotle and St. Thomas our mental lives take place where our lives take place, the world of first nature, the world of animals, plants, and inanimate beings. Our mental life is incomplete without its dwelling in that very world; there is none other. McDowell stops short of this because of his fear of "medieval superstition" and "enchantment," but that is prejudice, not argument.[11] "In a common medieval outlook, what we now see as the subject matter of natural science was conceived as filled with meaning, as if all of nature were a book of lessons for us."[12] This sweeping claim shows a lack of understanding of the complexity of those thousand or so years that we call "middle" or even worse "dark,"[13] not to mention a lack of self-reflection upon the usefulness for modern life of the symbol of a thoroughly deterministic natural science, that is, a lack of self-reflection upon the meaning for McDowell of the symbol of "first nature" as he employs it. The "medieval" appears in McDowell as simply another world (in the sense of a space of reasons) globally conceived, a third world which we choose not to inhabit, but that is useful as a foil against which to reassure ourselves of our progress and superiority.

Indeed, anyone who does not grant McDowell's account of first nature, an account he believes is forced upon the space of reasons by the very fact of modern science, is to be marginalized as lacking "a mark of intellectual progress."[14] It seems the freedom and spontaneity of the space of reasons is not quite so free or spontaneous. This concession and strict adherence to only one of the different ways of understanding the marvels of modern science is surprising from one who so strongly objects to the latent scientism of Putnam. A human animal who does not grant McDowell's account of first nature is not part of McDowell's world, his space of reasons. Since it is through education that McDowell believes one enters the space of reasons that constitutes being rational in our day and age, such a person presumably approaches being a sub-rational human animal. He is not yet qualified to enter the space of reasons, spontaneity, and freedom. He seems to approach the status of the similarly uneducated child whom McDowell explicitly excludes from his world since "human infants are mere animals, distinctive only in their potential."[15]

Not only is it the case that for St. Thomas our mental life is incomplete without its dwelling in the world of animals, plants, and inanimate objects, but perhaps even more striking is St. Thomas's position that the created world of animals, plants, and inanimate beings is itself incomplete until it is taken up into the mental life of the human person who dwells within it. This mutual indwelling is for St. Thomas the perfection of the created order, a more perfect image and likeness of God.[16] At the very least then, the individual that Aristotle and St. Thomas are talking about is not conceived of in the typically modern way that starts with the individual autonomous subject over against the world as an object, only to ask whether there is any such object-world for that subject-individual to be related to, and it is not McDowell's alternative, free-thinking individual in one world, eating and drinking in another, all the while lost in the cosmos.

Aristotle and St. Thomas took the linguistic expression of understanding to be the normal case, as when they discuss taking the nominal definition of a term from the ordinary use of the community as the first stage in the process of scientific understanding. This makes no sense if we are then to think that the rest of the scientific enterprise does not take place in a linguistic context.

But there are even more direct indications in St. Thomas's work of the appropriate place for the discussion of language and the social conditions of knowledge, indications that he has no desire to understand them in an "over-individualistic" way. In lesson 2 of the commentary on the *De interpretatione*, he argues that spoken language is necessary because man is "by nature a political and social animal," while written language is necessary "so that he might manifest his concepts to those who are distant according to place and to those who will come in future time."[17] In this context, he explicitly contrasts this social necessity for language with the counterfactual possibility that man might be a naturally solitary animal.

> If indeed a man were naturally a solitary animal, the passions of the soul, by which he is conformed to things themselves so as to have knowledge of them within himself, would suffice for him; but since a man is naturally a political and social animal, it was necessary that the conceptions of one man should be made known to another, which is accomplished through articulated sound; and so it was necessary that there should be significant articulated sound in order that men might live among one another. And hence those of diverse tongues are not able to live well among one another.[18]

While this point about the social necessity of language is something St. Thomas would have read in the Latin translation of Ammonius's commentary, St. Thomas adds the explicit counterfactual contrast with man taken as a naturally solitary animal. The social character of a man is not due to entering into an accidental aggregate, as one might think of the relation among apples in a bushel basket, nor is the social tacked on to the solitary. The solitary is contrary to fact, contrary to the natural fact of human existence.

St. Thomas's discussion helps us to better understand the conceptual relation among Aristotle's *De anima*, the *Nichomachean Ethics*, and the *Politics*. The temptation would be to think of what St. Thomas has to say about the soul in the *Commentary on the De anima* as pertaining to the solitary individual counterfactually mentioned here, while the social is something other, something added to a man's naturally solitary existence, and in need of treatment in thoroughly distinct works such as the *Ethics* and the *Politics*. But here it is useful to recall a difficulty about abstraction that St. Thomas often considers. The objection is made that to understand *Socrates' humanity* without considering the particular characteristics of its existence here and now, but rather as simply *rational* and *animal*, is to understand it falsely, since Socrates' humanity only exists with the particular characteristics of its existence here and now. Thus abstraction of its very nature leads to falsehood. St. Thomas's response is to distinguish between understanding some being without considering all the characteristics that pertain to it, and understanding some being as existing without those characteristics. The first is legitimate abstraction, while the second is false or vicious abstraction. The example he often gives is understanding a red apple as an apple while abstracting from its color. To understand a particular apple as the fruit of a plant with certain characteristic organic processes taking place within it, without at the same time considering it as red, is not to think of it falsely. It is simply to understand it in a certain way and not in another way, that is, abstractly but no less truly. This abstraction simply seems to be a condition of human understanding: we understand and talk about beings all the time, without either knowing or mentioning all their characteristics. However, to understand the apple as not-red when it is red would be a vicious abstraction and false. The first involves not considering a characteristic that a being has, while the second involves denying that the being has that characteristic.

The *De anima* and St. Thomas's commentary upon it are not considerations of a man in his solitary existence—that would involve a vicious abstraction. It would be an egregious mistake to think of them as considerations of a man in terms of what pertains to him as a solitary animal. Nothing pertains to him as a solitary animal, since he is not one; similarly nothing pertains to a red apple insofar as it is not red, since a red apple

Conclusion: Toward a More Perfect Form of Existence 283

that is not red is not an apple at all. The *De anima* is an abstract consideration of a man, a social and political animal, leaving out of consideration certain characteristics of his natural existence, such as those distinctive of his sociability and political character. It remains to be seen how the abstractions that occur in the *De anima* are related to those that occur in other sciences, like the *Nichomachean Ethics* and the *Politics,* as well as the wide range of abstract sciences which may consider human beings.

Now in the counterfactual case considered here in the passage from the *De interpretatione* commentary, St. Thomas is deliberately engaging in a vicious abstraction in order to tease out of it the implications for human existence of such an unnatural view of a man. The rhetorical effect of the counterfactual is to highlight by contrast the natural factual condition of human existence, namely, its social character exhibited in language.

Of course there is a sense in which any animal that engages in sexual reproduction is by nature a social animal. Any animal that seeks food in concert with other animals is by nature a social animal. But what St. Thomas is stressing is that the social character of a man goes all the way to the specific difference of the man from other animals, to the way in which reason expressed in language informs his animal activities and is itself an activity not engaged in by non-human beings. Once the pack of wolves kills the game, it is dog eat dog with respect to who will get the most. Even in instances of species where there is a more common sharing of the kill, it does not rise to the elaborate forms of preparation involved in human modes of cooking, and such species do not typically have elaborate feasts, the purpose of which is not simply to nourish the body but to play a larger role in a social life informed by reason and reasoned reflection upon what is "useful and hurtful, just and unjust, good and evil."[19] *Mutatis mutandis* for sexual reproduction. So we see the central significance of marriage feasts in human cultures, feasts that involve a public rite of union solidified by a public meal woven together by complex relations of reason, sociability, and custom. Spoken language is but one of the embodied reasonable forms of life that communicate these distinctively human passions of the animal soul, and only to the extent that it presupposes the others.

In the third chapter I examined a number of interpretations of Wittgenstein with an eye toward indicating the difficulty of identifying the target of his critique. Here it might be good to revisit briefly one of those interpretations to see how it might be friendlier to the Aristotelian-Thomistic tradition on just this point of the relation between language as a social and political form of life and the more basic animal forms of life. I want to emphasize one aspect of Colin McGinn's interpretation of Wittgenstein that accords very well with the reading of St. Thomas's Aristotelianism that I have tried to give in this work.

McGinn reacted negatively to Kripke's interpretation of Wittgenstein, according to which Wittgenstein believes that there are no facts about the individual language-user that constitute the meaning of a term used by that person. According to McGinn, Wittgenstein's positive view is one in which we grasp a rule not by interpreting it, but by the development of natural capacities for language use, which are not themselves divorced from the natural capacities we exercise in other "forms of life." Wittgenstein "does offer an account of the sort of thing understanding is: it is mastery of a technique, possession of a capacity, participation in a custom."[20] The technique or ability that constitutes understanding is the "ability to use signs." Thus, Wittgenstein's emphasis on actions not being interpretations—presumably these actions proceed for good or for ill from the natural capacities that have been developed in one's upbringing. McGinn believes that Wittgenstein is rejecting the modern rationalistic foundationalism that would require in our use of language a constant looking to some rule by reference to which our use counts as rational. Wittgenstein is substituting for it a form of anti-foundationalist naturalism. In the exercise of one's developed capacity for bravery, one does not consult a rule first to determine what it is rational to do. One acts in accord with the virtue. Thus, "what this shews. . ." is that "this other way of grasping of a rule" consists in the exercise of one's natural human ability to use signs.

Notice that the emphasis in McGinn is upon *understanding* as the natural human ability to use signs. However, lest the ability to use signs appear in Wittgenstein as a completely *sui generis* "form of life" divorced from the other natural capacities constitutive of human nature, McGinn adds in a footnote to his work that

> it is important that my use of language is interwoven with various kinds of non-linguistic activity in such a way as to fix what my words mean; so correctness of use will (partially) consist in how my linguistic actions fit in with my non-linguistic actions.[21]

One does not often read commentators on Wittgenstein remarking upon the continuity of language as a "form of life" with the other "forms of life" in which human beings engage. Without such a recognition, language as a form of life can appear as a practice untethered from the characteristically animal modes of human life, or at best with nothing more than a "foothold" in them. A philosophy that takes its starting point from this "foothold" can appear to be just another dualism that separates man's animal nature from his rational nature, or the animal part from the rational part of human nature, even when through language it projects the rational nature into publicly accessible acts in a "space of reasons."

St. Thomas would not agree that the intellect is nothing other than the ability to use signs, and that understanding a concept is nothing other than the ability to use linguistic signs appropriately. I think such an interpretation leads fairly quickly to the appearance of this *sui generis* form of life. The note that McGinn adds provides evidence that he fears this about his own interpretation. Nonetheless, this account of the intellect as the ability to use signs is Anthony Kenny's interpretation of St. Thomas in *Aquinas on Mind*.[22] Having provided that interpretation of intellect, Kenny proceeds to describe how according to St. Thomas conceptual content is produced by abstraction from phantasms. But he fails to show how or why that should result in a power that is to be primarily thought of as an ability to manipulate linguistic signs. By contrast, Haldane does not think that conceptual content can be acquired in the way Kenny suggests. Nonetheless, in his Wittgensteinian corrective to the "over-individualistic" account of mind that he finds in St. Thomas, he seems to share Kenny's view that the intellect should be understood as the ability to use linguistic signs.

> For Wittgenstein we learn to think as we learn to speak. The ability to structure experience is acquired through the learning of general terms. Alice is enabled to think *cat* by being taught the word 'cat' (or an equivalent). On this account, therefore, the concept is not innate, the child had to be taught it; and nor is it abstracted, she was not able to attend to cats *as cats* prior to being instructed in the use of the concept.[23]

Haldane wants to resist the position that concepts are acquired through more basic modes of experience than language learning, because of the abstractionist implications of such a position. Here Haldane has in mind Geach's critique of abstraction. For Haldane, the concept *cat* is not acquired by an active process of encountering cats on the part of the learner, but by a passive encounter with another human being who instructs the learner in the use of 'cat'. Once the concept is acquired in this fashion, the language learner can have "structured" experiences of cats—not just speakers who use 'cats'—because as Haldane argues in a number of other places the concept *cat* is formally identical with the substances that are cats. Until Alice has acquired the concepts through instruction, she cannot have the sorts of structured experiences of objects in the world that count as, and lead to, rational behavior toward those objects. Thus, a necessary condition for Alice's acting rationally toward cats is her having been introduced by another human being into what McDowell refers to as the space of reasons, having taken on what McDowell refers to as "second nature."[24]

Haldane objects elsewhere to McDowell's account of the social-cultural rationality of the space of reasons that it is obviously regressive—whence the rationality of the social-cultural milieu? But Haldane does not challenge the fundamental structure of McDowell's analysis. Rather he chides him for not raising, much less pursuing, the regress indicated by the rationality of the social.[25] Presupposing McDowell's analysis and pursuing this line of argument leads Haldane to another cosmological proof for the existence of God, in order to avoid an infinite regress of teachers. God is the "'Prime Thinker' or even, though metaphorically, [a] 'Prime Sayer'."[26] Haldane's picture suggests that God primordially sets up a correspondence, indeed an identity between the conceptual content of linguistic communities and the beings they encounter in the world, since the conceptual content is not drawn from the encounter with those beings, and yet its formal identity with them allows one to structure one's experience of those beings in rational ways.

I do not want to pursue here the question of whether Haldane's argument is successful as a cosmological argument for the existence of God. What concerns me here is the appearance of language as a *sui generis* form of life grafted upon the more basic human animal forms of life in which human beings encounter other worldly beings. This appears to be a different kind of Wittgensteinian "interweaving" of the linguistic form of life with the non-linguistic forms of life than the interweaving proposed by McGinn. McGinn emphasized that language is interwoven with non-linguistic activity because the latter fixes the meaning of the former, and "correctness of use will (partially) consist in how my linguistic actions fit in with my non-linguistic actions." In that sense the linguistic form of life is an outgrowth of the human life of animals. McGinn's account of Wittgenstein leaves open the door for a different account of abstraction than one sees criticized in Geach, and in Haldane after him. Linguistic activity is internal to the human form of animal life, and dependent upon it.

By contrast, Haldane "interweaves" language with non-linguistic activities in an external fashion. Language comes to those activities to render them rational. Alice can now "attend to cats *as cats* [having been] instructed in the use of the concept"; having been made rational by her teacher, she can now act rationally.[27] With apologies to McGinn, one might say that for Haldane it is important that my various kinds of non-linguistic acts are interwoven with the meanings of my linguistic acts in such a way as to fix the rationality of my non-linguistic acts; so correctness of non-linguistic actions will (partially) consist in how my non-linguistic actions fit in with my linguistic use.

To be fair, Haldane does think that the individual human, Alice, must have a potency for such a form of life, as does McDowell—you can't graft an oak branch onto a rock or real language use onto a parrot. There must be some "foothold." But the potency is a purely passive potency: if agency were associated with the human being in conceptual development, the account would be "over-individualistic." Nor can the linguistic form of life be a natural outgrowth or development of the more basic animal modes of specifically human life, and how a human animal through those modes or forms of life experiences other things; that would suggest for Haldane an "abstractionist" picture. The concept *cat* is indeed acquired through experience, not, however, through the experience of cats, but rather through the experience of one's teacher using 'cat'.

> In these terms one may say that Alice's intellect is receptive to, or potentially informed by, the concept *cat*, while the mind or intellect of James who has already mastered the use of the term is active with or actually informed by this concept. In teaching Alice the word, James imparts the concept and thereby actualizes her potentiality.[28]

Alice is not born with the ability to acquire concepts actively. Instead, she is born with "a second-order [power]; it is a power to acquire a (conceptual) power,"[29] an ability to acquire concepts passively.

There is an important difference here from St. Thomas, who posits in the individual human animal both an active ability and a passive capacity to acquire concepts in intellect, the active ability with the passive capacity that Haldane objects to as "over-individualistic." The active or agent intellect informs the passive intellect using more basic modes of experience as an instrument; the agency involved is intrinsic to human nature. By way of contrast, in Haldane's picture there is no active intellect in the learner to perform this activity. Rather, there is only a second-order passive intellect, that is, the passive potentiality to become an active potentiality. One might argue that in this scenario James forces the conception upon Alice, since in her ignorance she has no active part to play in acquiring it. "Through instruction Alice's hitherto unrealized potentiality is made actual through the activity of James."[30] In Haldane's picture, the agency involved in the interweaving of which McGinn speaks, the interweaving of the linguistic (rational) form of life with the non-linguistic (non-rational) forms of life, ultimately comes from outside human nature, even though human nature has a passive potency for it; thus the appeal to the "prime thinker and sayer." God both gives the first instruction and designs the human animal with the second-order passive capacity to be so instructed.

Haldane's God, like Descartes's before him, jumps in to fill The Gap between mind and world, though now the mind is not a private store of ideas, but a social store of words. Haldane suggests that this cosmological argument should bring to mind the passage of Genesis in which it is written, "Then God said, 'Let us make man in our image, after our likeness . . .' [then] out of the ground the Lord God formed every beast of the field and every bird of the air, and brought them to the man to see what he would call them; and whatever the man called every living creature that was its name."[31] Of course, Haldane only intends this as an illustrative image, but the Genesis story seems directly opposed to Haldane's argument. As the story goes, God does not teach Adam what the names of the beasts are. If anything, Adam instructs God as to their names! Doesn't this image suggest that God has created Adam with sufficient natural powers to act as an agent in his own learning?[32] Doesn't Adam's responsibility to name "every living creature" presuppose that Adam has sufficient experience of cows and cats to understand them in a conceptually "structured" and rational way that differentiates between them, in order that he not give the same name to cows and cats?

In Haldane's account rational agency and activity in human animals appears to be derived not from the nature of human animal life, but from an ideal of pure reason. In medieval discussions it is closer to Ibn Rushd's discussion of the agent intellect than St. Thomas's.[33] Language as a human *form of life* appears to be foreign to the more basic human *forms of life*, yet fortuitously "designed" to cut at the joints of reality so as to render rational from the outside those more basic forms of life. This guarantees the social character of knowledge and its acquisition, only at the expense of a shotgun wedding between language and the more basic human modes of animal life, a marriage destined for the sort of divorce that Quine captures when he writes that meaning is what essence becomes when it is divorced from being and wedded to the word.

For St. Thomas, whatever else it is, the intellect as a power of the soul is continuous with the other ways in which the human animal engages the world. My discussion of abstraction shows how it is not a form of "selective attention," not, that is, the *abstractionism* criticized by Geach. It shows how abstraction, nonetheless, presupposes and actively engages the more basic animal modes of encountering other things on the part of the human animal. We saw, in discussing Putnam's plurality of essences objection, that the human soul is for St. Thomas the fundamental metaphysical principle of unity in human acts. It is in virtue of the soul as formal principle of unity for the human substance that eating, sex, the pursuit of pleasure, and so on, are not simply the acts of an animal who happens also to have the passive potential to be made rational, but the acts of a *rational animal*. It is the intrinsic principle in virtue of which a human being not only reasons and understands apart from eating and sex, but also engages in reasonable eating, reasonable sex, the reasonable pursuit of pleasure, and so on.

Now consider what St. Thomas writes about language in his commentary on Aristotle's *De anima*. In human beings "nature . . . uses air that has been breathed in for the formation of articulated sounds, which is for the sake of a more perfect existence (*bene esse*)."[34] Earlier in this passage, St. Thomas mentions eating with the tongue and the dissipation of body heat through respiration as necessary activities for the mere conservation of the being of animals. By contrast, these activities then give weight to the central point of the passage, namely, the additional use of the tongue and respiration by rational animals in order to speak, a "more perfect" form of existence, more perfect than breathing and eating as such. When one considers this text in light of the formation of articulated sounds being necessary for the social and political nature of a man, it is evident that the "more perfect existence" referred to here is that very social and political existence made possible by language referred to in the *De interpretatione* commentary. Both find their proper context in Aristotle's statement in the *Politics* that "the city (*polis*) comes into existence, originating in the bare needs of life, and continuing in existence for the sake of a good life."[35]

Thus, in St. Thomas's commentary on the *De anima* itself, there is an indication that the full consideration of human nature does not end with a consideration of the soul in isolation in an "over-individualistic way," but rather that the consideration of the soul and its powers points to a more perfect natural mode of existence and a more complete examination of human nature in its social and political character. Against the background of McGinn's emphasis on the interweaving of language and other non-linguistic activities in human nature, as well as the technicalities of St. Thomas's account of the soul as the principle of unity for the various "forms of life," this discussion gives a certain non-dualistic and substantive coloring to Wittgenstein's celebrated and perhaps metaphorical statement that "my attitude towards him is an attitude towards a soul. I am not of the *opinion* that he has a soul."[36] The human soul is not an "inner" entity enclosed within a body that needs to discovered by inference. It is the actual unity of the bodily forms of life we recognize among one another as we live together, including but not limited to the linguistic form of life.

While discussing Putnam's plurality of essences objection, I showed that St. Thomas considers a view of the human person in which the individual would be informed by a plurality of souls, and thus a plurality of natures, the number of souls corresponding to and accounting for different groupings of vital activities. St. Thomas firmly rejects this view.

> Something is one simple thing only through one form through which it has being, since it is from the same principle that a thing is a being and is one thing. And so those things which are described by diverse forms are not one simple thing, as for example 'white man'. If, therefore, a man were to live on account of one form (the vegetative soul), and to be an animal on account of another form (the sensitive soul), and to be a man on account of still another form (the rational soul), it would follow that the man would not be one simple [substantial] thing.[37]

The implicit premise of St. Thomas's *modus tollens* here is this: *a man is one simple substantial thing,* that is, his vital activities of nutrition, growth, reproduction, sensation, and reason form a unity of life whose principle is the soul. Later, with the human soul explicitly in mind, he adds that

> diverse grades of perfection are found in material things, namely, to be, to live, to sense, and to understand. [In this order], what is added in the subsequent is always more perfect than the prior. Therefore, the form which only gives the first grade of perfection to matter is the most imperfect. But the form which gives [to matter] the first, second, third, and so on is the most perfect.[38]

"The form" here that gives "the first, second, third, and so on" is the human soul. Notice particularly, the "and so on," which follows upon and is more

Conclusion: Toward a More Perfect Form of Existence   291

perfect than the grade of perfection associated with sensation. To each of the sciences under which a man may fall there will be a definition of the man appropriate to that science expressing the perfection under consideration, for example, *changeable being, rational animal, ethical animal, social animal,* and *political animal*. But the plurality of definitions in the hierarchy of sciences, particularly animal and human sciences, does not imply a plurality of things defined, a plurality of natures and types of human soul. That is "the mistake of the Platonists." It is this point that Jacques Maritain captures particularly well in the French title of his *Degrees of Knowledge,* namely, *Distinguer pour Unir, ou Les degrés du savoir.*

If a man is by nature a political animal, it stands to reason that his political life, which necessarily involves communication, is the flower of his more basic vital activities or *forms of life* informed by reason. His political life is his flourishing, the "more perfect existence" that the individual naturally seeks, without which his individual existence is naturally incomplete and naturally less than perfect. "Less than perfect" is perhaps too mild a description of the counterfactual condition of the individual human existence apart from the social and political. Alasdair MacIntyre captures the extreme of this "person outside all traditions" when he writes that "to be outside all traditions is to be a stranger to enquiry; it is to be in a state of intellectual and moral destitution."[39] The nature that is animal in a man, that is rational in a man, and that is social and political in a man, is but one material nature whose principle is the human soul. The lower level activities or forms of life characteristic of animals generally, namely, growth, eating, sex, the pursuit of pleasure, and so on, are all drawn up into a higher and more perfect unity of life that participates in the specific human perfection of a reasonable social and political life. But that more perfect unity does not come from outside human nature as Haldane's account suggests; it is integral to and constitutive of human nature. Being rational, linguistic, and political are the specifically human ways of being an animal.

The *De anima* does not exclude the *Politics,* much less reduce it. Rather, it prepares the way for the discussion of man the talking, social, political animal. Any thorough non-reductive consideration of being human will recognize the seed of the political within men as individual human substances, but recognize its flower and perfection, its *telos* in their communities. Once one first recognizes how human language is rooted in understanding, which is an act of a person, that is, an act of a *subsistent individual of a rational nature,* the proper place for an examination of the essential social reality of language in human social and political sciences is evident.

This is confirmed by Aristotle's claim in the *Politics* and St. Thomas's commentary on it: speech proves that man is more of a political animal than "bees or any other gregarious animal," "since communication about [the useful and the hurtful, the just and the unjust, good and evil] makes a home and a city. Therefore, a man is by nature a domestic and civil animal."[40]

There will always be the temptation to find a distinction of things wherever one finds a difference in a thing. In the case at hand, if the Aristotelian emphasizes the difference between understanding expressed in speech and understanding *as such*, the temptation will be to think that he believes these are two different things, two different acts. Here, however, the Aristotelian emphasis upon the *developmental* character of *being*, that is, *potential being* rooted in prior *actual being*, comes to the fore. To say that understanding has not yet found its perfect expression is not equivalent to saying that it is not yet understanding at all, and it is certainly not equivalent to saying that its expression in less perfect modes of animal life is a *failure* of perfect understanding. For St. Thomas, to judge that 'X is imperfect or less than perfect' is not necessarily to judge that 'X is a failure or error'.[41] Perfection of being is achieved through the actualization of *potential* being (second nature) rooted in the actual being that one already is (first nature). What one already is makes possible the perfection of what one may become through one's acts.

When the Aristotelian distinguishes between understanding *as such* and the vocal expression of understanding, he is not necessarily distinguishing two things, that is, two acts. He is, in the first place, providing sufficient space for recognizing that understanding is expressed in all human action and not just in the manipulation of verbal or written symbols. He is also, in the second place, leaving sufficient conceptual space for a movement, that is, a development of the understanding expressed in all other modes of human action to the more perfect form of existence embodied in the expression of understanding in speech which is the fruit of understanding shared with the community. Thus, the Aristotelian is leaving enough room for Alice, through her experience of all the forms of life, to genuinely develop, transform, and even correct what she has gained from the community in learning language.

Anyone familiar with St. Thomas's writing, particularly his careful examination of authorities, his charitable concern for his students expressed at the beginning of the *Summa* "to set forth whatever is included in sacred doctrine as briefly and clearly as the matter itself may allow," and so on, can recognize the seriousness of what he writes about written language communicating to those distant in place and time. He is not merely stating a theoretical fact divorced from his own practice. The whole context of medieval learning intimately involved the written transmission of knowledge via authoritative texts, not the least of which was an authoritative text believed to be a salvific divine communication, one of whose books begins "in the beginning was the Word, and the Word was with God, and the Word was God."

This social reality is no less true of the *spoken* context of medieval university education, exhibiting as it did public disputation and real debate as its soul, a soul often lacking in our own contemporary universities and their panel discussions. Spoken words were the medium of communication in which this life took place. Here it might be good to recall that Aristotle is not only the author of the *De anima* and the *De interpretatione*, but also the *Topica*, *De Sophisticis Elenchis*, *Rhetorica*, and the *De poetica*. Strange as it may sound to our modern ears, St. Thomas includes even the last two among the works of logic in the preface to his commentary on the *Posterior Analytics*.

Every university master or student of the time would have had the earliest elements of his higher education built around a thorough study of the *Trivium* of *Logic*, *Grammar*, and *Rhetoric*. St. Thomas points out that people who do not share a common language find it difficult to live together. For a mendicant friar like St. Thomas, this spoken social context of the life of learning finds its focus and most authoritative embodiment in the office of the divine liturgy sung daily within the community, a developing oral tradition more than a thousand years old at the time, to which he would contribute and advance in the composition of the Feast of Corpus Christi. It is odd, to say the least, to suggest that philosophers and theologians trained by and living within that context were not aware of how social communities and their languages shape the acquisition of knowledge.

When St. Thomas attributes agency to the individual in the process of acquiring knowledge, it is not because he conceives of that process in an "over-individualistic" way, but rather to ensure that it is an act *done by* the person learning.[42] Here at least he would agree with Plato:

> Education isn't what some people declare it to be, namely, putting knowledge into souls that lack it, like putting sight into blind eyes.... Education is the craft concerned with doing this very thing, this turning around, and with how the soul can most easily and effectively be made to do it. It isn't the craft of putting sight into the soul. Education takes for granted that sight is there but that it isn't turned the right

way or looking where it ought to look, and it tries to redirect it appropriately.[43]

When St. Thomas turns to consider the relationship between teacher and student, he makes it clear that the student is not simply passive with respect to the knowledge that she acquires, but is in fact the principal agent of the acquisition rather than the teacher. As a secondary agent the teacher assists the learner by proposing for her consideration principles, arguments, and examples that the teacher has mastered.[44]

If learning is going to be the act of a person, it has to be the act of *that* person, and it presupposes that *that* person already has some tools for actively listening, no matter how rudimentary those tools may be. If it is *only* through language that conceptual functioning is acquired, as some would argue, and it is the child's language that is being learned, how is it that she actively attends to and appropriates for herself that language from her teacher? On the contrary, the evident explosive development of conceptual functioning acquired by the child in the learning of language would itself seem to presuppose an already existent conceptual functioning on her part.

What St. Thomas recognizes in his discussion of the teacher-student relationship is that unless the student is actively involved as an agent, indeed as the principal agent in the process, the knowledge imparted to her by the community represented by the teacher is never truly her own: she may parrot the words, but she will not learn the reality signified by them. Indeed, unless the student comes to the teacher with some already acquired knowledge, in terms of which she can actively attend to and appropriate the examples, principles, and arguments given to her in instruction, it is difficult to see how she could do anything but parrot back empty sounds. In our own age, we might appreciate St. Thomas's point in a more cynical and even sinister way: the teacher-student relationship risks involving nothing other than the imposition upon the student of an ideology, in terms of which she is unable to mount either an internal or an external critique of the teacher, since she is incapable of making her own what the teacher has shown her.

On the contrary, if knowledge is preserved and promoted within a community through language, it must be something that the individuals within that community actively acquire, promote, critique, and advance. A person is a *subsistent individual of a rational nature*. For St. Thomas, accepting this classical definition from Boethius means recognizing in a subsistent individual human the active capacity to engage in acts of reason. He does not deny that the exercise of that capacity is episodic, and that over time it develops habits which are influenced by the community to a greater or lesser degree, but it is a first-order capacity partially constitutive of one's being the kind of being one is. Its episodic exercise and its developmental character, shown in the habits it forms, do not tell against its fundamental presence in the subsistent individual human being any more than the fact that a child cannot exercise his capacity to reproduce until puberty tells against his having that capacity, and tells against his being an animal. In the human being the active capacity of intellect, in concert with the passive potency to receive the forms of things, is the highest capacity and as such gives its name to the specific difference of the human animal, namely *rational* animal.[45] To make the individual purely passive in learning is to treat her as less than a person. It is to grant her personhood only after she has entered into the politics of the language game.

I do not believe that Haldane would welcome the suggestion that his account renders the pre-linguistic human animal less than a person, but it is difficult to see how he avoids it. In his criticism of McDowell he mentions McDowell's statement to the effect that "human infants are mere animals, distinctive only in their potential."[46] He goes on to argue that he is uncomfortable with McDowell's account of first nature as lacking in natural teleologies and his "[reductionist] descriptions of natural science." Yet it is at this point that he fails to challenge the overall picture that McDowell has painted, as well as the portrait it provides of the pre-linguistic human animal. Rather, he presupposes its fundamentals, and reaffirms the "potential rationality of the infant" who is otherwise a "mere animal." Concerning McDowell's failure to address the issues raised by Haldane, the latter writes, "thus are left unexplained the actual rationality of the culture and the potential rationality of the infant."[47] Here Haldane appeals to God to explain the *presence* of this second-order "potential rationality": he refers the reader to his discussion in *Atheism and Theism,* where he had identified the "second-order" potential to become rational, to acquire, that is, the "first-order" power of rationality. He is proposing a solution to a problem from within the contours of McDowell's world, not a challenge to its presuppositions. Within that world, Haldane's problem with him is not the identification of human infants as "mere animals, distinctive only in their potential," but rather with McDowell's failure to explain whence comes the potential to be something more than mere animals.

Thus, Haldane's own solution to McDowell's failure does not overcome the difficulty I have been describing. He seems to agree with McDowell that the uneducated pre-linguistic human animal does not possess the power of reason but merely the potential to acquire the power of reason.

This pre-linguistic human animal is not a subsistent individual of a rational nature, but a subsistent individual of a non-rational nature that is potentially rational. She is not a person, but only potentially so. On the contrary, a "mere animal" is a mere abstraction according to St. Thomas, and a vicious one at that if it is projected back on to reality as in McDowell. Such a projection is, once again, the "error of the Platonists." McDowell's "mere animal" is unique in reality, a living organism that is a member of no species, who yet stands waiting to be granted admittance by the members of one particular kind of animal, the rational ones. As I have said, I do not think Haldane would welcome this result. The problem is that his proposed solution to McDowell does not avoid it.

For St. Thomas, a political society does not grant personhood; rather, a political society is constituted from and is the more perfect natural flourishing of the persons within it. On that account, it is not at all anachronistic that a *De anima* would have something to say about how the individual human being, the human person, acquires knowledge, engages in acts of reason, and communicates through language. The intellectual and moral destitution that MacIntyre describes is the destitution of a person, not a "mere animal." It is our fault, not Aristotle's or St. Thomas's, if we do not then move on to the *Politica*, *Rhetorica*, and *De poetica*, not to mention the *Summa Theologiae*. What Aristotle and St. Thomas recognized was that we come together across time and space in our common desire for knowledge, understanding, and wisdom, and that language is the handmaiden of that unity. We name as we know, and we all bring something to the table to offer and share, no matter how great or meager. Not only does St. Thomas say the importance of language, in his life he shows it, and his actions speak for themselves.

It is true that thought and understanding have often been conceived in modern philosophy as if they are private acts hermetically sealed off from our more basic physical acts, as well as from our social and political relationships, and that language is the remedy for that privacy and loneliness. But to insist upon the priority of understanding over linguistic expression is not to retreat to that modern privacy. I have been insisting that for the Thomistic-Aristotelian, conceptual activity is displayed in all human acts, non-linguistic as much as linguistic. That, I take it, is the point behind the famous distinction in St. Thomas between human acts and the acts of a human.[48] In a different genre and more dramatically, Andre Dubus captures St. Thomas's point well:

> We can bring our human, distracted love into focus with an act that doesn't need words, an act which dramatizes for us what we are to-

gether. The act itself can be anything: five beaten and scrambled eggs, two glasses of wine, running beside each other in rhythm with the pace and breath of the beloved. They are all parts of that loveliest of sacraments between man and woman, that passionate harmony of flesh whose breath and dance and murmur says: we are, we are, we *are*.[49]

Similarly, we can lie with our bodies as much as with our words, and perhaps more successfully so, since there it is perhaps least expected. Here St. Thomas the theologian might well point to the passage from John, "If anyone says, 'I love God,' yet hates his brother, he is a liar." Presumably the hatred expressed here that renders what is said a "lie" is not simply a hatred expressed in words alone. The words are a lie because of the ways in which they are interwoven with the more fundamental deeds. This is the crux of the argument for maintaining that understanding cannot be completely identified with the capacity for the manipulation of vocal or written signs, namely, the truth of our actions as much as of our words. No matter how much instruction may help, talking about eating rationally does not give Alice the ability to do so. She has the ability to do so, and must form the appropriate habits for doing so. If she is lucky she will find a good teacher.

Analysis must of necessity consider apart what may be one in reality. Emphasis upon one aspect of the analysis may mislead one into thinking that it stands alone, apart from the other. I have not meant to suggest that the more basic rational forms of human animal life take part in strict isolation from the linguistic form of life. Quite the contrary, I have been trying to insist upon their continuity. But not just their continuity. All these forms of life are the warp and woof of the complex fabric of a human life, the life of a rational, social, political, talking *animal*.

Against the background of the emphasis in Aristotle and St. Thomas on the social and political nature of a human animal as a more perfect form of natural existence, in which the individual desires by nature to know but also desires by nature to communicate the "passions of the soul" to another, I would be inclined to suggest that keeping our thoughts private is the exception to the rule. It is more appropriate to say "I can keep my thoughts to myself, if I choose to," rather than "my thoughts are private and I can choose to communicate them." Typically, the times we *deliberately* try to keep our thoughts private are when we ask someone to try and guess them, as a way of proving that they *are* private, or when we fear the repercussions of speaking openly. Even then we seem to engage in a silent dialogue between ourselves and an imagined audience. It takes an effort to keep our thoughts to ourselves. Thought is not by nature private, but rather public. But the public is not just the educated or even uneducated speaking public. A better, and non-reductive understanding of the sciences under which human beings fall presents an opportunity for developing an authentic philosophical anthropology. Here, perhaps, one can see a deep affinity between *Thomist Realism* and a certain understanding of Wittgenstein and the *Linguistic Turn*. For the Thomistic Aristotelian, conceptual functioning on the part of the human animal is naturally ordered toward expression in non-linguistic and linguistic acts alike, forming through these linguistic and non-linguistic acts social and political communities. But like any human act, we can deliberately choose to cut word and deed short of their completion or fulfillment. When we do so, we choose not to flourish as human beings. We choose a less perfect form of existence.

Notes

Introduction

1. See John Haldane's new argument for the existence of God in J.J. Haldane and J.J.C. Smart, *Atheism and Theism* (Oxford: Blackwell, 1996), 106–16.

2. Here I depart from other authors who attribute to St. Thomas the view that the system of concepts and judgments that involve them constitutes a mental language. See, for example, Janet Coleman, "MacIntyre and Aquinas," in *After MacIntyre* (Notre Dame, Ind.: University of Notre Dame Press, 1994), 68. See also Robert Pasnau, "Aquinas on Thought's Linguistic Nature," *The Monist: Analytical Thomism* 80, no. 4 (October 1997): 558–75.

3. Josef Pieper, *The Silence of St. Thomas* (New York: Pantheon Books, 1957).

4. St. Thomas Aquinas, *In Libros Posteriorum Analyticorum Expositio* (Turin: Marietti, 1955), *Prooemium* 4 (hereafter, *CPA*). Reference to particular books, such as *Liber I*, will be a Roman numeral while lessons, such as lectio 2, will be an Arabic numeral, and sections within a lecture will have the Marietti numbering from the beginning of the text. Thus a complete reference to the text might look like this: *CPA* I.2.15. I will follow the same scheme for citing other Marietti commentaries. Where the text of the citation from the Marietti may differ from the 1989 Leonine edition, I will use the Leonine.

5. Text of Aristotle in 16a3–9 in St. Thomas Aquinas, *Expositio Libri Peryermenias* (Rome: Leonine Commission, 1989); hereafter *CPH*.

6. Jerry Fodor, "Mental Representation: An Introduction," in *Scientific Inquiry in Philosophical Perspective*, ed. N. Rescher (Washington, D.C.: University Press of America, 1987), 114.

7. Michael Dummett, "Frege, Gottlob," in *The Encyclopedia of Philosophy*, ed. Paul Edwards (New York: Macmillan, 1967), vol. 3, p. 228.

8. Richard Rorty, *The Linguistic Turn* (Chicago: University of Chicago Press, 1992), 374.

9. Hilary Putnam, "Language and Philosophy," in *Mind, Language, and Reality*, Philosophical Papers 2 (Cambridge: Cambridge University Press, 1975), 8, 14.

10. Ibid., 14.

11. Ian Hacking, *Why Does Language Matter to Philosophy?* (Cambridge: Cambridge University Press, 1975), 52. See Rorty's difficulties with this passage from Hacking in Rorty, *Linguistic Turn*, 364.

12. Plato, *Cratylus*, in *The Collected Dialogues of Plato*, ed. Edith Hamilton and Huntington Cairns (Princeton, N.J.: Princeton University Press, 1961), 438d.

13. Ludwig Wittgenstein, *Philosophical Investigations*, trans. G. E. M. Anscombe (New York: Macmillan, 1953), §201, p. 81.

14. Rorty, *Linguistic Turn*, 365.

15. John Searle, *Expression and Meaning: Studies in the Theory of Speech Acts* (Cambridge: Cambridge University Press, 1979), p. xi, and Searle, *The Rediscovery of Mind* (Cambridge, Mass.: MIT Press, 1992), p. xi.

16. See my "Verbum Mentis: Theological or Philosophical Doctrine?" *Proceedings of the American Catholic Philosophical Association* 74 (2001). Here it suffices to say that my reason for denying that the *verbum mentis* is part of St. Thomas's account is its absence from St. Thomas's latest and most extensive philosophical discussions, namely the *Commentary on the De anima*, the *Commentary on the De interpretatione*, and Questions 75–93 of the first part of the *Summa Theologiae*. Of course my denial that the *verbum mentis* serves any philosophical purpose in St. Thomas does not imply that it serves no theological purpose.

17. "nature ... utitur aere respirato ad formationem vocis, quod est ad *bene esse*" (*Commentary on the De anima* II.18.473, hereafter *CDA*).

Chapter 1. Aristotle's Semantic Triangle in St. Thomas

1. *Aristotle's Categories and Propositions*, trans. Hippocrates Apostle (Grinnell, Iowa: Peripatetic Press, 1980), 97 n. 1. The use of 'proposition' here does not have the typically contemporary sense of an abstract proposition. I will use "enunciation" from here on, in order to relate the discussion more closely to St. Thomas.

2. *De interpretatione* 16a1–2. At 16b27 we learn that a sentence is a significant articulated sound, some part of which is separately significant, though not as an affirmation or negation. At 17a2–5 we learn that a proposition or enunciation is a statement that is either true or false. At 17a9 we learn that propositions or enunciations are divided into affirmations and denials, and conjunctions of these. At 17a25–26 we learn that an affirmation is an enunciation "of one thing of another," while a negation is an enunciation taking one thing from another.

3. See *Aristotle's Categories and Propositions*, trans. Apostle, 98 n. 3.

4. *On Interpretation*, trans. E. M. Edghill, in *The Works of Aristotle*, Great Books of the Western World (Chicago: Encyclopedia Britannica, 1952), 25.

5. *CPH* I.2.22.

6. *CPH* proem.1.

7. *CPA* proem.4.

8. Latin text of Aristotle in St. Thomas Aquinas, *Sentencia Libri De Anima* (Rome: Leonine Commission, 1984) III.6.430a26–b5, b26–32. See St. Thomas's *Commentary*, III.11.

9. *De interpretatione* 16a9–18.

10. *Aristotle's Categories and De Interpretatione*, trans. J.L. Ackrill (Oxford: Clarendon, 1963), note to 16a9–18, p. 114.

11. *CPA* proem.4.

12. *CPH* proem.1.

13. Ibid.

14. Paul Vincent Spade warns against identifying "signification" with "meaning" in contemporary semantical discussions, despite the similarity of the ordinary language overtones of the two words. Signification is what he calls, in a difficult phrase, a "psychologico-causal" term. His point is framed with Boethius's translation of the *De interpretatione* as the background, in particular: "[Verbs] spoken in isolation are names and signify something. For he who speaks [them] establishes an understanding and he who hears [them] rests" ("The Semantics of Terms," in *The Cambridge History of Later Medieval Philosophy*, ed. Norman Kretzmann, Anthony Kenny, and Jan Pinborg [Cambridge: Cambridge University Press, 1982], 188–96, at 188). Spade writes that "'to signify' something was 'to establish an understanding' of it." Consequently, he identifies signification with a "species of the causal relation." E.J. Ashworth adds that "signification is also closely associated with being a sign, that is, with representing or making known something beyond itself" (see "Signification and Modes of Signifying in Thirteenth-Century Logic: A Preface to Aquinas on Analogy," *Medieval Philosophy and Theology* 1 [1991]: 39–67, at 44). Their emphasis on the cognitive and causal character of the discussion of signification is on target. However, care should be taken with respect to Spade's animadversion, since his term "psychologico-causal" is far from clear. Just what sense of "cause" is at play in his warning?

15. *CPH* I.2.15.

16. *Commentary on the Sentences of Peter Lombard*, III Sententia 6.1.3, quoted in Gyula Klima, "The Semantic Principles Underlying Saint Thomas Aquinas's Metaphysics of Being," *Medieval Philosophy and Theology* 5 (March 1996). I am indebted to Klima's analysis for the formal notion of *significata*. There are of course further distinctions to be made with respect to the supposition and signification of concrete and abstract terms. See section 56, pp. 15–20 of Klima's essay.

17. For purposes of simplicity, I am restricting 'signification$_2$' to *res extra animam*. Concepts may well be signified$_2$, but how that occurs by an act of reflection will only be evident after the discussion of chapter 7.

18. Walker Percy, *The Message in the Bottle* (New York: Farrar, Strauss, Giroux, 1975). Percy himself is exploring and developing a fruitful interaction of C.S. Peirce's "thirdness" and Jacques Maritain's and John of St. Thomas's "esse intentionale" and "relation of intentional identity."

19. "Speakers are supposed to be able to direct their mental attention *to* concepts by means of something akin to perception, and, if A and B are different concepts, then attending to A and attending to B are different mental states. . . . [t]he mental state of the speaker determines which concept he is attending to, and thereby determines what it is he refers to" (Hilary Putnam, *Representation and Reality* [Cambridge, Mass.: MIT Press, 1988], chapter 2, 129 n. 1). I am tempted to write here "have come before our conscious attention," to more closely parallel Putnam's description. However, I think "conscious attention" is a little too loaded an expression.

20. *CPH* I.2.19.

21. Some qualification has to be made on this last statement in virtue of the analogous use of 'universal' predicated of the nature that exists in *res extra animam*, but that may be abstracted from such *res*. By abstraction the intention of universality accrues to it. It is the nature *in anima* with the intention of universality that is properly called a universal, while the nature itself is called universal because of this accident that accrues to it as understood.

22. *CDA* III.8.717–18.

23. St. Thomas Aquinas, *Opusculum De Ente et Essentia* [Turin: Marietti, 1957], 4.1. The *De ente* is an early work, preceding St. Thomas's becoming Master at Paris, dated between 1252 and 1256. See James Weisheipl, O.P., *Friar Thomas D'Aquino* (Washington, D.C.: Catholic University of America Press, 1983), 78–79, 386–87. See also St. Thomas Aquinas, *Quaestiones Quodlibetales* (Turin: Marietti, 1949), 8.8.1, *respondeo*) (hereafter *QDL*); *Summa Theologiae* (Turin: Marietti, 1948), I.85.3, ad 1, 2, and 4 (hereafter *ST*) and *CDA* I.1.13.

24. *ST* I.85.1, ad 1.

25. Joseph Owens, "Common Nature: A Point of Comparison between Thomistic and Scotistic Metaphysics," in *Inquiries into Medieval Philosophy*, ed. James Ross (Westport, Conn.: Greenwood, 1971), 192.

26. This contrast does not consider cases in which something is known from its effects, in which case the principle of knowing is not formally the same as the principle of being. See St. Thomas Aquinas, *Quaestiones Disputatae De Veritate* (Turin: Marietti, 1948), 3.3, ad 7 (hereafter *QDV*).

27. St. Thomas Aquinas, *Quaestiones Disputatae de Potentia* (Turin: Marietti, 1949), q. 8, a. 1, *respondeo* (hereafter *QDP*). See also *CDA* III.8.700.

28. *QDL* 8.1.1. See also Weisheipl, *Friar Thomas D'Aquino*, 78–79, 386–87.

29. Avicenna, quoted in Owens, "Common Nature," 187. My summary of Avicenna is heavily dependent upon Owens's analysis.

30. Ibid., 1C, 190.

31. See Owens, "Common Nature." See also St. Thomas's discussion of *unum* as a transcendental characteristic of being in *ST* I.11.1, and especially *QDV* 1.1, *respondeo*, and 21.1, *respondeo*.

32. See *De ente* 4.2. For an excellent discussion of the conundrums involved in the "Socrates est species" paralogism, and St. Thomas's *De ente et essen-*

tia, see Gyula Klima, "Socrates Est Species," in *Argumentationstheorie: Scholastische Forschungen zu den logischen und semantischen Regeln korrekten Folgerns*, ed. Klaus Jacobi (Leiden: E.J. Brill, 1993), 491–506.

33. See Ralph McInerny, "Being and Predication," in his *Being and Predication: Thomistic Interpretations* (Washington, D.C.: Catholic University of America Press, 1986), 173–228.

34. John Haldane finds St. Thomas's analysis in the *De ente et essentia* inadequate to the extent that he believes it cannot account for what is common between individuals of the same kind, or what is common between token occurrences of the same type concept. See "Forms of Thought," in *The Philosophy of Roderick Chisholm*, ed. Lewis Edwin Hahn (Chicago: Open Court, 1997), 149–70.

35. *De ente* 4.1.

36. *QDV* 2.5, ad 17; *ST* I.76.2, ad 4; and *CDA* III.13.789.

37. *De ente* 4.2. Also *CDA* II.12.380.

38. Ibid.

39. *De ente* 4.2.

40. Ibid.

41. Of course in St. Thomas "conceptio" has a broader use than I am employing here. It can also refer to the terminus of the second operation (*CPH* I.4.44). See Robert W. Schmidt, S.J., *The Domain of Logic According to Saint Thomas Aquinas* (The Hague: Martinus Nijhoff, 1966), 120.

42. The use of 'concept' here is different from Frege's use of 'concept' (*Begriff* in the German) in "Concept and Object," and those who follow him in this usage. Frege was adamant that by 'concept' he did not have a mental act, or in his way of putting it a "psychological" sense, in mind. See "On Concept and Object" in *Translations from the Philosophical Writings of Gottlob Frege*, ed. Peter Geach and Max Black (Oxford: Basil Blackwell, 1970), 42. However his understanding of mental acts and their "subjective" character differs markedly from St. Thomas's. It would be interesting, however, to pursue in another place parallels with Frege's distinction between "properties" and "marks of an object" ("On Concept and Object," 51–52). He writes "in my way of speaking, a thing can be at once a property and a mark, but not of the same thing." We may say "Socrates is a rational animal," or we may say equivalently that "Socrates is a man." For Frege, *man*, *rational*, and *animal* are properties of Socrates. But *rational* and *animal* are "marks" of the property *man*. They are not properties of *man* taken as an object. In St. Thomas, *rational* and *animal* are elements of the nature of *man considered absolutely*, and as such are predicated along with it (*convenire*) when the nature is predicated of a man like Socrates. But they are not predicated of the nature *absolutely considered*, as if it were an object or thing having those characteristics. In fact, in David Wiggins's application of the Fregean terminology, a *concept*, distinguished from a *conception* (Fregean *sense*) begins to look an awful lot like St. Thomas's nature *absolutely considered*. See David Wiggins, "Putnam's Doctrine of Natural Kind Words and Frege's Doctrines of Sense, Reference, and Extension:

Can They Cohere?" in *Reading Putnam*, ed. Peter Clark and Bob Hale (Oxford: Basil Blackwell, 1994), 201–15, but esp. 214 n. 10.

43. *Concept* can be synonymous with *ratio intellecta*. See Schmidt, *The Domain of Logic*, 110, esp. n. 56, and 111: "The *ratio intellecta* or intelligible aspect of the thing which is understood, considered *as* understood, is the intention or concept taken objectively: 'Intellectus enim intelligendo concipit et format intentionem, sive rationem intellectam, quae est interius verbum.'" The quote is from St. Thomas, *Summa Contra Gentiles* IV.11 (hereafter *SCG*).

44. *ST* I.85.2, *respondeo* and I–II.93.1, ad 2.

45. *ST* I.77.3, ad 3.

46. *SCG* I.33.

47. *CPH* I.2.14.

48. Ibid., 19.

49. Ibid.

50. See *CPA* I.4.3 and also II.6.468. Aquinas is referring to *Topics* II at 110a13–23, where Aristotle writes, ". . . you should define what kind of things should be called as most men call them, and what should not. For this is useful both for establishing and for overthrowing a view: e.g. you should say that we ought to use our terms to mean the same things as most people mean by them, but when we ask what kind of things are or are not of such and such a kind, we should not here go with the multitude: e.g. it is right to call 'healthy' whatever tends to produce health, as do most men: but in saying whether the object before us tends to produce health or not, we should adopt the language no longer of the multitude but of the doctor" (in *The Works of Aristotle*, Great Books of the Western World, ed. Robert Maynard Hutchins [Chicago: Encyclopaedia Britannica, 1952]).

51. See Saul Kripke, *Naming and Necessity* (Cambridge, Mass.: Harvard University Press, 1980), passim but especially 138. See Putnam, *Mind, Language, and Reality*, in particular chapter 8 ("Is Semantics Possible"), chapter 11 ("Explanation and Reference"), and chapter 12 ("The Meaning of 'Meaning'"). For an alternative that recognizes "speaking-sensitive semantics," see Putnam's remarks and references in *The Threefold Cord* (New York: Columbia University Press, 2000), 87–91, 124–25.

52. This thesis could be elaborated rigorously in terms of the usual distinctions concerning word types and token occurrences of them. However I do not think such rigor would significantly add to the point made.

53. See the worries of Wittgenstein and Dummett about such meaning-conferring acts in Wittgenstein, *Philosophical Investigations*, §§ 329, 334, and Michael Dummett, *Origins of Analytical Philosophy* (Cambridge, Mass.: Harvard University Press, 1994), 44, 49–50. Compare John Haldane, "The Life of Signs," *Review of Metaphysics* 47 (March 1994): 469.

54. Haldane, "Life of Signs," 454.

55. *CPH* I.2.19.

56. *CDA* II.7.681.
57. *Textus Aristotelis,* 16a5–7 in *CPH.*
58. *CPH* I.2.21.
59. *QDV* 2.5, ad 17. For a discussion of how 'likeness' or 'image' is and is *not* being used here, see Ralph McInerny, *The Logic of Analogy* (The Hague: Martinus Nijhoff, 1961), ch. 4, "The Signification of Names," 51–52.
60. *ST* I.76.2, ad 4.

Chapter 2. Three Rival Versions of Aristotle

1. Norman Kretzmann, "Aristotle on Spoken Sound Significant by Convention," in *Ancient Logic and Its Modern Interpretations,* ed. J. Corcoran (Dordrecht: D. Reidel, 1974), 4.
2. Ibid., 5.
3. Norman Kretzmann, "Semantics, History of," in *The Encyclopedia of Philosophy,* ed. Paul Edwards (New York: Macmillan, 1967), vol. 7, p. 362.
4. Kretzmann, "Aristotle on Spoken Sound," 4.
5. Ibid., 3.
6. Ibid., 10–12.
7. Ibid., 18.
8. Ibid., 14.
9. Ibid., 7, 19.
10. For Kretzmann's analysis of the debate about the Greek, see ibid., 18 n. 4.
11. Ibid., 18 n. 2.
12. *Aristotle's Categories and De Interpretatione,* trans. Ackrill, 16a3–9, p. 43.
13. Kretzmann, "Aristotle on Spoken Sound," 5, 7. See also 19 n. 8.
14. Ibid., 18 n. 6.
15. *Aristotle: On Interpretation. Commentary by St. Thomas and Cajetan,* trans. Jean T. Oesterle (Milwaukee: Marquette University Press, 1962), 23 n. 10.
16. Kretzmann, "Aristotle on Spoken Sound," 5.
17. Kretzmann, "Semantics, History of," 367.
18. Kretzmann, "Aristotle on Spoken Sound," 8 n. 4.
19. Ibid., 4–5.
20. John Magee, *Boethius on Signification and Mind* (Leiden: E.J. Brill, 1989), 32–48.
21. Kretzmann, "Aristotle on Spoken Sound," 5.
22. Ibid., 8.
23. Ibid.
24. Ibid., 6.
25. Kretzmann's thesis has found at least one detractor in Magee's *Boethius on Signification and Mind,* which provides a detailed paleographical examination, both internal and external to the text, and a critique of Kretzmann's claims about

the use of σύμβολα and σημεῖα in ancient Greek texts (see pp. 36–48). The upshot of his critique is that Kretzmann is wrong to suppose that they have distinct meanings in the text.

26. Kretzmann, "Aristotle on Spoken Sound," 7–8.
27. Ibid., 8, passim.
28. Ibid., 5.
29. Ibid., 8.
30. *Aristotle's Categories and De Interpretatione,* trans. Ackrill, 16a26–28.
31. Kretzmann, "Aristotle on Spoken Sound," 16.
32. Ibid., 8.
33. Ibid.
34. Aristotle, *Metaphysics* IV.2.
35. C. S. Peirce as quoted in William Alston, *Philosophy of Language* (Englewood Cliffs, N.J.: Princeton University Press, 1964), 55.
36. *Aristotelis Categoriae et Liber De Interpretatione,* ed. L. Minio-Paluello (Oxford: Clarendon Press, 1949).
37. *Aristotle's Categories and De Interpretatione,* trans. Ackrill, p. 49.
38. *Aristotle's Categories and Propositions,* trans. Apostle.
39. *The Categories; On Interpretation,* trans. Harold P. Cooke, and *Prior Analytics,* trans. Hugh Tredennick, Loeb Classical Library (Cambridge, Mass.: Harvard University Press, 1983).
40. *Categories and De Interpretatione,* trans. Ackrill.
41. Kretzmann, "Aristotle on Spoken Sound," 18.
42. See Magee, *Boethius on Signification,* 43–45.
43. "Dicendum quod secundum Philosophum, I Perihermeneias, *voces sunt signa intellectum, et intellectus sunt rerum* similitudines; et sic patet quod voces referuntur ad res significandas mediante conceptione intellectus" (*Summa Theologiae,* trans. Herbert McCabe, O. P. [Blackfriars, 1964]) I.13.1.
44. Bernard Lonergan, *Verbum: Word and Idea in Aquinas,* ed. David B. Burrell, C.S.C. (Notre Dame, Ind.: University of Notre Dame Press, 1967), 2.
45. Magee argues that ταῦτα actually refers back to both vocal utterances and written marks. In that case, Magee believes that Kretzmann's argument falls apart because there is no way in which written marks could be seen as the natural effects of mental impressions, related to the latter in a relationship of constant concurrence (*Boethius on Signification and Mind,* 43–45). But see my comments later in the body of the text about why it appears that written marks should count as the natural effects of mental impressions, in the sense of *natural* at play in Kretzmann's analysis when he claims that spoken words are natural effects of mental impressions. For the sake of argument let us grant Kretzmann that it refers only to the vocal utterances.
46. *De Interpretatione* in *Aristotelis Latinus,* vol. 2.1 (Leiden: E. J. Brill, 1965), p. 5.
47. Of course *primum* could be a nominative or accusative singular adjective, but in the context that seems ruled out. Magee notes that Moerbeke may

have been translating an abbreviated form of πρώτων, probably the result of a shortening of the final omega in the genitive plural (*Boethius on Signification and Mind*, 28).

48. *De Interpretatione* in *Aristotelis Latinus*, vol. 2.2, p. 41.

49. G. Verbeke, "Deux Traductions de Moerbeke," in *Ammonius: Commentaire sur Le Peri Hermeneias d'Aristote, traduction de Guillaume de Moerbeke* (Louvain: Universaires de Louvain, 1961), p. xc.

50. *Ammonius In Aristotelis De Interpretatione Commentarius*, ed. Adolfus Busse, in *Commentaria in Aristotelem Graeca* (Berlin, 1895), 24.

51. Verbeke, *Ammonius: Commentaire sur Le Peri Hermeneias*, 45–46.

52. Verbeke's paraphrase of Busse, "Deux Traductions de Moerbeke," p. lxxvi.

53. Ibid., p. lxxxvi.

54. Magee, *Boethius on Signification*, 26.

55. Kretzmann, "Aristotle on Spoken Sound," 18 n. 4.

56. *De Interpretatione* in *Aristotelis Latinus*, p. 41.

57. Verbeke, *Ammonius: Commentaire sur Le Peri Hermeneias*, 45.

58. Ibid., 45–46. Compare my translation of the Latin Ammonius with David Blanck's recent translation of Ammonius directly from the Greek: "he says: 'Yet those of which, as the first ones, these are signs', where 'these' are what is in the vocal sound, i.e. names and verbs, thus 'those of which, as the first ones, these are signs' (he means thoughts; for things are also signified by them, not immediately, however, but by means of thoughts; however, thoughts are not signified by means of other items, but first and immediately" (*Ammonius: On Aristotle's On Interpretation 1–8*, trans. David Blanck [Ithaca. N.Y.: Cornell University Press, 1996], 33). By the way, Blanck translates the genitive plural πρώτως. He notes that Busse's text has the adverb. But against Busse and for the adjective he cites E. Montanari, *La sezione linguistica del Peri Hermeneias di Aristotele*, vol. 1 (Florence 1984), 126–32 and vol. 2 (1988), 45–57, as well as Brunschwig in the 13th *Symposium Aristotelicum*, 37–40. See Blanck, 142 n. 90.

59. Ibid., 32.

60. Magee, *Boethius on Signification*, 29.

61. Ibid.

62. Boethius, *Commentaries on Aristotle's De Interpretatione* (New York: Garland, 1987), II.33.27.

63. Ibid., vol. 1, p. 40.

64. Recall that I have not claimed that Ammonius explicitly uses the terms 'secondary' or 'second', but rather that they are implicit in the account he gives.

65. John Locke, *An Essay Concerning Human Understanding* (Oxford: Clarendon Press, 1975) III.2.2.

66. E.J. Ashworth suggests that Locke had near contemporary scholastics at Oxford in mind in these sorts of texts. See E.J. Ashworth, "Locke on Language," *Canadian Journal of Philosophy* 14, no. 1 (March 1984): 45–73, and "'Do Words Signify Ideas or Things?' The Scholastic Sources of Locke's Theory of Language," *Journal of the History of Philosophy* 19 (July 1981): 299.

67. For ease of reference, as well as continuity with the method of citations used to this point by the other authors, I will use the Marietti paragraph divisions, despite the fact that I am translating the Leonine edition. I should point out that in passage 13 I have translated 'immediate' as 'without mediation'. At first sight it might seem more natural to simply transliterate it as 'immediately'. However, I hope the rest of the chapter will make clear why I think 'without mediation' is a better translation.

68. *Ammonius*, 34–35.

69. *CPH* I.2.20.

70. See *Commentaries on Aristotle's De Interpretatione*, vol. II, p. 41. See also Magee's discussion of this text, *Boethius on Signification*, 73–74.

71. *CPH* I.2.20.

72. 16a9–18. Compare also St. Thomas's commentary at III.11 on the *De anima* passage with the commentary on the passage from the *Peri hermeneias*.

73. 430a26–29, in *Aristotle: De Anima*, trans. D. W. Hamlyn (Oxford: Clarendon Press, 1993).

74. For the Greek text of *De anima*, see Loeb Classical Editions, vol. 8 (Cambridge, Mass.: Harvard University Press, 1986), 432a10–12, p. 180.

75. 432a10–12, trans. Hamlyn. See also St. Thomas's commentary at III.13.

76. See Magee, *Boethius on Signfication*, 105–116.

77. In *Boethius on Signification*, 19, Magee cites M. T. Larkin, *Language in the Philosophy of Aristotle* (The Hague: Janua Linguarum, 1971), 22, 34, and Joseph Owens, *The Doctrine of Being in the Aristotelian Metaphysics*, 3d. ed. (Toronto: Pontifical Institute of Medieval Studies, 1978), 120.

78. I say "implicitly" because St. Thomas does not explicitly contrast his interpretation of *primorum* with Boethius's or Ammonius's. In other instances, he is more than happy to mention how his interpretation differs from various predecessors. Why not here? I'm not sure, but I have an unsubstantiated speculation. Typically in this commentary St. Thomas will mention predecessors when there is a dispute among them. He will review their differing positions and responses to one another. Boethius's reply to Andronicus's objection is a case in point. Then he will provide his interpretation as a resolution of the dispute. In this case, my speculation is that he recognizes that there is no dispute for him to settle. His interpretation simply differs from the uniform interpretation of his predecessors.

79. *CPH* I.3.24.

80. Most other recent commentators probably find the presence of the initial text as perplexing as Ackrill does, as opposed to the transitional character that St. Thomas sees in it. Ackrill writes, "There are grave weaknesses in Aristotle's theory of meaning. Fortunately, the notion that utterances are symbols of affections in the soul and that these are likenesses of things does not have a decisive influence on the rest of the *De interpretatione*. For example, Aristotle does not often appeal to psychological experiences or facts to explain or support what

he says about names, verbs, statements, &c.; most of what he says is independent of the special theory about words, thoughts, and things. Aristotle's main and official discussion of thinking (to which he—or an editor—here refers us) is in *De anima* III 3–8" (Ackrill's note to 16a3, in *Aristotle's Categories and De Interpretatione*, p. 113).

81. There is little question that he was familiar with Boethius's commentaries. Magee suggests, however, that Moerbeke's translation of Ammonius may not have been finished in time for St. Thomas to use it (*Boethius on Signification*, 18 n. 30). Magee's suggestion, however, seems to be belied by the text of St. Thomas, where in addition to Alexander, Porphyry, and Boethius, he refers to Ammonius a number of times; perhaps his acquaintance with Ammonius was not direct, but rather through other authors.

82. *CPH* I.2.12.

83. Ibid., I.2.15.

84. Of course I am using the English equivalents of the Latin terms, but the distinction arises in the Latin from which the English is derived.

85. Cf. G. E. M. Anscombe, "A Theory of Language?" in *Perspectives on the Philosophy of Wittgenstein*, ed. Irving Block (Cambridge, Mass.: MIT Press, 1981), 155.

86. *ST* I.85.2.

Chapter 3. Language and Mental Representationalism

1. Michael Dummett, *Frege: Philosophy of Language*, 2d. ed. (Cambridge, Mass.: Harvard University Press, 1981), 3–4, 637–38.

2. *Aristotle's Categories and De Interpretatione*, trans. Ackrill, ch. 1, note to 16a3.

3. Robert Sokolowski, "Exorcising Concepts," *Review of Metaphysics* 40 (March 1987): 456.

4. John Haldane, "Reid, Scholasticism and Current Philosophy of Mind," in *The Philosophy of Thomas Reid*, ed. M. Dalgarno and E. Matthews (Dordrecht: Kluwer, 1989), 287.

5. See Alexander Broadie, "Medieval Notions and the Theory of Ideas," in *Proceedings of the Aristotelian Society*, vol. 87 (London: Aristotelian Society, 1987), 153–67.

6. Bertrand Russell, "Knowledge by Acquaintance and by Description," *Proceedings of the Aristotelian Society* (London: Aristotelian Society, 1910–11): 119.

7. W. V. O. Quine, "Speaking of Objects," *Ontological Relativity and Other Essays* (New York: Columbia University Press, 1969), 1.

8. Dummett, *Origins of Analytic Philosophy*, 39–40.

9. Rene Descartes, *Ouvres de Descartes*, ed. Charles Adam and Paul Tannery, vol. 7 (Paris: J. Vrin, 1964), 56.

10. Ibid., 42.

11. Ibid., 37.
12. Locke, *Essay*, II.1.1.
13. Ibid., IV.1.1.
14. Ibid., IV.4.3.
15. Ibid.
16. "[I]t is not in the Power of the most exalted Wit, or enlarged Understanding, by any quickness or variety of Thought, to *invent or frame one new simple Idea* in the mind, not taken in by the ways before mentioned [sensation or reflection]: nor can any force of the Understanding, *destroy those that are there*. . . . I would have any one try to fancy any Taste, which had never affected his Palate; or frame the *Idea* of a Scent, he had never smelt: And when he can do this, I will also conclude, that a blind Man hath *Ideas* of Colours, and a deaf Man true distinct Notions of Sounds" (ibid., II.2.2).
17. Ibid., IV.4.4.
18. Ibid., IV.4.5.
19. Ibid., IV.4.12.
20. See, e.g., ibid., II.24.1, 25.6, 29.8; IV.7.16 and 11.1.
21. Ibid., II.31.6.
22. Ibid., III.3–7; V.6.
23. "The simple *Ideas* where of we make our complex ones Substances, are all of them (bating only the Figure and Bulk of some sorts) Powers; which being Relations to other Substances, we can never be sure that we know all the Powers, that are in any one Body, till we have tried what Changes it is fitted to give to, or receive from other Substances, in their several ways of application: which being impossible to be tried upon any one Body, much less upon all, it is impossible we should have adequate *Ideas* of any Substance, made up of a Collection of all its Properties" (ibid., II.31.8).
24. Ibid., V.17.8.
25. Ibid., II.11.17.
26. "The understanding seems to me not to have the least glimmering of any ideas which it doth not receive from one of these two. *External* objects furnish the mind with the ideas of sensible qualities, which are all those different perceptions they produce in us; and *the mind* furnishes the understanding with ideas of its own operations" (ibid., II.1.5). "For since the things the mind contemplates are none of them besides itself, present to the understanding, it is necessary that something else, as a sign or representation of the thing it considers, should be present to it; and these are ideas" (ibid., IV.21.4).
27. For the following summary, see ibid., II.1.2–5 and 8.7–10.
28. Ibid., II.8.15.
29. Ibid., III.2.2.
30. Ibid., III.3.11.
31. See Ashworth, "Locke on Language," and "'Do Words Signify Ideas or Things?'"

32. Locke, *Essay*, II.3.6.

33. However, see later *CPH* I.10.6.

34. Locke, *Essay*, III.2.2.

35. See Ashworth, "Locke on Language," and "'Do Words Signify Ideas or Things?'"

36. For Kretzmann, "the basic semantic relation in Locke's account of language is that of a word used by some speaker as a proper name for some idea in that speaker's mind. It seems to follow from this doctrine that as long as one does use words in this (the only approved) way, one cannot misuse them; and Locke does sometimes suggest that in the early chapters of Book III (see, for instance, 3.2.3). Those chapters indeed present a classic formulation of what Wittgenstein was later to criticize as the notion of a "private language" ("Semantics, History of," 380). For Brykman, see Genevieve Brykman, "Locke on Private Language," in *Minds, Ideas, and Objects: Essays on the Theory of Representation in Modern Philosophy*, ed. Phillip D. Cummins and Guenter Zoeller, North American Kant Society Studies in Philosophy (Atascadero, Calif.: Ridgeview, 1992), 125–34.

37. Antoine Arnauld, *On True and False Ideas*, trans. Stephen Gaukroger (Manchester: Manchester University Press, 1990), 67. See also Steven Nadler, *Arnauld and the Cartesian Philosophy of Ideas* (Princeton, N.J.: Princeton University Press, 1989), 159–65.

38. Locke, *Essay*, III.2.1–2 and IV.21.4.

39. "[N]ature . . . utitur aere respirato ad formationem vocis, quod est ad bene esse" (*CDA* II.18.473).

40. Locke, *Essay*, III.2.8.

41. Ibid., III.2.5.

42. Ibid., III.4.2.

43. George Berkeley, *The Principles of Human Knowledge* (London: Penguin Books, 1988), 59.

44. "Here are *Essences* and *Properties*, but all upon supposition of a Sort, or general abstract *Idea*, which is considered as immutable: but there is no individual parcel of Matter, to which any of these Qualities are so annexed, as to be *essential* to it, or inseparable from it. That which is *essential*, belongs to it as a Condition, whereby it is of this or that Sort: But take away the consideration of its being ranked under the name of some abstract *Idea*, and then there is nothing necessary to it, nothing inseparable from it. Indeed, as the *real Essence* of Substances, we only suppose their Being, without precisely knowing what they are: But that which annexes them still to the *Species*, is the nominal Essence, of which they are the supposed foundation and cause" (Locke, *Essay*, III.6.6). See also section 8.

45. Berkeley, *Principles of Human Knowledge*, 73.

46. Ibid., 55. Putnam nonetheless believes this is "the heart of Berkeley's argument."

47. Ibid., 85.

48. Ibid., 39.

49. "I do not deny absolutely that there are general ideas, but only that there are any *abstract general ideas*. . . . An idea, which considered in itself is particular, becomes general, by being made to represent or stand for all other particular ideas of the same sort" (ibid., 43). See also pp. 42–47.

50. Ibid., 43. Consider also "*universality*, so far as I can comprehend, not consisting in the absolute, positive nature or conception of anything, but in the relation it bears to the particulars signified or represented by it: by virtue whereof it is that things, names, or notions, being in their own nature *particular*, are rendered *universal*" (ibid., 45).

51. Locke, *Essay*, III.5.20, passim.

52. Hume chided Locke for applying the term "idea" to all the mind's perceptions, writing that Locke "perverted" the term from its original sense. However the dispute is a terminological one, as Hume specifies that ideas are a species of perception less vibrant and forceful, though similar to the original perceptions that are impressions of sensation.

53. David Hume, *A Treatise of Human Nature* (Oxford: Clarendon Press, 1967), I.4.2.47. I am not considering in this exegesis how his views may or may not have changed in the *Enquiries*.

54. Ibid., I.4.5.15.

55. Ibid., I.4.2.53.

56. Ibid., I.1.7.6.

57. "As to those impressions, which arise from the sense, their ultimate cause is . . . perfectly inexplicable by human reason, and it will always be impossible to decide with certainty, whether they arise immediately from the object, or are produced by the creative power of the mind, or are derived from the Author of our being" (ibid., I.1.1.7).

58. Ibid., I.4.6.4. For "monstrous offspring," see I.4.2.52. For "palliative" see I.4.2.46.

59. Ibid., I.1.7, passim.

60. Ibid.

61. "But, however absurd this doctrine might appear to the unlearned, who consider the existence of the objects of sense as the most evident of all truths, and what no man in his senses can doubt, the philosophers who had been accustomed to consider ideas as the immediate objects of all thought, had no title to view this doctrine of Berkeley in so unfavourable light" (Thomas Reid, *Inquiry into the Human Mind*, in *Philosophical Works*, vol. 1 [Hidlesheim: Georg Olms Verlagsbuchhandlung, 1967], 281a). For Reid's discussion of the theory of ideas, see Roger D. Gallie, *Thomas Reid and "The Way of Ideas"* (Dordrecht: Kluwer, 1989). See also Haldane, "Reid, Scholasticism and Current Philosophy of Mind," 285–301. For criticism of Reid's treatment of Descartes and the Cartesians, see Nadler, *Arnauld and the Cartesian Philosophy of Ideas*.

62. Reid, *Inquiry into the Human Mind*, 293a–b.

63. John Stuart Mill, *A System of Logic, Ratiocinative and Inductive: Being a Connected View of the Principles of Evidence and the Method of Scientific Investigation* (New York: Longmans, 1906).

64. Edmund Husserl, *Logical Investigations*, trans. J.N. Findlay (New York: Humanities Press, 1970), I.4.30. For a discussion of the original interaction and common concerns of Husserl and Frege, see J. Mohanty, *Husserl and Frege* (Bloomington: Indiana University Press, 1982); Herbert Spiegelberg, *The Phenomenological Movement* (The Hague: Martinus Nijhoff, 1982), 84–100; Barry Smith, "On the Origins of Analytic Philosophy,"*Graezer philosophische Studien* 35 (1985): 153–73; Dummett, *Origins of Analytic Philosophy*; Robert Sokolowski "Husserl and Frege," *Journal of Philosophy* 84, no. 10 (1987): 521–28; Barry Smith "Frege and Husserl: The Ontology of Reference," *Journal of the British Society for Phenomenology* 9, no. 2 (May 1978): 111–25; Richard Cobb-Stevens, *Husserl and Analytic Philosophy* (Dordrecht: Kluwer, 1990). This last is particularly interesting for the stress it places upon seeing Husserl as continuing the Aristotelian tradition.

65. Husserl, *Logical Investigations*, II.2.10. Also "now we completely see the deceptive confusions in Locke's train of thought. From the obvious truth that each universal name has its own peculiar universal meaning, he passes on to assert that a *general idea* corresponds to every general name, which idea is for him simply a *separate intuitive presentation*." And, "we tend naturally to turn our gaze among logical phenomena to whatever has primary intuitive palpability; we are then misled into taking the inner pictures which are found to accompany our names as the meanings of those names. If we become clear, however, that a meaning is merely what we mean, or what we understand by an expression, we cannot maintain such a conception" (ibid., II.3.15).

66. Ibid., II.5.34.

67. Sokolowski, "Exorcising Concepts," 456. See also Smith, "Frege and Husserl."

68. Introduction to Gottlob Frege, *Foundations of Arithmetic*, trans. J.L. Austin (Evanston, Ill.: Northwestern University Press, 1980), p. x.

69. Gottlob Frege, "The Thought: A Logical Inquiry," trans. A.M. and Marcell Quinton, *Mind* 65, no. 259 (July 1956): 306.

70. Colin McGinn, *Wittgenstein on Meaning* (Oxford: Basil Blackwell, 1984), 98–100.

71. Ibid., 100.

72. See also Dummett's argument with McDowell and Evans. McDowell and Evans assert that allowing for terms that have senses but no reference in fact leads to skepticism with respect to the reference of terms generally. See John McDowell, "On the Sense and Reference of a Proper Name," *Mind* 86 (1977): 159–85. Dummett summarizes the point this way, "[I]f we criticize that strand in Frege's notion of sense in accordance with which senses are themselves objects,

however, we are, once more, concerned with whether they are rightly conceived as objects of *apprehension*, rather than whether they can be objects of *reference*: It was precisely because Frege failed to distinguish between these two senses of the word 'object', or these two ingredients in our notion of an object, that Russell came to aim off target, in his criticism of Frege.... For Evans and McDowell, by treating the mode of presentation as if it were itself an object of apprehension, we are detaching it from the object which is the referent ... if it is only by the medium of senses, as thought-components, that we apprehend objects, we appear to be threatened by the same danger as threatened the empiricists who saw us as enclosed in a world of mental images and sense-impressions, the inner world of ideas of which Frege spoke in 'Der Gedanke'. How can we know that we ever do reach the object, or that there really is any object, if a sense always interposes between us and it, a sense that carries no guarantee of any corresponding referent?" (Michael Dummett, *The Interpretation of Frege's Philosophy* [Cambridge, Mass.: Harvard University Press, 1981], 132–33).

73. Wittgenstein does not address explicitly Descartes, Locke, Berkeley, and Hume. The impetus for my reading of the *Investigations* this way came from reading Haldane, "The Life of Signs."

74. Wittgenstein, *Philosophical Investigations*, §693.

75. "At the first glance the proposition [Satz]—say as it stands printed on paper—does not seem to be a picture of the reality of which it treats. But nor does the musical score appear at first sight to be a picture of a musical piece; nor does our phonetic spelling (letters) seem to be a picture of our spoken language. And yet these symbolisms prove to be pictures—*even in the ordinary sense of the word*—of what they represent" (Ludwig Wittgenstein, *Tractatus Logico-Philosophicus*, trans. C. K. Ogden [London: Routledge and Kegan Paul, 1986], §4.011. Emphasis added).

It is important to recognize that according to Wittgenstein in the *Tractatus*, sentences are not pictures of objects, but rather of *facts* [§2.1], which facts are themselves composed of *states of affairs*. Even more it is important to recognize that he did not conceive of the picturing relation generally along sensual lines, despite his phrase "even in the ordinary sense of the word [pictures]." A sentence pictures a fact because it shares the very same logical structure as the fact, not because its sensual appearance is like the sensual appearance of the fact it pictures—it pictures the fact's logical structure. For more on the *Tractatus* and picturing, see Justus Hartnack, *Wittgenstein and Modern Philosophy*, 2d. ed. (Notre Dame, Ind.: University of Notre Dame Press, 1986). See also G. E. M. Anscombe, *An Introduction to Wittgenstein's Tractatus* (Philadelphia: University of Pennsylvania Press, 1971), chs. 3–5.

76. "On Sense and Reference," 56–57. I have not been able to find Frege actually using the phrase "sense determines reference," though admittedly I am making use of translations, and have confined my search to "Sense and Reference," "The Thought," "Concept and Object," and *The Foundations of Mathematics*.

Those who use this phrase in the context of Frege typically cite the passage in which he says that the sense of a term is the mode of presentation of its reference. See Michael Dummett, *Frege: Philosophy of Language* (Cambridge, Mass.: Harvard University Press, 1981), ch. 5, "Sense and Reference," 93–97. See also Hilary Putnam "The Meaning of 'Meaning'," in *Mind, Language, and Reality*. (Cambridge: Cambridge University Press, 1975), 215–71.

77. Wittgenstein, *Philosophical Investigations*, §140.

78. Haldane, "The Life of Signs," 459. Compare Hume's account in the *Treatise*, "Abstract ideas are, therefore, in themselves individual, however they may become general in their representation. The image in the mind is only that of a particular object, though the application of it in our reasoning be the same as if it were universal," and " a particular idea becomes general by being annex'd to a general term; that is, to a term, which from a customary conjunction has a relation to many other particular ideas, and readily recalls them in the imagination" (Hume, *Treatise*, 1.1.7.6 and 10).

79. Wittgenstein, *Philosophical Investigations*, §663.

80. Here I am indebted to Michael Letteney for helping me think through this cautionary point.

81. See John Searle, *Intentionality: An Essay in the Philosophy of Mind* (Cambridge: Cambridge University Press, 1988), ch. 1, The Nature of Intentional States, esp. 13–14, 21–22, esp. p. 21). Searle speaks in the idiom of "Intentionality." See, however, Haldane, "The Life of Signs," 459, using the idiom of reference. See also John McDowell, "Meaning and Intentionality in Wittgenstein's Later Philosophy," *Midwest Studies in Philosophy* 17 (1992): 40–52, esp. 42, and Jerry Fodor, *The Language of Thought* (Cambridge, Mass.: Harvard University Press, 1975), passim, but especially chapter 2, "Private Language, Public Languages."

82. Wittgenstein, *Philosophical Investigations*, §198.

83. Ibid., §329. For a discussion of the inadequacies of identifying "meaning" with actual use in accord with established rules of use, and nothing more, see Haldane's "The Life of Signs," esp. Part III.

84. McDowell, "Meaning and Intentionality in Wittgenstein," quoting Saul Kripke, *Wittgenstein on Rules and Private Language: An Elementary Exposition* (Cambridge, Mass.: Harvard University Press, 1982), 107.

85. Kripke, *Wittgenstein on Rules and Private Language*, 95.

86. Ludwig Wittgenstein, *The Blue and Brown Books* (New York: Harper & Row, 1958), 93.

87. McDowell, "Meaning and Intentionality in Wittgenstein," 45.

88. Locke, *Essay*, III.6.35.

89. Wittgenstein, *Philosophical Investigations*, §85.

90. Wittgenstein, *The Brown Book*, p. 97.

91. McDowell, "Meaning and Intentionality in Wittgenstein," 46. See also his "Putnam on Mind and Meaning," in *Philosophical Topics* 20, no. 1 (1992): 35–48.

92. Searle, *Intentionality*, 7–22, esp., "Conditions of satisfaction are those conditions which, as determined by the Intentional content, must obtain if the state is to be satisfied. For this reason the *specification* of the content is already a *specification* of the conditions of satisfaction" (p. 13).
93. Wittgenstein, *Philosophical Investigations*, §201.
94. McDowell, "Meaning and Intentionality in Wittgenstein," 49–51.
95. See McDowell, "Putnam on Mind and Meaning," passim.
96. McGinn, *Wittgenstein on Meaning*, 68–69.
97. Ibid., 70.

Chapter 4. The Language of Thought

1. Fodor, "Mental Representation," 105–106, 118.
2. "[H]ere is the first respect in which cognitive science departs from classical versions of mental representation theory; where they had a mental picture gallery, we have a language of thought" (ibid., 123).
3. Jerry Fodor, *Psychosemantics* (Cambridge, Mass.: MIT Press, 1987), 98.
4. Fodor, *The Language of Thought*, 70.
5. Ibid., 4.
6. Ibid., 43.
7. Jerry Fodor, *A Theory of Content* (Cambridge, Mass.: MIT Press, 1990), 20.
8. Fodor, *Psychosemantics*, 13.
9. Fodor, *A Theory of Content*, 20.
10. Fodor, *Psychosemantics*, 18.
11. Ibid., 145, original emphasis.
12. Ibid., 17.
13. "Computers show us how to connect semantical with causal properties *for symbols*. . . . You connect the causal properties of a symbol with its semantic properties via its syntax. The syntax of a symbol is one of its second order physical properties" (Colin McGinn, *Mental Content* [Oxford: Basil Blackwell, 1989], 22). See also Fodor, *Psychosemantics*, 17–19.
14. Ibid., 18–20.
15. Cf. ibid., 44.
16. Here I will be paraphrasing Fodor, in particular Appendix 1 of *Psychosemantics* and sections of *The Language of Thought*.
17. Fodor, *A Theory of Content*, 167.
18. Fodor does not discuss states of affairs in this context, but I am extending it to them. I am using something like the principle that *a proposition, p = 'A is B', is true if and only if the state of affairs of 'A's being B' obtains*.
19. Cf. Fodor, *A Theory of Content*, 168.
20. "The question we're arguing about isn't . . . whether mental states have a semantics. Roughly, it's whether they have a syntax. Or, if you prefer, it's

whether they have a *combinatorial* semantics: the kind of semantics in which there are (relatively) complex expressions whose content is determined, in some regular way, by the content of their relatively simple parts" (Fodor, *Psychosemantics*, 138).

21. Cf. ibid., 151.
22. Fodor, *The Language of Thought*, 63–64.
23. See ibid., 100–103, and *A Theory of Content*, 53–57. "[I]f, for example, Chomsky is right . . . then learning a first language involves constructing grammars consonant with some innately specified system of language universals and testing those grammars against a corpus of observed utterances in some order fixed by an innate simplicity metric. And, of course, there must be a language in which the universals, the candidate grammars, and the observed utterances are represented. And, of course, this language cannot be a natural language since, by hypothesis, it is his first language that the child is learning" (*The Language of Thought*, 58).
24. Ibid., 59 n. 4.
25. See G. E. M. Anscombe, "Report on Analysis 'Problem' No. 10: 'It is impossible to be told anyone's name'," *Analysis* 17, no. 3 (1957): 49–53.
26. Fodor, *The Language of Thought*, 80. I am paraphrasing chapter 2 of *The Language of Thought*.
27. Ibid., 65.
28. Ibid., 66.
29. Ibid., 67.
30. Ibid., 97.
31. Cf. ibid., 9–26, esp. 25. See also *A Theory of Content*, 138–41.
32. Fodor, *A Theory of Content*, 58. Later, Fodor gives the following for two events being "covered" by a law: *Covering principle*: If an event $e1$ causes an event $e2$, then there are properties F, G such that:

1. $e1$ instantiates F
2. $e2$ instantiates G,

and

3. "F instantiations are sufficient for G instantiations" is a causal Law.

"When a pair of events bears this relation to a law, I'll say that the individuals are each *covered* or *subsumed* by that law and I'll say that the law *projects* the properties in virtue of which the individuals are subsumed by it. Notice that when an individual is covered by a law, it will always have some property in virtue of which the law subsumes it. If, for example, the covering law is that Fs cause Gs, then individuals that get covered by this law do so either in virtue of being Fs (in case they are subsumed by its antecedent) or in virtue of being Gs (in case they are subsumed by its consequent)" (142–43).

33. "For expository convenience, I shall usually assume that sciences are about events in at least the sense that it is the occurrence of events that makes the laws of a science true. Nothing, however, hangs on this assumption" (Fodor, *The Language of Thought*, 10 n. 8).

34. Ibid., 93.

35. Fodor, *Psychosemantics*, 99.

36. Fodor, *A Theory of Content*, 91.

37. This is an ontological dependence. Later Fodor tells us that he is not sure how to get a "normative" dependence out of this relation. That is, he does not know how to get from the ontological dependence to the thesis that 'cow' misrepresents cats. He writes that "there's no obvious reason why the fact that one way of using a symbol is asymmetrically dependent on another implies that we should prefer the second way of using it to the first. It seems, not just here but also in the general case, that ontological priority is normatively neutral, Plato to the contrary notwithstanding" (*A Theory of Content*, 128).

38. Ibid., 92–93.

39. I will not discuss here a third element in Fodor's account, namely that it is not enough to have the causal relations, but that there must be actual instances of cows causing 'cow' tokens in the language of thought. This third element is involved in his efforts to overcome Putnam's Twin Earth problems. See Fodor, *A Theory of Content*, 121–22. But it also introduces difficulties with unexemplified representations like "unicorn," see 123–24.

40. Fodor, *Psychosemantics*, 45.

41. The background to the embarrassment is the thesis that identity of intension determines identity of extension.

42. Fodor, *Psychosemantics*, 48, original emphasis.

43. Ibid., 51.

44. Ibid., 50.

45. Ibid., 51.

46. My example, but see ibid., 50–51.

Chapter 5. Hilary Putnam's Criticism of Aristotelian Accounts of Language and Mental Representationalism

1. This despite what Putnam writes in "How Old is the Mind": "If there is one value which a historical survey of what has been thought on these matters can have, it is to caution us against thinking that it is obvious even what the questions are. For part of what we have seen in this survey is that each previous period in the history of Western thought had a quite different idea of what such a term as *mind* or *soul* might stand for, and a correspondingly different idea of what the puzzles were that we should be trying to solve." This was originally published in *Exploring the Concept of Mind*, ed. Richard M. Caplan (Iowa City: University of Iowa Press,

1986). At least from time of publication it predates some of the materials I will be using to discuss his account of mental representation.

2. Putnam, *Representation and Reality*, 19.
3. Ibid.
4. Ibid.
5. Ibid.
6. Putnam, "The Meaning of 'Meaning'," 218.
7. Putnam, *Representation and Reality*, 20.
8. Ibid., 21.
9. Ibid., 6–7, 21.
10. Kretzmann, "Aristotle on Spoken Sound," 15.
11. Ibid., 5.
12. Putnam, *Representation and Reality*, 6.
13. Ibid., 19.
14. Ibid., 20.
15. See Michael Hallett, "Putnam and the Skolem Paradox," in *Reading Putnam*, ed. Peter Clark and Bob Hale (Oxford: Basil Blackwell, 1994), 66–97, for a detailed and in-depth discussion, as well as a bibliography of Putnam's use of the Lowenheim-Skolem Theorem.
16. Recently a number of classic discussions of Putnam's thought experiment have been reprinted in a volume entitled *The Twin Earth Chronicles: Twenty Years of Reflection on Hilary Putnam's "The Meaning of 'Meaning'*," ed. Andrew Pess and Sanford Goldberg (Armonk: M.E. Sharpe, 1996).
17. In fact, bats are featherless bipeds, so that pair does not have identical extension. But that just underscores the point, since we believe they have different extension because of the existence of bats, not because of their different intension. It could have turned out that the extension was identical, and there would still be the difference of intension.
18. W.V.O. Quine, "On What There Is," in *From a Logical Point of View* (Cambridge, Mass.: Harvard University Press, 1980), 11–12.
19. Ibid.
20. W.V.O. Quine, "Two Dogmas of Empiricism," in *From a Logical Point of View* (Cambridge, Mass.: Harvard University Press, 1980), 22.
21. Ibid.
22. Putnam, "The Meaning of 'Meaning'," 219.
23. Ibid., 220.
24. Ibid.
25. In "The Meaning of 'Meaning'" (p. 219), Putnam actually uses two different propositions than 2) and 3), which we have gotten from *Representation and Reality*. In the former, which is the earlier work, he uses 2') that knowing the meaning of a term is just a matter of being in a certain psychological state, and 3') that the meaning of a term determines its extension.
26. Ibid., 222.

27. McGinn, *Mental Content*, 31.

28. The example in Twin Earth is the term 'water', and the substances H_2O and XYZ, but as Putnam indicates it could be real world examples involving gold and brass, and the term 'gold'. 'Elms' and 'beeches' also function as examples. Also his "brain in a vat" can be taken to be a variant. The latter is in a way an updating of Descartes' evil demon thought experiment.

29. Locke, *Essay*, III.3–7 and IV.6.

30. Putnam points out somewhere that pure gold is white. The yellowish color in ordinary samples is due to the presence of copper.

31. In both *Representation and Reality* (e.g., p. 38) and "The Meaning of 'Meaning'" (e.g., p. 234), Putnam associates his intentions with Kripke's in *Naming and Necessity*.

32. Putnam, *Representation and Reality*, 38.

33. Putnam, "The Meaning of 'Meaning'," 238.

34. See Kripke, *Naming and Necessity*, 128–29.

35. Putnam, "The Meaning of 'Meaning'," 244.

36. Ibid., 241.

37. Ibid., 232.

38. For his attack on *causal representation*, see Hilary Putnam, "Aristotle after Wittgenstein," in *Modern Thinkers and Ancient Thinkers*, ed. Robert W. Sharples (Boulder, Colo.: Westview Press, 1993). See also his Gifford Lectures, *Renewing Philosophy* (Cambridge, Mass.: Harvard University Press, 1992), 34–59, and *Representation and Reality*, 43–56.

39. Donald Davidson, "The Individuation of Events," in *Actions and Events* (Oxford: Clarendon Press, 1980).

40. H. L. A. Hart and A. M. Honoré, *Causation in the Law* (Oxford: Clarendon Press, 1959).

41. Putnam, "Aristotle after Wittgenstein," 124.

42. Fodor, *Psychosemantics*, 47, original emphasis.

43. Ibid., original emphasis.

44. Putnam, *Representation and Reality*, 129 n. 1.

45. Ibid., 19.

46. Ibid., 39.

47. Hilary Putnam, *Reason, Truth, and History* (Cambridge: Cambridge University Press, 1981), 21, 66–67.

48. Putnam, *Representation and Reality*, 21–22.

49. Putnam, *Reason, Truth, and History*, 5.

50. Ibid., 2.

51. See McGinn's discussion of Putnam's "twin earth" cases in *Mental Content*, 30–36.

52. Ibid., 2.

53. See Searle, *Intentionality*, 200–208.

54. Putnam, "The Meaning of 'Meaning'," 220.

Chapter 6. The Third Thing Thesis

1. Putnam, "Aristotle after Wittgenstein," 125.
2. John McDowell, "Singular Thought and the Extent of Inner Space," in *Subject, Thought, and Context*, ed. P. Pettit and J. McDowell (Oxford: Clarendon Press, 1986), 137–68, emphasis added.
3. Walker Percy, *Love in the Ruins* (New York: Avon Books, 1978), 181.
4. Searle, *Intentionality*, 230. It is part of the burden of his work to say what "representation" is.
5. See Putnam, *Reason, Truth, and History*, ch. 1.
6. Donald Davidson, "A Coherence Theory of Truth and Knowledge," in *Truth and Interpretation: Perspectives on the Philosophy of Donald Davidson*, ed. Ernest LePore (Oxford: Basil Blackwell, 1986), 312, emphasis added.
7. John McDowell, *Mind and World* (Cambridge, Mass.: Harvard University Press, 1994), passim.
8. Ibid., 84.
9. Ibid., 109.
10. Ibid., 95.
11. "It has long been conventional to attack dualism—indeed to regard it as some kind of superstition. But it is little realized that the principal objections to dualism apply equally to the kind of materialism which for some is now philosophical orthodoxy" (David Braine, *The Human Person: Animal and Spirit* [Notre Dame, Ind.: University of Notre Dame Press, 1992], 23 and passim).
12. Aristotle, *Categories and Propositions*, trans. Apostle, 1a22.
13. Aristotle, *Physics*, trans. Hippocrates G. Apostle (Bloomington: Indiana University Press, 1969).
14. St. Thomas Aquinas, *In Octo Libros Physicorum Aristotelis Expositio* (Turin: Marietti, 1954) IV.4.435. From here on, referred to as *CLP*.
15. QDV 2.3, ad 1, and 8.7, ad *sed contra* 2.
16. SCG I.46.
17. SCG I.46. See also QDV 5.6, *respondeo*, and QDP 9.5.
18. QDV 2.5, ad 17, and 10.4.
19. Exceptions of course must be made when the knower and his accidents become the thing known.
20. QDV 14.8, ad 5.
21. QDV 1.4, *respondeo*.
22. QDV 3.3, ad 17.
23. Here I do not want to get into issues of *per accidens* versus *per se* existential dependence. I write "actively" since a chair might depend upon me for its existence, to the extent that it depended upon me to make it. But now it persists in its own existence, and does not depend upon mine. On the other hand, my typing at this moment actively depends upon me for its existence, since it persists in my existence as my act.

24. John Haldane, "Brentano's Problem," *Graezer Philosophische Studien* 35 (1989): 1–32, at 25. Though not always agreeing with them, in what follows I am indebted to Haldane and Searle in *Intentionality* for clearing away many obstacles in my own thought on this point.

25. See *CDA* III.7.690. See also *ST* I–II.17.5, ad 2.

26. *ST* I.85.2, ad 3.

27. *ST* I.76.3

28. Frege, "On Concept and Object," 46.

29. This is perhaps an overly felicitous example for my case, since it seems correct to think of the pitch as somehow "interposed between" the pitcher and the batter.

30. The seed of the example comes from Roderick Chisholm's criticism of one way in which Donald Davidson tries to explicate the formal structure of sentences that make reference to events, in "States of Affairs Again," *Nous* 5 (1971): 182. I am, however, making a different point.

31. Wittgenstein, *The Blue and Brown Book: The Blue Book*, 1.

32. See *ST* I.14.2, *respondeo* as well as I.56.1, *respondeo*.

33. See *SCG* I.47, 51, and 52; II.55, 59. *ST* I.14.2. *QDV* 16.1, ad 13.

34. *ST* I.14.2, *respondeo*.

35. *Compendium Theologiae*, cap. LXXXIII. *Sancti Thomae Aquinatis Doctoris Angelici Ordinis Praedicatorum: Opera Omnia secundum impressionem Petri Fiaccadori Parmae 1852–1873* (New York: Musurgia, 1950).

36. *CDA* III.8.700 (cap. II, 18–19).

37. I have added "form" here for emphasis; but this is justified, since form just is the actuality.

38. Anscombe, "A Theory of Language?" 155.

39. *Commentary on Aristotle's Metaphysics* VIII.5.1767 (hereafter *CMA*). See also *Quaestio disputata De Spiritualibus Creaturis* (Turin: Marietti, 1949),1.3, *respondeo* (hereafter *QSC*).

40. *Commentary on the Sentences of Peter Lombard*, II Sent. 3.3.1c and 35.1.1, ad 3. These texts were brought to my attention by Robert Pasnau's *Theories of Cognition in the Later Middle Ages* (Cambridge: Cambridge University Press, 1996). I am wary of using the *Sentences Commentary* for two reasons. First, there is no critical edition of it. Second, though I am equally wary of the trope that generally sets Augustine against Aristotle in St. Thomas's work, I think the earlier works do in fact show a pronounced Augustinian flavor on cognitive issues that is much more strongly mediated by Aristotelian analyses in the later work. In this particular instance I see no difficulty with using this text, since I think it is supported by the larger argument about the unity of form and matter that I make preceding it.

41. *CMA* IX.8.1865.

42. *De anima*, II.1 412a11.

43. *ST* I.79.6, ad 3.

44. Pasnau, *Theories of Cognition*, 190.//
45. "Omnis cognito est secundum aliquam formam quae est in cognoscente principium cognitionis. Forma autem huiusmodi dupliciter potest considerari: uno modo secundum esse quod habet in cognoscente, alio modo secundum respectum quem habet ad rem cuius est similitudo. Secundum quidem primum respectum facit cognoscentem actu cognoscere, sed secundum respectum secundum determinat cognitionem ad aliquod cognoscibile determinatum" (QDV 10.4c). See Pasnau's translation, *Theories of Cognition*, 105.
46. Pasnau, *Theories of Cognition*, 113, 163, 171, 176, 190, 198, 211.
47. CDA II.23.225–31 [sec. 547]. See Pasnau's translation, *Theories of Cognition*, 127.
48. Pasnau, *Theories of Cognition*, 127.
49. Ibid., 126.
50. Ibid., 190.
51. Ibid., 2.
52. To be fair, Pasnau briefly considers the possibility that the *species* is in fact one with the cognitive power, only to quickly dismiss it (*Theories of Cognition*, 191). He does this on the basis of the two texts from St. Thomas's *Sentences Commentary* that I mentioned above. Those texts indicate that the *species* forms a unity with the cognitive power in the act of understanding. He suggests that Ockham may not have a more parsimonious account because Aquinas "says on many occasions [that the intelligible species] is a form of intellect." But here Pasnau summarily dismisses the texts, preferring to continue to read Aquinas through Ockham's eyes. He claims that here St. Thomas is inconsistent since he will argue that St. Thomas treats *species* as objects of apprehension. I will consider that claim in the next chapter.
53. Pasnau, *Theories of Cognition*, 14.
54. QDL 8.2.1c. Quoted in Pasnau, *Theories of Cognition*, 127.
55. See *Theories of Cognition*, 163 and 198 for instances of Pasnau taking 'source' as evidence for the agency of the species.
56. Roy J. Deferrari, *A Latin-English Dictionary of St. Thomas* (Boston: Daughters of St. Paul, 1986), 838.
57. CMA V.I.756–762. It is important to note that 'extrinsic' in the context of my argument with Pasnau does not necessarily mean extrinsic to the being that cognizes. The agent intellect, for example, is extrinsic to the act of cognition, but it is *in anima*. Here extrinsic simply means extrinsic to what it causes. The heart, while part of the same being, the organism, is extrinsic to the brain when it causes blood to flow to it. Thus Pasnau's position is that while it is *in anima* the *species* is extrinsic to the act, while I am arguing that it is not.
58. SCG I.26.
59. SCG II.58. See also IV.81.
60. ST I.77.6, *respondeo*. See also QDP 3.9, ad 9; QSC, q. 1, a.1, ad 9.

61. *ST* 1.75.5, ad 3; *QDP* 5.1, *respondeo*.

62. *QDL* I.6, ad 2.

63. J.L. Austin, *Sense and Sensibilia*, reconstructed from the manuscript notes by G.J. Warnock (Oxford: Oxford University Press, 1964), 2.

64. "exprimunt ipsius modum, qui nomine ipsius entis non exprimitur," "aliquis specialis modus entis," and "modus generaliter consequens omne ens" (*QDV* 1.1, *respondeo*).

65. *De ente* 1.1. See also, *SCG* III.8; Aristotle's *Metaphysics* IV.7 1017a8; *CMA* V.9.

66. Compare John Searle, "an intentional state has a representative content, but it is not about or directed at its representative content. Part of the difficulty here derives from "about," which has both an extensional and an intensional-with-an-s-reading. In one sense (the intensional-with-an-s), the statement or belief that the King of France is bald is about the King of France, but in that sense it does not follow that there is some object which they are about. In another sense (the extensional) there is no object which they are about because there is no King of France. On my account it is crucial to distinguish between the *content* of a belief (i.e., a proposition) and the *objects* of a belief (i.e., ordinary objects)" (*Intentionality*, 17).

67. "... being is said absolutely and primarily of substances, thereafter and in a certain way of accidents" (*De ente* 2.1). See also, "It is true that this name 'being', insofar as it means a thing to which [participated] being is appropriate, signifies the essence of the thing, and is divided among the ten categories; nevertheless it is not said univocally since the same intelligible character is not appropriate to each; rather it said *per se* of substance, and in another way of the others" (*QDL* II.2.1, *respondeo*).

68. *CMA* VII.4.1331. Generally, on the analogous senses of *ens* see *CMA* IV.1.

69. *QDV* 1.1.

70. "So beyond *being*, which is the first conception of the intellect, *one* adds something merely of reason, namely *negation;* for it says that *being* is as it were *undivided*. But *true* and *good* are said positively; hence they only add a mere relation of reason" (*QDV* 21.1, *respondeo*).

71. *CMA* IV.2.553.

72. *De ente* 1.2.

73. *CPA* II.6.461.

74. St. Thomas Aquinas, *Expositio Super Job Ad Litteram* (Rome: Leonine Commission, 196), cap. 39, 116–19.

75. *De ente* 5.3.

76. Joseph Bobik, *Aquinas on Being and Essence: A Translation and Interpretation* (Notre Dame, Ind.: University of Notre Dame Press, 1965), 164.

77. "sciendum est enim quod hoc nomen Homo, imponitur a quiddate, sive a natura hominis; et hoc nomen Res imponitur a quiddiate tantum; hoc vero nomen Ens, imponitur ab actu essendi: et hoc nomen Unum, ab ordine vel indi-

visione. Est enim unum ens indivisum. Idem autem est quod habet essentiam et quiddiatem per illam essentiam, et quod est in se indivisum. Unde ista tria, res, ens, unum, significant omnino idem, sed secundum diversas rationes" (CMA IV.2.553).

78. "... cum negatio, quae in ratione unius includitur, sit negatio in subjecto (alias non ens, unum dici posset)" (CMA IV.3.565).

79. Ibid. Here it is clear from the context that by "of such a nature as to be the subject of an affirmation" he has in mind the sorts of affirmations made of ens_r that we make when we attribute properties to them. It is not the sort of affirmation spoken of in the *De ente* that applies as much to fictional beings as to dogs.

80. QDV 1.1.

81. For an account of how, after and in contrast to St. Thomas, *ens* as *possible or actual* began to supplant *ens* as *actual* as the subject matter of Metaphysics, see Lilli Alanen and Simo Knuutilla, "Modality in Descartes and His Predecessors," in *Modern Modalities*, ed. Simo Knuutilla (Dordrecht: Kluwer, 1988), 1–69.

82. Of course taking into account St. Thomas's example of the phoenix above, one might allow for the application of *res* in that instance, just as St. Thomas allows for an application of *essentia* or *quiddity*.

83. De ente 2.1.

84. CMA VII.4.1137.

85. Of course there is no need to write ens_r here, as e.g., "ens_r in re." "Ens_r in re" would be redundant, since *in re* specifies the sense of *ens* at play.

86. De ente 4.2.

87. This way of putting it is restricted to the context at hand, which involves questions of substantial or *per se* unity. It should not be taken to imply that any unity not found in the nature of things is consequently a being of reason. A *per accidens* unity is not a *being of reason*, as in the example I give in the text of my skin color and my human nature.

88. Dallas Willard, while not suggesting that *Psychologism* was true as an account of logic, suggests that there was a kernel of truth in what prompted it, namely the fact that we all think that logic ought to express normative relations that our acts of reasoning ought to exemplify. See Willard's "The Paradox of Logical Psychologism: Husserl's Way Out," in *Readings on Edmund Husserl's Logical Investigations*, ed. J. N. Mohanty (The Hague: Martinus Nijhoff, 1977), 45.

89. See Hilary Putnam, *The Many Faces of Realism* (LaSalle, Ill.: Open Court, 1987), Lecture I, "Is There Still Anything To Say About Reality And Truth?" 3–21.

90. Ibid., 19, original emphasis.

Chapter 7. The Introspectibility Thesis

1. ST I.84.
2. CDA III.8.717. See also, ST I.85.1, *respondeo*.

3. Putnam, *Representation and Reality*, 129 n. 1.

4. Ibid., 19.

5. *ST* I.14.4, respondeo.

6. *CDA* III.8.717. See also *ST* I.84.7.

7. *CDA* II.12.378.

8. Aquinas, *Commentary on Aristotle's De Anima*, trans. Foster and Silvester Humphries.

9. An objectual interpretation of the quantifiers "supports [the] conviction that the values of the quantifiers are existents; hence, everything that is exists, and the particular quantifier ∃, which expresses membership in the domain of discourse, is *the* essential quantifier. Merely possible and impossible objects are by definition not real and cannot be the values of quantification." Compare this with a "substitutional" interpretation of the quantifiers, in which "an existential quantification holds if there is a constant whose substitution for the variable of quantification would render the matrix true" (Hector-Neri Castaneda, "Quine's Experiment with Intensional Objects and His Existentialist Quantified Modal Logic," in *On Quine*, ed. Paolo Leonardi and Marco Santambrogio [Cambridge: Cambridge University Press, 1995], 140–63).

10. Frege, "On Concept and Object," 48. Of course this itself involves the translation of the German '*Gegenstand*' as 'object'. Fregean scholars can debate the adequacy of that translation.

11. Gottlob Frege, "On Sense and Reference," in *Translations from the Philosophical Writings of Gottlob Frege*, trans. Peter Geach and Max Black (Oxford: Basil Blackwell, 1970), 61.

12. Ibid., 57. Allowing that definite descriptions are proper names is the locus of disagreement between Frege and Russell on logically proper names and definite descriptions. See Michael Dummet's analysis of their disagreement in *Interpretation of Frege's Philosophy*, 128–39.

13. Frege, "On Concept and Object," 47–48.

14. Frege, "On Sense and Reference," 62.

15. Frege, "On Concept and Object," 51.

16. Smith, "Frege and Husserl."

17. Dummett, *Interpretation of Frege's Philosophy*, 321.

18. Quine, "Speaking of Objects," 1.

19. For example, Wittgenstein, *Tractatus*, §3.3411; Kripke, *Naming and Necessity*, esp. 57, on objects designated by descriptions, 96 on a name associated with an object by "baptism," and 107 on the object which is both Cicero and Tully; Quine, "On What There Is," esp. 18–19 on mathematical and physical objects, and "Two Dogmas of Empiricism," esp. 30 in *From a Logical Point of View*; and Putnam, "The Trail of the Human Serpent Is Over All," esp. 18–19, in *The Many Faces of Realism*, on the bearing of conceptual relativity on what is to count as an "object" in a world.

20. Wittgenstein, *The Brown Book*, 89.

21. Perhaps the most concise statement is in *ST* I.77.5, *respondeo*. The whole *respondeo* recapitulates the *De anima* discussion.

22. *CDA* III.7.675 and *ST* I.79.2.

23. *ST* I.84.2; *CMA* IX.10.1894; *CDA* II.6.308.

24. *CDA* II.13.393.

25. By making this comparison of sense to intellect, I do not mean to ignore the fact that the *obiecta* of sense powers are as they stand actually sensible, while the *obiectum* of the intellect is as it stands only potentially intelligible and has to be rendered actually intelligible in the process of abstraction involving the agent intellect. Nor do I intend to ignore the fact that the way in which "alteration" is said of the sense powers and the intellect when they "receive" the sensible or intelligible form, that "alteration" and "reception" are not said in their proper senses, but in extended senses. See *CDA* II.11.365–66.

26. "Sed quia omne esse est secundum aliquam formam, oportet quod esse sensibile sit secundum formam sensibilem et esse intelligibile secundum formam intelligibilem" (*CDA* II.5.286).

27. *ST* I.45.4, ad 1.

28. *CDA* III.8.718.

29. *CDA* II.13.393. See also III.1.575–81, on how the sensible object of one faculty may be the indirect object of another faculty, because the diverse senses bear upon the "same sensible thing." Honey, a sweet thing, may well be seen, and thus something sweet is seen, but its sweetness is not seen as such.

30. *ST* I.77.4, ad 3.

31. *ST* I.85.2, *respondeo*.

32. Cf. *CMA* I.10–17, especially 15.227.

33. One can also know the cause through the effect, which results from demonstration *quia*. But *scientia* is supposed to move from demonstration *quia* to demonstration *propter quid*.

34. *Metaphysics*, I 987b13, trans. W. D. Ross, in *The Basic Works of Aristotle*, ed. Richard McKeon (New York: Random House, 1941).

35. Ibid., 991a21–22, trans. Ross.

36. *CMA* I.15.227.

37. *ST* I.85.2

38. Compare Jacobs's and Zeis's claims about the importance of an Aristotelian-Thomistic [A/T] theory of cognition. Against "the narrow boundaries around the theory of knowledge which have been posted since the seventeenth century," they write that "we believe that the recognition of a need for an extension of these boundaries is a good sign, but so far, the contemporary discussion has not taken into consideration the unique contribution which an A/T theory of cognition can provide, and until it does, *our cognitive ability to go out of our minds will remain a mystery*" (Jonathan Jacobs and John Zeis, "Form and Cognition: How to Go Out of Your Mind," in *The Monist: Analytical Thomism* 80, no. 4 (October, 1997): 554, emphasis added).

39. *ST* I.85.2, *respondeo*.
40. Ibid.
41. *CDA* III.8.718–19.
42. *CDA* III.8.718.
43. *ST* I.87.4, *respondeo*.
44. *ST* I.84.5, *respondeo*.
45. *ST* I.14.5, *respondeo*.

46. It is interesting to note that John of St. Thomas and many of those who follow his interpretation of St. Thomas compare the concept to a mirror in which the object is seen (*Cursus Theologicus*, Editio Monachorum Solesmensium [Tornaci: Desclee et Soc., 1930] Disp.21, a.1). The concept is what John of St. Thomas calls a formal sign. John Frederick Peifer in his *The Concept in Thomism* (New York: Bookman, 1952) provides a classic exposition of this line of interpretation. See esp. p. 192 for John of St. Thomas on the mirror image.

47. Cf. *ST* I.77.1 on whether the essence of the soul is its power; and I.79.2 on whether the intellect is a passive power.

48. *ST* I.85.1, ad 3. This passage refers to the likeness as a representation. I translate 'phantasm' by the neutral 'sense experience' for fear of reintroducing the problems at the level of sensation. For an interpretation of Aristotle that denies he is committed to treating *phantasmata* as mental images, see Martha Nussbaum's interpretive essay in her translation of Aristotle's *De motu animalium* (Princeton, N.J.: Princeton University Press, 1978), 221–69.

49. *QDV* 10.6, ad 7.
50. See also *QDL* VIII.2.1, *respondeo*. For primary and secondary causality, see *QDP* III, esp. a. 7; *ST* I.45 especially articles 2 and 5, I.105.5: *SCG* III.67.
51. *QDV* 10.6, ad 7.
52. *ST* I.84.8, 85.1, ad 3 and 4.
53. Peter Geach, *Mental Acts* (London: Routledge & Kegan Paul, 1957), 18. Cf. chs. 6–11.
54. Ibid., see Geach's Appendix on Aquinas and Abstractionism, 130–31.
55. *ST* I.85.3, *respondeo*.
56. *CLP* I.1.11.
57. *ST* I.84.7, *respondeo*.
58. Ibid.
59. *ST* I.85.7, ad 2.
60. *ST* I.84.6, ad 3; *QDV* 10.6, ad 5.
61. *ST* I.84.6, ad 2.

62. Even if, as a matter of empirical fact this particular claim may not be true of *every* child, the point holds. Anyone who has been around children learning to speak is familiar with the phenomenon of *interesting* mistakes. By an interesting mistake I mean a use of a word that is not quite right and yet is not simply wrong. My son never called anyone but me 'papa'. For a long time, however, he did refer to the lids of the pots he played with on the kitchen floor as 'doors'. That is an interesting mistake.

63. Cf. *ST* I.1.1, *respondeo*.
64. *ST* I.85.2, *respondeo*.
65. *ST* I.84.3, *respondeo*.
66. *CDA* III.9.725. See this lesson throughout on how the possible intellect is intelligible, as well as *ST* I.87.
67. Cf. Haldane, "The Life of Signs."
68. "Et ideo id quod primo cogniscitur ab intellectu humano est huiusmodi objectum; et secundario cognoscitur objectum; et per actum cognoscitur ipse intellectus, cuius est perfectio ipsum intelligere. Et ideo Philosophus dicit quod objecta praecognoscuntur actibus, et actus potentiis" (*ST* I.87. 3, *respondeo*). I have translated "the nature of material things" for "objectum," because the passage immediately before the one quoted makes it clear that that is what 'objectum' refers to.
69. Ibid., I.87.4, ad 2.
70. *CDA* II.6.308.
71. See in particular *CDA* II.24.
72. It is necessary to avoid reading 'spiritual' in a misleading fashion, as if 'spiritual' meant 'ghostlike' or 'wispy' or 'airy'. St. Thomas often uses 'intentional' and 'immaterial' as synonyms for this 'spiritual reception'.
73. *ST* I.78.3. See also *CDA* II.14.418, as well as I.4.43, 2.179; II.12.377, 24.552–53; III.13.789.
74. See *ST* I.84.1.
75. For the sense in which the actuality of a natural power is and is not an accident of the thing, see *ST* I.77.1, ad 5.
76. *ST* I.77.1, *respondeo*.
77. *ST* I.75.1, ad 2.
78. Of course for Aristotle and St. Thomas, with the exception of intellectual cognition, this generally goes for all the animals in different ways appropriate to their natures.
79. Pasnau, *Theories of Cognition*, 195–97.
80. Ibid., 209, emphasis added.
81. Ibid., 216–19.
82. McDowell, "Putnam on Mind and Meaning," 44.

Chapter 8. The Internalist Thesis and St. Thomas's "Externalism"

1. *CPH* I.2.21.
2. See for just one example, *ST* I.76.2, ad 1.
3. McGinn, *Mental Content*, 3.
4. Ibid., 2. The terminology is in fact fairly standard in contemporary philosophy, but McGinn paraphrases the sense of the terms particularly well for this discussion.
5. This is said of *first intentions;* the story of *second intentions* is of course different, but sufficiently familiar to support the claim that they too have their

remote foundation in *res extra animam*. See McInerny, *The Logic of Analogy*, 41–48, esp. 63. See also Schmidt, *The Domain of Logic*, 117–21.

6. Putnam, "Aristotle after Wittgenstein," 126.
7. *CDA* II.12.379.
8. *ST* I.87.1, *respondeo*.
9. Ibid.
10. Ibid.
11. Ibid., ad 1.
12. For recent discussions that I believe disagree with me on this point, see Haldane, especially "Mind-World Identity Theory and the Anti-Realist Challenge," in *Reality, Representation, and Projection*, ed. J. Haldane and C. Wright (New York: Oxford University Press, 1993), 15–37; "Brentano's Problem"; "Putnam on Intentionality," *Philosophy and Phenomenological Research* 52 (1992): 671–82; and "Forms of Thought." See also Jacobs and Zeis, "Form and Cognition: How to Go Out of Your Mind."
13. *QDV* 1.3, *respondeo*.
14. Cf. *ST* I.85.5, ad 3.
15. The discussion is much more complicated than this, in particular in instances when the negation is true, or when non-existent subjects are involved, and so on.
16. See *ST* I.85.6, *respondeo*. See *ST* I.89.1 for an argument from St. Thomas to the effect that it is befitting the perfection of the universe, as well as the clarity of human understanding, that man be joined to a body, and receive his knowledge from sensible things.
17. I am being deliberately vague here, in order to avoid getting entangled in a metaphysical discussion of divine practical knowledge. It's just an example. See *QDV* 3–4, and *ST* I.14–16.
18. Cf. *QDV* 2.1, ad 6, *ST* I.14.8, ad 3 and I.45.4.
19. For a discussion of *passio* as an analogous name, see Ralph McInerny, *Studies in Analogy* (The Hague: Martinus Nijhoff, 1968), 30–33. The primary text that McInerny has in mind is our text from the *Peri hermeneias* "passiones animae" and St. Thomas's commentary.
20. What I am interested in here is the passive "moment" in the active receptivity of intellect. See *CDA* II–III, and *ST* I.75–89.
21. "It is difficult in this manner, however, to acquire [a whatness of] an object if we do not know that the object exists; and the cause of the difficulty, as stated before, is the fact that we may not know whether the object exists or not, except by way of an accident" (93b28–35, *Aristotle's Posterior Analytics*, trans. Hippocrates G. Apostle [Grinnell, Iowa: Peripatetic Press, 1981]).
22. "A single science is one whose domain is a single genus, viz. all the subjects constituted out of the primary entities of the genus—i.e. the parts of this total subject—and their essential properties" (*Posterior Analytics*, 87a38–b4).
23. *CPH* I.2.20.

24. See Ralph McInerny's response to Ackrill in *The Logic of Analogy*.
25. *CPH* I.2.21.
26. The exception being, as St. Thomas notes, when something is known through its effects.
27. Geach, *Mental Acts*, 12–14. Harm J. M. J. Goris's work *Free Creatures of an Eternal God* (Nijmegen: Stichting Thomasfonds, 1997) points out that St. Thomas does not always call the first act an act of simple apprehension. This work is useful for detailing the differences of terminology that characterize St. Thomas's work. See especially chapter 5.
28. Geach, *Mental Acts*, 12–14, passim, for all the passages.
29. "Constituit enim qui dicit intellectum, et qui audit quiescit," in *Textus Aristotelis*, 16b20–21, *CPH*.
30. "Currit." From the context, St. Thomas seems to have in mind an instance in which I say 'he runs' where the context does not make it clear who is referred to by 'he'. Of course in Latin this is more pronounced by the grammatical absence of the pronoun in the sentence.
31. *CPH* I.5.68.
32. Ibid.
33. "The ability to express a judgment in words thus presupposes a number of capacities, previously acquired, for intelligently using the several words and phrases that make up the sentence" (Geach, *Mental Acts*, 12). I am indebted to Ralph McInerny for helping me to understand this point.
34. Plato already knew this. See *Sophist*, 262c–263b, in *The Collected Dialogues of Plato*, ed. Edith Hamilton and Huntington Cairns (Princeton, N.J.: Princeton University Press, 1961).
35. Dummett, *Frege: Philosophy of Language*, 3–4.
36. McGinn, *Mental Content*, 7.
37. Ibid.
38. See Putnam, "The Meaning of 'Meaning'," 227–29, as well as *Representation and Reality*, 22–26.
39. Of course this applies to the recognition of identical accidental features in different subjects as well.
40. Putnam, "Aristotle after Wittgenstein," 124.
41. Ibid.
42. Cf. ibid., 78.
43. He does recognize that the criterion of set membership used by any particular science may not provide sharp boundaries for extraordinary cases. Cf. ibid., 77. But that lack of sharp boundaries is itself caught up with what the sciences consider essential. See discussion below.
44. Ibid., 124–29. My emphasis.
45. Ibid., 77.
46. Ibid., 74.
47. Putnam, *Mind, Language, and Reality*, 14.

48. *ST* I.76.3.
49. *ST* I.76.3, ad 4.
50. *ST* I.76.3, *respondeo*.
51. *CMA* VIII.3.1725.
52. It is not necessary to take 'already' here as a temporal adverb.
53. One way in which this identity of Socrates and his nature shows up is in St. Thomas's account of what proper names signify, *CPH* I.10.124 and 132.
54. St. Thomas Aquinas, In Symbolum Apostolorum Expositio in *Opuscula Theologica*, ed. R. M. Spiazzi (Turin: Marietti, 1954), par. 864.
55. For an argument why in this life we can never have a final comprehensive grasp of the essence of anything, see Joseph Pieper's *The Silence of St. Thomas*. Pieper's argument is that we can never have a final comprehensive understanding of the essence of any creature, because we cannot have a final comprehensive grasp of the cause of its being, which is God.
56. *ST* I.85.5, *respondeo*.
57. Ibid.
58. *De ente* 4.2.
59. Josef Pieper, *Guide to Thomas Aquinas*, trans. Richard and Clara Winston (New York: Pantheon, 1962), 160.
60. The Dewey Lectures appear as "Sense, Nonsense, and the Senses: An Inquiry into the Power of the Human Mind," *Journal of Philosophy* 91, no. 9 (September 1994): 445–517.

Chapter 9. Conclusion

1. *CPH* I.2.19.
2. Putnam, *Representation and Reality*, 21–22.
3. Putnam, *Reason, Truth, and History*, 5.
4. McDowell, "Putnam on Mind and Meaning," 43.
5. Ibid., 42.
6. See, for example, Larkin, *Language in the Philosophy of Aristotle*, 96, and John Poinsot, *Tractatus De Signis* (Berkeley: University of California Press, 1985), 246.
7. Haldane and Smart, *Atheism and Theism*, 114.
8. McDowell, "Putnam on Mind and Meaning," 40.
9. Ibid., 45.
10. See ibid., 85 and 95. See McDowell's comparisons throughout the book of the "experiences" of animals and the experiences properly so called of those endowed with reason.
11. "I found among . . . scholars that what lay behind the arrogant contempt for philosophy was the bad aftereffect of—a philosopher to whom they now denied allegiance on the whole without, however, having broken the spell of

his cutting evaluation of other philosophers—with the result of an over-all irritation with all philosophy" (Friedrich Nietzsche, "We Scholars," in *Beyond Good and Evil*, trans. Walter Kaufmann [New York: Vintage Books, 1966], 122).

12. McDowell, *Mind and World*, 71.

13. For a very good reading of the complexity of medieval approaches to the "meaningfulness" of the natural world, symbolic or otherwise, see Umberto Eco, *Art and Beauty in the Middle Ages* (New Haven, Conn.: Yale University Press, 1986).

14. McDowell, *Mind and World*, 71.

15. Ibid., 123.

16. Cf. *QDV* 2.2, *respondeo*.

17. *CPH* I.2.12, trans. Oesterle.

18. Ibid.

19. I have no intention of making an empirical claim about the extent to which the eating habits of other species of animals serve social functions other than nourishing the individual and biologically preserving the species. What I have in mind are things like baking cakes to commemorate birthdays, or elaborate feasts celebrating the public proclamation of a permanent sexual union, or closer to home, the obligatory drinks and dinner following a talk by an invited speaker in an academic forum.

20. McGinn, *Wittgenstein on Meaning*, 70.

21. Ibid., 88 n. 34.

22. Cf. Anthony Kenny, *Aquinas on Mind* (London: Routledge, 1993), 15, 18.

23. Haldane and Smart, *Atheism and Theism*, 114. By abstraction here, Haldane has in mind the "selective attention" described and criticized by Geach in *Mental Acts*.

24. This is not to suggest that Haldane would agree with everything McDowell has to say about the space of reasons, and second nature. But I find the parallel useful. And as I will presently show, in this area at least, Haldane agrees with almost everything McDowell has to say about the "space of reasons" and the language user's introduction into it.

25. John Haldane, "Rational and Other Animals," in *Verstehen and Humane Understanding*, ed. Anthony O'Hear (Cambridge: Cambridge University Press, 1996), 24–26.

26. My claim that Haldane presupposes McDowell's analysis arises from his criticism of McDowell in "Rational and Other Animals." There as a solution to the problem raised by McDowell, he refers the reader to his treatment in "The Mystery of Emergence," *Proceedings of the Aristotelian Society* 96 (1996), and the discussion in Haldane and Smart, *Atheism and Theism*. Haldane considers a number of issues I raise here in the second edition of J.J.C. Smart and J.J. Haldane, *Atheism and Theism*, 2nd. ed. (Oxford: Blackwell, 2002), ch. 6.

27. Haldane and Smart, *Atheism and Theism*, 114.

28. Ibid.

29. Ibid., 115.

30. Ibid.

31. Genesis 1:26; 2:19, as quoted in Haldane and Smart, *Atheism and Theism*, 116.

32. Were this discussion to go beyond the biblical image, it would be necessary to examine St. Thomas's complicated account of secondary causality in created agents, which does not exclude but rather presupposes God's primary and concurrent causality. However that primary and concurrent causality is not properly understood as an "external" agency with respect to the secondary causality of created agents, as Haldane's analysis suggests. See *ST* I.25.1–6, I.95.1–8, and *QDP* 3.7. For an excellent discussion of St. Thomas's account of divine concurrence in the secondary causality of created agents see Alfred Freddoso, "Medieval Aristotelianism and the Case against Secondary Causation in Nature," in *Divine and Human Action: Essays in the Metaphysics of Theism*, ed. Thomas V. Morris (Ithaca, N.Y.: Cornell University Press, 1988), 74–118, and "God's General Concurrence with Secondary Causes: Pitfalls and Prospects," *American Catholic Philosophical Quarterly* 67 (1994): 131–56.

33. *Averrois Cordubensis Commentarium Magnum in Aristotelis De anima Libros*, ed. F. S. Crawford (Cambridge, Mass.: Medieval Academy of America, 1953). For Aquinas's engagement of this theme from Ibn Rushd, see Ralph McInerny, *Aquinas against the Averroists* (West Lafayette, Ind.: Purdue University Press, 1993). McInerny provides both the Latin text of Aquinas and a translation as well as a very useful commentary for understanding the issues involved.

34. In *CDA* II.18.473.

35. *Politics* 1252b28–29, trans. Benjamin Jowett, in *The Basic Works of Aristotle*, ed. Richard McKeon (New York: Random House, 1941). Cf. Thomas Aquinas, *Expositionem in VIII Lib. Politicorum* (Turin: Marietti, 1951), I.1.

36. Wittgenstein, *Investigations*, 178. iv. I write these comments in full awareness of McGinn's own fears about dualistic tendencies in Wittgenstein's attitudes toward "psycho-physiological correspondence." See *Wittgenstein on Meaning*, 91–117.

37. *ST* I.76.3, *respondeo*.

38. *ST* I.76.4 ad, *respondeo*.

39. Alasdair MacIntyre, *Whose Justice? Which Rationality?* (Notre Dame, Ind.: University of Notre Dame Press, 1988), 367.

40. *Expositionem in VIII Lib. Politicorum*, I.1.

41. *QDV* 2.2, *respondeo*.

42. *ST* I.79.4, *respondeo*.

43. Plato, *The Republic*, trans. G. M. A. Grube (Indianapolis: Hackett, 1992), VII518d, p. 190.

44. Cf. *ST* I.117.1, *respondeo*, and *QDV* 11, passim.

45. "[S]imilarly, the intellectual soul is at times called by the name 'intellect', as from its highest power, as it is said in I *de Anima*, that *intellect is a substance*.

And also in this way Augustine says that mind is spirit or essence" (*ST* I.79.1, ad 1). "[W]here several [powers] exist in the soul, each is a part; but the soul itself is named after the principal part, whether sensitive or intellectual" (*CDA* II.4.270). See also *Metaphysics* VIII, chs. 2–3, and Aquinas's commentary VIII.12–3 on definition.

    46. Haldane, "Rational and Other Animals," 26.
    47. Ibid., 26.
    48. *ST* I–II.1.1.
    49. Andre Dubus, "On Charon's Wharf," from *Broken Vessels* (Boston: David R. Godine, 1991), 82.

BIBLIOGRAPHY

Alanen, Lilli, and Simo Knuutilla. "Modality in Descartes and His Predecessors." In *Modern Modalities*, edited by Simo Knuutilla, 1–69. Dordrecht: Kluwer, 1988.

Alston, William. *Philosophy of Language*. Englewood Cliffs, N.J.: Princeton University Press, 1964.

Ammonius. *Ammonius In Aristotelis De Interpretatione Commentarius*. Edited by Adolfus Busse. Commentaria in Aristotelem Graeca. Berlin, 1895.

———. *Ammonius: On Aristotle's On Interpretation 1–8*. Translated by David Blanck. Ithaca, N.Y.: Cornell University Press, 1996.

Anderson, Robert D. "Medieval Speculative Grammar: A Study of the Modistae." Dissertation, University of Notre Dame, 1989.

Anscombe, G. E. M. *An Introduction to Wittgenstein's Tractatus*. Philadelphia: University of Pennsylvania Press, 1971.

———. "Report on Analysis 'Problem' No. 10: 'It is impossible to be told anyone's name'." *Analysis* 17, no. 3 (1957).

———. "A Theory of Language?" In *Perspectives on the Philosophy of Wittgenstein*, edited by Irving Block. Cambridge, Mass.: MIT Press, 1981.

Aquinas, St. Thomas. *Commentary on Aristotle's De anima*. Translated by Kenelm Foster, O. P. and Silvester Humphries, O. P. New Haven, Conn.: Yale University Press, 1951.

———. *Commentary on Aristotle's Metaphysics*. Translated by John P. Rowan. Chicago: Regnery 1961.

———. *Compendium Theologiae*. In *Sancti Thomae Aquinatis Doctoris Angelici Ordinis Praedicatorum: Opera Omnia secundum impressionem Petri Fiaccadori Parmae 1852–1873*. New York: Musurgia, 1950.

———. *Expositio Libri Peryermenias*. Rome: Leonine Commission, 1989.

———. *Expositio super Job ad litteram*. Rome: Leonine Commission, 1965.

———. *Expositionem in VIII Lib. Politicorum*. Turin: Marietti, 1951.

———. *In Libros Posteriorum Analyticorum Expositio*. Turin: Marietti, 1955.

———. *In Metaphysicam Aristotelis Commentaria*. Turin: Marietti, 1926.

———. *In Octo Libros Physicorum Aristotelis Expositio.* Turin: Marietti, 1954.
———. *In Symbolum Apostolorum Expositio.* In *Opuscula Theologica,* edited by R. M. Spiazzi. Turin: Marietti, 1954.
———. *Opusculum De Ente et Essentia.* Turin: Marietti, 1957.
———. *Quaestio disputata De Spiritualibus Creaturis.* Turin: Marietti, 1949.
———. *Quaestiones disputatae De malo, Aquinas.* Rome: Leonine Commission, 1982.
———. *Quaestiones disputatae De potentia.* Turin: Marietti, 1949.
———. *Quaestiones disputatae De veritate.* Turin: Marietti, 1948.
———. *Quaestiones Quodlibetales.* Turin: Marietti, 1948.
———. *Sentencia Libri De anima.* Rome: Leonine Commission, 1984.
———. *Summa Theologiae.* Ottawa: Garden City Press, 1941.
———. *Summa Theologiae.* Turin: Marietti, 1948.
Aristotle. *Aristotle's Categories and De Interpretatione.* Translated by J. L. Ackrill. Oxford: Clarendon Press, 1963.
———. *Aristotle's Categories and Propositions.* Translated by Hippocrates G. Apostle. Grinnell, Iowa: Peripatetic Press, 1980.
———. *Aristotle: De Anima.* Translated by D. W. Hamlyn. Oxford: Clarendon Press, 1993.
———. *Aristotle's Posterior Analytics.* Translated by Hippocrates G. Apostle. Grinnell, Iowa: Peripatetic Press, 1981.
———. *Aristotle: On Interpretation. Commentary by St. Thomas and Cajetan.* Translated by Jean T. Oesterle. Milwaukee: Marquette University Press, 1962.
———. *Aristotelis Categoriae et Liber De Interpretatione.* Edited by L. Minio-Paluello. Oxford: Clarendon Press, 1949.
———. *The Basic Works of Aristotle.* Edited by Richard McKeon. New York: Random House, 1941.
———. *De anima.* Greek text, Loeb Classical Editions. Vol. 8. Cambridge, Mass.: Harvard University Press, 1986.
———. *De Interpretatione.* Translated by Moerbeke. *Aristotelis Latinus.* Vol. 2.2. Leiden: E. J. Brill, 1965.
———. *De Interpretatione.* Translated by Boethius. *Aristotelis Latinus.* Vol. 2.1. Leiden: E. J. Brill, 1965.
———. *Metaphysics.* Translated by Hippocrates G. Apostle. Bloomington: Indiana University Press, 1966.
———. *Physics.* Translated by Hippocrates G. Apostle. Bloomington: Indiana University Press, 1969.
———. *The Works of Aristotle.* Edited by Robert Maynard Hutchins. Great Books of the Western World. Chicago: Encyclopedia Britannica, 1952.
Arnauld, Antoine. *On True and False Ideas.* Translated by Stephen Gaukroger. Manchester: Manchester University Press, 1990.
Ashworth, E. J. "Do Words Signify Ideas or Things? The Scholastic Sources of Locke's Theory of Language." *Journal of the History of Philosophy* 19 (July 1981): 299–326.

———. "Locke on Language." *Canadian Journal of Philosophy* 14, no. 1 (March, 1984): 45–74.

———. "Signification and Modes of Signifying in Thirteenth-Century Logic: A Preface to Aquinas on Analogy." *Medieval Philosophy and Theology* 1 (1991): 39–67.

———. *The Tradition of Medieval Logic and Speculative Grammar from Anselm to the End of the Seventeenth Century: A Bibliography from 1836 Onwards.* Toronto: Pontifical Institute of Medieval Studies, 1978.

Atherton, Margaret. "The Inessentiality of Lockean Essences." *Canadian Journal of Philosophy* 14 (1984): 277–94.

Augustine. *The Teacher. The Fathers of the Church*, vol. 59. Washington, D.C.: Catholic University of America Press, 1968.

Austin, J.L. *Sense and Sensibilia.* Reconstructed from the manuscript notes by G.J. Warnock. Oxford: Oxford University Press, 1964.

Ayers, Michael. "Locke's Logical Atomism." In *Rationalism, Empiricism, & Idealism*, edited by Anthony Kenny, 6–22. Oxford: Clarendon Press, 1986.

Barnes, Jonathan. "Proof and the Syllogism." In *Aristotle on Science: The "Posterior Analytics."* "Proceedings of the Eighth Symposium Aristotelicum, 1978. Padova: Editrice Antenore, 1981.

Berkeley, George. *Principles of Human Knowledge.* London: Penguin Books, 1988.

Bobik, Joseph. *Aquinas on Being and Essence: A Translation and Interpretation.* Notre Dame, Ind.: University of Notre Dame Press, 1965.

Boethius. *Commentaries on Aristotle's De Interpretatione.* Vols. 1–2. New York: Garland, 1987.

Bolton, Martha Brandt. "The Idea-Theoretic Basis of Locke's Anti-Essentialist Doctrine." In *Minds, Ideas, and Objects: Essays on the Theory of Representation in Modern Philosophy*, edited by Phillip D. Cummins and Guenter Zoeller, 85–96. North American Kant Society Studies in Philosophy. Atascadero, Calif.: Ridgeview, 1992.

Braine, David. *The Human Person: Animal and Spirit.* Notre Dame, Ind.: University of Notre Dame Press, 1992.

Broadie, Alexander. *Introduction to Medieval Logic.* Oxford: Clarendon Press, 1993.

———. "Medieval Notions and the Theory of Ideas." *Proceedings of the Aristotelian Society*, vol. 87, 153–67. London: Aristotelian Society, 1987.

Brykman, Genevieve. "Locke on Private Language." In *Minds, Ideas, and Objects: Essays on the Theory of Representation in Modern Philosophy*, edited by Phillip D. Cummins and Guenter Zoeller, 125–34. North American Kant Society Studies in Philosophy. Atascadero, Calif.: Ridgeview, 1992.

Burnyeat, Miles F. "Wittgenstein and Augustine's *De Magistro.*" *Proceedings of the Aristotelian Society: Supplementary Volume* 61 (1987): 1–24.

Caplan, Richard M., ed. *Exploring the Concept of Mind.* Iowa City: University of Iowa Press, 1986.

Castaneda, Hector-Neri. "Quine's Experiment with Intensional Objects and His Existentialist Quantified Modal Logic." In *On Quine*, edited by Paolo

Leonardi and Marco Santambrogio, 140–63. Cambridge: Cambridge University Press, 1995.

Chisholm, Roderick. "States of Affairs Again." *Nous* 5 (1971).

Cobb-Stevens, Richard. *Husserl and Analytic Philosophy.* Dordrecht: Kluwer, 1990.

Cohen, Marc S. "The Credibility of Aristotle's Philosophy of Mind." In *Aristotle Today: Essays on Aristotle's Ideal of Science,* edited by Mohan Matthen, 103–25. Edmonton: Academic, 1987.

Coleman, Janet "MacIntyre and Aquinas." In *After MacIntyre: Critical Perspectives on the Work of Alasdair MacIntyre,* edited by John Horton and Susan Mendus. Notre Dame, Ind.: University of Notre Dame Press, 1994.

Colish, Marsha. *The Mirror of Language: A Study in the Medieval Theory of Knowledge.* Lincoln, Neb.: University of Nebraska Press, 1983.

Dalcourt, Gerald J. "Poinsot and the Mental Imagery Debate." *The Modern Schoolman* 72 (1994): 1–12.

Davidson, Donald. "A Coherence Theory of Truth and Knowledge." In *Truth and Interpretation: Perspectives on the Philosophy of Donald Davidson,* edited by Ernest Lepore. Oxford: Basil Blackwell, 1986.

———. "The Individuation of Events." In *Actions and Events.* Oxford: Clarendon Press, 1980.

Deferrari, Roy J. *A Latin-English Dictionary of St. Thomas.* Boston: Daughters of St. Paul, 1986.

Dennett, Daniel. *Brainstorms.* Montgomery, Vt.: Bradford Books, 1978.

Descartes, Rene. *Meditations of First Philosophy.* Translated by John Cottingham and Robert Stoothoff. New York: Cambridge University Press, 1988.

———. *Ouvres de Descartes.* Edited by Charles Adam and Paul Tannery. Vol. 7. Paris: J. Vrin, 1964.

Dubus, Andre. "On Charon's Wharf." From *Broken Vessels.* Boston: David R. Godine, 1991.

Dummett, Michael. "Frege, Gottlob." In *The Encyclopedia of Philosophy,* edited by Paul Edwards. New York: Macmillan, 1967.

———. *Frege: Philosophy of Language.* 2d ed. Cambridge, Mass.: Harvard University Press, 1981.

———. *The Interpretation of Frege's Philosophy.* Cambridge, Mass.: Harvard University Press, 1981.

———. *Origins of Analytical Philosophy.* Cambridge, Mass.: Harvard University Press, 1994.

Eco, Umberto. *Art and Beauty in the Middle Ages.* New Haven, Conn.: Yale University Press, 1986.

Fodor, Jerry. *The Language of Thought.* Cambridge, Mass.: Harvard University Press, 1975.

———. "Mental Representation: An Introduction." In *Scientific Inquiry in Philosophical Perspective,* edited by N. Rescher. Washington, D. C.: University Press of America, 1987.

———. *Psychological Explanation*. New York: Random House, 1968.
———. *Psychosemantics*. Cambridge, Mass.: MIT Press, 1987.
———. "Something of the State of the Art." Introduction to *Representations: Philosophical Essays on the Foundations of Cognitive Science*, edited by Jerry Fodor. Cambridge, Mass.: MIT Press, 1981.
———. *A Theory of Content*. Cambridge, Mass.: MIT Press, 1990.
Freddoso, Alfred. "God's General Concurrence with Secondary Causes: Pitfalls and Prospects." *American Catholic Philosophical Quarterly* 67 (1994): 131–56.
———. "Medieval Aristotelianism and the Case against Secondary Causation in Nature." In *Divine and Human Action: Essays in the Metaphysics of Theism*, edited by Thomas V. Morris, 74–118. Ithaca, N.Y.: Cornell University Press, 1988.
Frege, Gottlob. *The Basic Laws of Arithmetic*. Translated by Montgomery Furth. Berkeley: University of California Press, 1964.
———. "On Concept and Object." In *Translations from the Philosophical Writings of Gottlob Frege*, translated by Peter Geach and Max Black. Oxford: Basil Blackwell, 1970.
———. *Foundations of Arithmetic*. Translated by J.L. Austin. Evanston, Ill.: Northwestern University Press, 1980.
———. "On Sense and Reference." In *Translations from the Philosophical Writings of Gottlob Frege*, translated by Peter Geach and Max Black. Oxford: Basil Blackwell, 1970.
———. "The Thought: A Logical Inquiry." Translated by A. M. and Marcell Quinton. *Mind* 65, no. 259 (July 1956).
Gallie, Roger D. *Thomas Reid and "The Way of Ideas."* Dordrecht: Kluwer, 1989.
Geach, Peter. *Mental Acts*. London: Routledge & Kegan Paul, 1957.
Geach, Peter, and Max Black, eds. *Translations from the Philosophical Writings of Gottlob Frege*. Oxford: Basil Blackwell, 1970.
Gilson, Etienne. *Thomist Realism and the Critique of Knowledge*. Translated by Mark A. Wauck. San Francisco: Ignatius Press, 1986.
Goris, Harm J. M. J. *Free Creatures of an Eternal God*. Nijmegen: Stichting Thomasfonds, 1997.
Grene, Marjorie. *A Portrait of Aristotle*. Chicago: University of Chicago Press, 1963.
Hacking, Ian. *Why Does Language Matter to Philosophy?* Cambridge: Cambridge University Press, 1975.
Haldane, John. "Brentano's Problem." *Graezer Philosophische Studien* 35 (1989).
———. "Forms of Thought." In *The Philosophy of Roderick Chisholm*, edited by Lewis Edwin Hahn. Chicago: Open Court, 1997.
———. "The Life of Signs." *Review of Metaphysics* 47 (March 1994).
———. "Mind-World Identity Theory and the Anti-Realist Challenge." In *Reality, Representation, and Projection*, edited by J. Haldane and C. Wright, 15–37. New York: Oxford University Press, 1993.
———. "The Mystery of Emergence." *Proceedings of the Aristotelian Society* 96 (1996).

———. "Putnam on Intentionality." *Philosophy and Phenomenological Research* 52 (1992): 671–82.

———. "Rational and Other Animals." In *Verstehen and Humane Understanding*, edited by Anthony O'Hear. Cambridge: Cambridge University Press, 1996.

———. "Reid, Scholasticism, and Current Philosophy of Mind." In *The Philosophy of Thomas Reid*, edited by M. Dalgarno and E. Matthews. Dordrecht: Kluwer, 1989.

Haldane, J.J., and J.J.C. Smart. *Atheism and Theism*. Oxford: Blackwell, 1996.

Hallett, Michael. "Putnam and the Skolem Paradox." In *Reading Putnam*, edited by Peter Clark and Bob Hale, 66–97. Oxford: Basil Blackwell, 1994.

Hart, H.L.A., and A.M. Honoré. *Causation in the Law*. Oxford: Clarendon Press, 1959.

Hartnack, Justus. *Wittgenstein and Modern Philosophy*. 2d ed. Notre Dame, Ind.: University of Notre Dame Press, 1986.

Hume, David. *A Treatise of Human Nature*. Oxford: Clarendon Press, 1967.

Husserl, Edmund. *Logical Investigations*. Translated by J.N. Findlay. Vol. 1. New York: Humanities Press, 1970.

Jacobs, Jonathan, and John Zeis. "Form and Cognition: How to Go Out of Your Mind." *The Monist: Analytical Thomism* 80, no. 4 (October, 1997).

Jaeger, Werner. *Aristotle*. 2d ed. Oxford: Clarendon Press, 1948.

Kenny, Anthony. *Aquinas on Mind*. London: Routledge, 1993.

Klima, Gyula. "On Being and Essence in St. Thomas Aquinas's Metaphysics and Philosophy of Science." In *Knowledge and the Sciences in Medieval Philosophy: Proceedings of the Eighth International Congress of Medieval Philosophy*, edited by Simo Knuuttila, Reijo Tyorinoja, and Sten Ebbesen, 210–21. Helsinki: Luther Agricola Society, 1990.

———. "The Semantic Principles Underlying Saint Thomas Aquinas's Metaphysics of Being." *Medieval Philosophy and Theology* 5 (March 1996).

———. "Socrates Est Species." In *Argumentationstheorie: Scholastische Forschungen zu den logischen und semantischen Regeln korrekten Folgerns*, edited by Klaus Jacobi, 491–506. Leiden: E.J. Brill, 1993.

Kretzmann, Norman. "Aristotle on Spoken Sound Significant by Convention." In *Ancient Logic and Its Modern Interpretations*, edited by J. Corcoran. Dordrecht-Holland: D. Reidel, 1974.

———. "Semantics, History of." In *The Encyclopedia of Philosophy*, edited by Paul Edwards. New York: Macmillan, 1967.

Kretzmann, Norman, Anthony Kenny, and Jan Pinborg, eds. *The Cambridge History of Later Medieval Philosophy*. Cambridge: Cambridge University Press, 1982.

Kripke, Saul. *Naming and Necessity*. Cambridge, Mass.: Harvard University Press, 1980.

———. *Wittgenstein on Rules and Private Language: An Elementary Exposition*. Cambridge, Mass.: Harvard University Press, 1982.

Larkin, M.T. *Language in the Philosophy of Aristotle*. The Hague: Janua Linguarum, 1971.

Lewis, C. S. *The Discarded Image*. Cambridge: Cambridge University Press, 1967.
Locke, John. *An Essay Concerning Human Understanding*. Oxford: Clarendon Press, 1975.
Lonergan, Bernard, S.J. *Verbum: Word and Idea in Aquinas*. Edited by David Burrell, C.S.C. Notre Dame, Ind.: University of Notre Dame Press, 1967.
MacIntyre, Alasdair. *Whose Justice? Which Rationality?* Notre Dame, Ind.: University of Notre Dame Press, 1988.
Mackie, J.L. "Representational Theories of Perception." In *Problems from Locke*, 37–71. Oxford: Clarendon Press, 1976.
Magee, John. *Boethius on Signification and Mind*. Leiden: E.J. Brill, 1989.
Martin, C.J.F. *Introduction to Medieval Philosophy*. Edinburgh: Edinburgh University Press, 1996.
Matthen, Mohan, ed. *Aristotle Today: Essays on Aristotle's Ideal of Science*. Edmonton: Academic, 1987.
McDowell, John. "Meaning and Intentionality in Wittgenstein's Later Philosophy." *Midwest Studies in Philosophy* 17 (1992): 40–52.
———. *Mind and World*. Cambridge: Cambridge University Press, 1994.
———. "On the Sense and Reference of a Proper Name." *Mind* 86 (1977): 159–85.
———. "Putnam on Mind and Meaning." *Philosophical Topics* 20, no.1 (1992).
———. "Singular Thought and the Extent of Inner Space." In *Subject, Thought, and Context*, edited by P. Pettit and J. McDowell. Oxford: Clarendon Press, 1986.
McGinn, Colin. *Mental Content*. Oxford: Basil Blackwell, 1989.
———. *Wittgenstein on Meaning*. Oxford: Basil Blackwell, 1984.
McInerny, Ralph. *Aquinas against the Averroists*. West Lafayette, Ind.: Purdue University Press, 1993.
———. "Being and Predication." In *Being and Predication: Thomistic Interpretations*, 173–228. Washington, D.C.: Catholic University of America Press, 1986.
———. *The Logic of Analogy*. The Hague: Martinus Nijhoff, 1961.
———. *Studies in Analogy*. The Hague: Martinus Nijhoff, 1968.
Mill, John Stuart. *A System of Logic, Ratiocinative and Inductive: Being a Connected View of the Principles of Evidence and the Method of Scientific Investigation*. New York: Longmans, 1906.
Mohanty, J. *Husserl and Frege*. Bloomington: Indiana University Press, 1982.
Montaigne, Michel de. "An Apology for Raymond Sebond." In *The Essays of Michel De Montainge*, translated by M.A. Screech. London: Allen Lane, 1991.
Nadler, Stephen. *Arnauld and the Cartesian Philosophy of Ideas*. Princeton, N.J.: Princeton University Press, 1989.
Nietzsche, Friedrich. "We Scholars." In *Beyond Good and Evil*, translated by Walter Kaufmann. New York: Vintage Books, 1966.
Nussbaum, Martha. Introduction to Aristotle's *De motu animalium*. Princeton, N.J.: Princeton University Press, 1978.
O'Callaghan, John P. "Verbum Mentis: Theological or Philosophical Doctrine?" *Proceedings of the American Catholic Philosophical Society* 74 (2001).

Owens, Joseph. "Common Nature: A Point of Comparison between Thomistic and Scotistic Metaphysics." In *Inquiries into Medieval Philosophy*, edited by James Ross. Westport, Conn.: Greenwood, 1971.

———. *The Doctrine of Being in the Aristotelian Metaphysics*. 3d ed. Toronto: Pontifical Institute of Medieval Studies, 1978.

Pannacio, Claude. "From Mental Word to Mental Language." *Philosophical Topics* 20, no. 2 (1992).

Pasnau, Robert. "Aquinas on Thought's Linguistic Nature." *The Monist: Analytical Thomism* 80, no. 4 (October 1997): 558–75.

———. *Cognitive Theory in the Later Middle Ages*. Cambridge: Cambridge University Press, 1996.

Peifer, John Frederick. *The Concept in Thomism*. New York: Bookman, 1952.

Percy, Walker. *Love in the Ruins*. New York: Avon Books, 1971.

———. *The Message in the Bottle*. New York: Farrar, Strauss, Giroux, 1975.

Pess, Andrew, and Sanford Goldberg, eds. *The Twin Earth Chronicles: Twenty Years of Reflection on Hilary Putnam's "The Meaning of 'Meaning'."* Armonk: M. E. Sharpe, 1996.

Pieper, Josef. *Guide to Thomas Aquinas*. Translated by Richard and Clara Winston. New York: Pantheon, 1962.

———. *The Silence of St. Thomas*. New York: Pantheon Books, 1957.

Plato. *Cratylus*. In *The Collected Dialogues of Plato*, edited by Edith Hamilton and Huntington Cairns. Princeton, N.J.: Princeton University Press, 1961.

———. *The Republic*. Translated by G. M. A. Grube. Indianapolis: Hackett, 1992.

———. *Sophist*. In *The Collected Dialogues of Plato*, edited by Edith Hamilton and Huntington Cairns. Princeton, N.J.: Princeton University Press, 1961.

Poinsot, John. *Tractatus De Signis*. Berkeley: University of California Press, 1985.

Putnam, Hilary. "Aristotle after Wittgenstein." In *Modern Thinkers and Ancient Thinkers*, edited by Robert W. Sharples. Boulder, Colo.: Westview Press, 1993.

———. *The Many Faces of Realism*. LaSalle, Ill.: Open Court, 1987.

———. "The Meaning of 'Meaning'." In *Mind, Language, and Reality*. Cambridge: Cambridge University Press, 1975.

———. *Mind, Language, and Reality*. Philosophical Papers 2. Cambridge: Cambridge University Press, 1975.

———. *Reason, Truth, and History*. Cambridge: Cambridge University Press, 1981.

———. *Renewing Philosophy, The Gifford Lectures*. Cambridge, Mass.: Harvard University Press, 1992.

———. *Representation and Reality*. Cambridge, Mass.: MIT Press, 1988.

———. "Sense, Nonsense, and the Senses: An Inquiry Into the Powers of the Human Mind." *Journal of Philosophy* 91, no. 9 (September 1994): 445–517.

———. *The Threefold Cord*. New York: Columbia University Press, 2000.

———. *Words and Life*. Cambridge, Mass.: Harvard University Press, 1994.

Putnam, Hilary, and Martha Nussbaum. "Changing Aristotle's Mind." In *Essays on Aristotle's De Anima*, edited by Martha C. Nussbaum and Amelie Oksenberg Rorty. Oxford: Clarendon Press, 1992.

Quine, W. V. O. *From a Logical Point of View.* Cambridge, Mass.: Harvard University Press, 1980.

———. "Speaking of Objects." In *Ontological Relativity and Other Essays.* New York: Columbia University Press, 1969.

Reale, Giovanni. *The Concept of First Philosophy and the Unity of the Metaphysics of Aristotle.* Edited and translated from the third edition by John R. Catan. Albany, N. Y.: SUNY Press, 1980.

Reid, Thomas. *Inquiry into the Human Mind.* In *Philosophical Works,* vol. 1, with notes and supplementary dissertations by Sir William Hamilton. Hidlesheim: Georg Olms Verlagsbuchhandlung, 1967.

Rorty, Richard. *The Linguistic Turn.* Chicago: University of Chicago Press, 1992.

Russell, Bertrand. "Knowledge by Acquaintance and by Description." *Proceedings of the Aristotelian Society.* London: Aristotelian Society, 1910–11.

Schmidt, Robert W., S. J. *The Domain of Logic According to Saint Thomas Aquinas.* The Hague: Martinus Nijhoff, 1966.

Schwartz, Stephen, ed. Introduction to *Naming, Necessity, and Natural Kinds.* Ithaca, N. Y.: Cornell University Press, 1977.

Searle, John. *Expression and Meaning: Studies in the Theory of Speech Acts.* Cambridge: Cambridge University Press, 1979.

———. *Intentionality: An Essay in the Philosophy of Mind.* Cambridge: Cambridge University Press, 1988.

———. *The Rediscovery of Mind.* Cambridge, Mass.: MIT Press, 1992.

Smith, Barry. "Frege and Husserl: The Ontology of Reference." *Journal of the British Society for Phenomenology* 9, no. 2 (May 1978): 111–25.

———. "On the Origins of Analytic Philosophy." *Graezer philosophische Studien* 35 (1985): 153–73.

Sokolowski, Robert. "Exorcising Concepts." *Review of Metaphysics* 40 (1987).

———. "Husserl and Frege." *Journal of Philosophy* 84, no. 10 (1987): 521–28.

Solmsen, Friedrich. *Die Entwicklung de aristolischen Logik und Rhetorik.* Berlin, 1929.

Sorabji, Richard. *Necessity, Cause, and Blame: Perspectives on Aristotle's Theory.* Ithaca, N. Y.: Cornell University Press, 1980.

Spade, Paul Vincent. "The Semantics of Terms." In *The Cambridge History of Later Medieval Philosophy,* edited by Norman Kretzmann, Anthony Kenny, and Jan Pinborg. Cambridge: Cambridge University Press, 1982.

Spiegelberg, Herbert. "Augustine in Wittgenstein: A Case Study in Philosophical Stimulation." *Journal of the History of Philosophy* 17 (1979): 327.

———. *The Phenomenological Movement.* The Hague: Martinus Nijhoff, 1982.

Verbeke, G. *Ammonius: Commentaire sur Le Peri Hermeneias d'Aristote, Traduction de Guillaume de Moerbeke.* Louvain: Universitaires de Louvain, 1961.

Weisheipl, James, O. P. *Friar Thomas D'Aquino.* Washington, D. C.: Catholic University of America Press, 1983.

Wiggins, David. "Putnam's Doctrine of Natural Kind Words and Frege's Doctrines of Sense, Reference, and Extension: Can They Cohere?" In *Reading Putnam,* edited by Peter Clark and Bob Hale, 201–215. Oxford: Basil Blackwell, 1994.

Willard, Dallas. "The Paradox of Logical Psychologism: Husserl's Way Out." In *Readings on Edmund Husserl's Logical Investigations*, edited by J. N. Mohanty. The Hague: Martinus Nijhoff, 1977.

Wittgenstein, Ludwig. *The Blue and Brown Books.* New York: Harper & Row, 1958.

———. *Philosophical Investigations.* Translated by G. E. M. Anscombe. New York: Macmillan, 1953.

———. *Tractatus Logico-Philosophicus.* Translated by C. K. Ogden. London: Routledge & Kegan Paul, 1986.

Wolfe, Tom. *The Painted Word.* New York: Farrar, Strauss, Giroux, 1975.

INDEX

abstraction
  Aquinas on, 20, 21, 37, 73, 166–68, 170–71, 184, 202–3, 218–24, 227, 282–83, 285, 289, 296, 302n.21, 327n.25
  Berkeley on, 94–95, 312n.49
  Geach on, 221, 285, 286, 289
  Haldane on, 285–86
  Locke on, 94, 98, 99, 106–7, 116, 311n.44
  See also essences; intellect; intelligible species
Ackrill, J. L.
  on *De interpretatione*, 16, 18–19, 22, 80, 308n.80
  translation of *De interpretatione*, 43–44, 46, 54, 57, 64, 69
Ammonius, on *De interpretatione*, 58–59, 60–65, 66–67, 69–71, 72–74, 76, 77, 89, 91, 281, 307n.58, 308n.78, 309n.81
Andronicus, 66
Anscombe, Elizabeth, 171–72
Apostle, Hippocrates
  on *De Interpretatione*, 15, 16, 22, 300n.1
  translation of *De Interpretatione*, 54, 56
Aquinas, St. Thomas
  on abstraction, 20, 21, 37, 73, 166–68, 170–71, 184, 202–3, 218–24, 227, 282–83, 285, 289, 296, 302n.21, 327n.25
  on actuality and potentiality, 172–73, 174–75, 229–31, 292
  vs. Ammonius, 66–67, 69–71, 72–74, 76, 77, 308n.78, 309n.81
  on Aristotle's logical works, 3–5
  on Aristotle's semantic triangle, 72–76, 147, 236, 247, 276, 286
  vs. Avicenna, 28–29, 184, 188
  vs. Boethius, 66–67, 68, 69–71, 72–74, 76, 77, 308n.78, 309n.81
  on concepts, 31–39, 67–77, 87–88, 165–82, 189–90, 194, 200, 221–24, 236, 237–56, 272–74, 277–78, 304n.43, 331n.27
  on *De anima*, 16–17, 18, 19, 25–26, 30, 68, 71, 91, 202, 205–6, 207, 208, 211–12, 215, 237, 251, 256, 282–83, 289–91, 296, 300n.16, 334n.45
  on education, 293–94
  on efficient causes, 177–79, 182, 232, 233–35
  on *entia rationis*, 190–94, 195, 325n.87
  on enunciations, 16–17, 30–31, 71–72, 166–67, 169, 300n.1
  on essences, 4–5, 16–17, 19, 26–31, 37–39, 68–69, 184–87, 188–90, 191–92, 193, 194, 199–203, 206–8, 212, 215, 225, 228, 238–41, 244–46, 251, 257–74, 290–91, 303n.42, 324n.67, 325n.82
  on formal causes, 179–80, 181–82, 217, 232, 233–34, 235
  on form vs. matter, 164–65
  vs. Frege, 303n.42
  on God's causality, 334n.32
  on God's understanding, 200–201, 218
  on human acts vs. acts of a human, 296–97
  on human rationality, 290–91, 294–95, 296, 334n.45
  on human soul, 205–6, 242, 267–69, 289–91, 334n.45
  on immanent acts, 169, 173–74, 214–15

347

Aquinas, St. Thomas (cont.)
  on intrinsic vs. extrinsic principles, 179–80, 181, 182, 323n.57
  on introspection, 224–27, 241–43
  vs. Kretzmann, 69–70, 76
  on language and perfection, 13, 91, 280–81, 289–93, 297–98
  on linguistic convention, 33–34, 276, 278–79, 285
  vs. Locke, 87–89, 91–92
  on matter as principle of individuation, 238
  on *modus significandi* of general words, 19–22, 32, 38, 72, 73–74, 76, 87–88, 89
  on naming as we know, 3, 76–77, 246, 296
  on natures considered absolutely, 26–31, 37–39, 191–92, 212, 238–41, 303n.42
  on nominal definitions, 250, 251–52, 255–56, 281
  on passions of the soul, 12, 16, 19–20, 21–22, 33, 36–37, 55–56, 65–67, 68–71, 72–76, 208–13, 224–32, 237–49, 281, 297
  on perfection, 13, 91, 280–81, 289–93, 296, 297–98
  on Plato, 20, 22, 73, 83, 167, 199–200, 201, 210–11, 212–13, 215, 265, 266, 268, 269, 291, 296
  on powers/faculties of soul, 205–6
  on *res extra animam*, 12, 19–21, 88–89, 164–92, 199–208, 210, 214–18, 227–28, 236, 237–39
  on *res in anima* vs. *res extra animam*, 26–28, 164–92, 190, 210–12
  on *res significata*, 20–21, 38, 73
  on science, 34–36, 39, 210–13, 221–22, 255, 266, 267–68, 271–74, 281, 291
  on sense experience, 177–79, 206–8, 214–15, 216–17, 219–23, 224, 228–29, 327n.25, 328nn.36, 48
  on signification, 12, 19–21, 22–26, 31–36, 65–67, 70–77, 88, 276, 301n.14
  on similitude/likeness, 25, 89, 165, 173, 208, 214, 217–18, 225, 227–32, 236, 244, 246, 248, 257, 276, 278
  and skepticism, 213, 241–42, 258, 273
  on social necessity of language, 278–83, 289, 291–92, 297–98
  on *species in medio*, 177–79
  on transient acts, 169, 173–74, 214–15

  on truth and falsity, 4, 26, 67–68, 71, 167–68, 244–45, 250–51, 254, 263, 275, 297
  on unity and being, 28–31, 268–70, 289, 290–91
  on universality, 32, 39, 87–88, 202–3, 217–18, 221–24, 238, 243, 252–53, 272–73, 302n.21
  on *verbum mentis*, 12, 300n.16
  on written language, 292–93
  See also Aquinas, St. Thomas, works of; being; intellect; intelligible characters; intelligible species; signification; Thomistic-Aristotelian tradition
Aquinas, St. Thomas, works of
  Commentary on the De Anima, 18, 26, 30, 68, 91, 202, 208, 211–12, 215, 256, 282–83, 289–91, 300n.16
  Commentary on the De Interpretatione, 4–5, 12, 15–22, 25, 34, 37, 55, 65–77, 88, 250–52, 268, 276, 281–82, 289, 300n.16
  Commentary on the Metaphysics, 173–74, 184, 186–87, 190–91, 209, 211
  Commentary on the Physics, 222
  Commentary on the Posterior Analytics, 4–5, 17, 19, 185, 242, 293
  Commentary on the Sentences of Peter Lombard, 173, 322n.40, 323n.52
  De ente et essentia, 26–31, 39, 165, 186, 191, 192, 201–2, 212, 238–39, 240, 256, 303n.34, 324n.67, 325n.79
  De Principiis Naturae, 180
  Quaestiones Disputatae de Veritate, 184, 186, 219–20, 244–45, 257
  Quaestiones Quodlibetales, 324n.67
  Summa Contra Gentiles, 304n.43
  Summa Theologiae, 25, 26, 30, 55–56, 76, 166–67, 181, 199–200, 205–6, 208, 210–11, 222, 229, 235, 256, 296, 300n.16
Aristotle
  on αρχή, 179–80
  on causes, 178, 181, 211, 228, 235
  on essences, 4–5, 16–17, 18, 19
  on formal vs. efficient causes, 178, 181
  on habits, 175
  on immanent vs. transient activity, 214
  on the intellect's composing and dividing, 4–5, 16–18, 19
  on language, 91

Index 349

on linguistic convention, 5, 6, 42, 47,
  49–51, 138, 276
logical works of, 3–5
on passions of the soul, 5–6, 15–16,
  19–20, 31, 36–37, 39, 65–67, 68,
  138–40, 276
  vs. Plato, 20, 22, 73, 83, 154, 167,
  199–200, 201, 210–11, 256
on science, 34–36, 39, 161–63, 210–11,
  250, 281, 330nn.21, 22
on soul, 205–6, 237, 267–68
on truth and falsity, 17–18, 26, 67–68, 70,
  71, 300n.2
on understanding of indivisibles (essences),
  4–5, 16–17, 18, 19
on words/articulated sounds, 5–6, 15–16,
  17–18, 19–20, 238, 268, 276, 289
  See also Aristotle, works of; Aristotle's
    De interpretatione; Aristotle's semantic
    triangle; Thomistic-Aristotelian tradition
Aristotle, works of
  Categories, 4–5, 69, 72, 164–65
  De Anima, 16–19, 25–26, 41, 68, 70, 71,
    237, 251, 282–83, 289–91, 293, 296,
    309n.80, 334n.45
  De Sophisticis Elenchis, 293
  Metaphysics, 79, 209, 211, 214, 269
  Nichomachean Ethics, 282
  Physics, 164–65
  Poetics, 4, 293, 296
  Politics, 282, 289, 291–92, 296
  Posterior Analytics, 4–5, 22, 273–74,
    330nn.21, 22
  Rhetoric, 4, 293, 296
  Topica, 293, 304n.50
  See also Aristotle; Aristotle's De interpretatione; Aristotle's semantic triangle
Aristotle's De interpretatione, 4–5, 41–77,
  79–81, 293
  Ackrill on, 16, 18–19, 22, 80, 308n.80
  Ackrill translation of, 43–44, 46, 54, 57,
    64, 69
  affirmations and denials in, 15, 300n.2
  Ammonius on, 58–59, 60–65, 66–67,
    69–71, 72–74, 76, 77, 89, 91, 281,
    307n.58, 308n.78, 309n.81
  Apostle on, 15, 16, 22, 300n.1
  Apostle translation of, 54, 56
  Bekker text of, 54–56, 58
  Edghill translation of, 16, 55, 56, 73
  enunciations in, 5, 15, 300nn.1, 2

Greek texts of, 54–58, 70
Latin translations, 12, 15–22, 43–44,
  46–47, 57–60, 62, 63, 80, 86, 301n.14
Magee on, 45, 49, 55, 59, 61–62, 65, 70,
  305n.25, 306nn.45, 47, 309n.81
Minio-Paluello text of, 54, 56, 57, 59, 70
nouns and verbs in, 5, 15, 17–18, 71–72
πρώτων in, 43–45, 54, 56–60, 61–62,
  64–65, 68, 70–71, 306n.47
πρώτως in, 43–56, 57–60, 61–62, 64–65,
  70–71, 80
reference to De Anima in, 16–19, 25–26,
  41, 42–43
signs (σημεῖα) in, 41–42, 43–55, 56,
  57–58, 63, 69, 76, 77, 80, 138–40,
  305n.25
statements and sentences in, 15, 300n.2
symbols (σύμβολα) in, 41–42, 43–55, 56,
  57–58, 63, 69, 76, 77, 80, 138–40,
  305n.25
truth and falsity in, 17–18, 70, 71, 300n.2
  See also Aristotle, works of; Aristotle's
    semantic triangle
Aristotle's semantic triangle, 5–8, 15–22, 253,
  278–79
  Aquinas on, 72–76, 147, 236, 247, 276, 286
  criticisms of, 6–8, 44–45, 54, 79–81, 136,
    239, 241, 246
  Kretzmann on, 5, 12, 41–58, 60, 61, 63,
    64, 67, 73, 77
  and mental representationalism, 5–7, 8, 39,
    53, 64, 77, 79–81, 153–55
  passions of the soul in, 5–6, 15–16, 31, 39,
    41–77, 239
  Putnam on, 6, 8, 12, 135–36, 140–41,
    145–46, 239, 246, 277
  things in, 5–6, 15–16, 39, 41–77
  words/articulated sounds in, 5–6, 15–16,
    31, 39, 41–77
  See also Aristotle's De interpretatione;
    signification; similitude/likeness
Arnauld, Antoine, 90
Ashworth, E. J., 89, 301n.14
Augustine, St., 3, 322n.40, 335n.45
Austin, J. L., 182
Avicenna, 28–29, 184, 188, 218

behaviorism, 2, 118, 122
being
  act of being (actus essendi), 30, 184, 188,
    194, 270–71, 289

being (cont.)
  in anima, 26–31, 36, 39, 164–94, 212, 238–41
  and the categories, 183–84
  ens as existence (ens$_e$), 183–84, 185–90, 191, 192, 193–94, 199, 218, 251, 324n.67, 325n.79, 325n.85
  ens as propositional being (ens$_p$), 183–84, 185–91, 192–94, 195, 251
  ens rationis, 190–94, 195
  independent vs. dependent existence, 165
  relationship to essence, 28–30, 184–87, 188–90, 192, 193, 194
  relationship between res and ens, 182–98, 325n.85
  in singularibus/res extra animam, 26–29, 30, 39, 164–65, 212, 238–41
  and unity, 28–31, 268–70, 290–91
Bekker, Immanuel, 54–58
Berkeley, George
  on abstract ideas, 94–95, 312n.49
  on Descartes, 92–93
  vs. Hume, 96, 97–98, 99
  on ideas, 92–96, 97–98, 99–100, 101, 112, 157, 278
  on language, 12, 278
  vs. Locke, 92–96, 99
  on signification, 94–96, 99, 278, 312n.50
  on spirits, 93–94, 96
  on universality, 95, 97–98, 312n.50
  Wittgenstein on, 12
Blanck, David, 307n.58
Bobik, Joseph, 186
Boethius, 19, 250, 294
  vs. Aquinas, 66–67, 68, 69–71, 72–74, 76, 77, 308n.78, 309n.81
  on De interpretatione, 61–65, 66–67, 68, 69, 69–71, 72–74, 76, 77, 86, 89, 91, 308n.78, 309n.81
  Kretzmann on, 12, 43–44, 46–47, 57–58, 60, 63, 80
  Latin translation of De interpretatione, 12, 15–22, 43–44, 46–47, 57–60, 62, 63, 80, 86, 301n.14
  vs. Locke, 64, 65, 77, 86, 89, 91
Bolzano, Bernard, 204
Braine, David, 163–64, 321n.11
British Empiricism, 6–7, 12, 79–80, 236, 277
  and Fodor, 113–14, 115–17, 125, 126, 132, 133

See also Berkeley, George; Hume, David; Locke, John
Broadie, Alexander, 81
Brykman, Genevieve, 90
Busse, Adolfus, 59, 307n.58

Carnap, Rudolf, 6, 136, 145
causes
  Aquinas on efficient causes, 177–79, 182, 232, 233–35
  Aquinas on formal causes, 179–80, 181–82, 217, 232, 233–34, 235
  Aristotle on, 178, 181, 211, 228, 235
  Aristotle on formal vs. efficient causes, 178, 181
  Fodor on causal relation between mental representations and external world, 114, 116–17, 124–28, 129–33, 137, 155, 248, 317n.32, 318nn.37, 39
  Fodor on causal relation between thoughts, 116–20, 121, 152–53
  intelligible species as efficient cause, 176–82, 232–35
  intelligible species as formal cause, 165, 176–82, 217, 232, 233–35, 247–49
Chomsky, Noam, 6, 122, 317n.23
concepts
  as acts of intellect, 31–32, 67–77, 88, 168–75, 176, 184, 189–90, 194, 200, 251–56, 257, 304n.43, 331n.27
  as acts of simple apprehension, 251–55, 331n.27
  Aquinas on, 31–39, 67–77, 87–88, 165–82, 189–90, 194, 200, 221–24, 236, 237–56, 272–74, 277–78, 304n.43, 331n.27
  as capacities for linguistic expression, 253–54
  as complex, 67–72, 249–53, 255–56, 257
  definition of, 31–32, 250, 251–52
  developmental view of, 3, 221–24, 250, 252–53, 272–74
  of fictional beings, 250, 251–52, 255
  as formally identical with res extra animam, 167, 170–71, 202, 212, 235, 237–49, 265
  Frege on, 11, 168, 303n.42
  Haldane on, 166, 168–69, 285–89
  as having both propositional and existential being, 184

identity of, 37–39, 238, 243, 251, 256, 257, 265
and introspection, 224–27, 241–42
John of St. Thomas on, 328n.46
and nominalization, 168–70
as passions of the soul (*passiones animae*), 19–20, 21–22, 31–32, 36–37, 60–61, 62–63, 66, 73–76, 88, 89, 165, 208, 217, 227–32, 236, 237–49, 250–51, 276
Putnam on, 194–98, 200, 237, 257–72, 258, 262–63
as simple, 67–72, 249–54, 256, 257
See also ideas; mental representationalism; passions of the soul
convention, linguistic, 24, 33–36, 37, 76
Aristotle on, 5, 6, 42, 47, 49–51, 138, 276
Cooke, Harold, 54
Crossett, John, 56

Davidson, Donald, 152, 161
Deferrari, Roy J., 179
definite descriptions, 11, 204, 326n.12
Descartes, René, 83, 161, 164, 242
Berkeley on, 92–93
on ideas, 90, 92–93, 160, 236, 288
Reid on, 99–100
Dewey, John, 274
Dubus, Andre, 296–97
Dummett, Michael, 6–7, 102, 204, 205, 254, 278, 313n.72
on ideas, 6–7, 79–80, 82–83

Edghill, E. M., translation of *De interpretatione*, 16, 55, 56, 73
essences
as apprehended by first act of intellect, 4–5, 16–17, 19–21, 26–31, 68–69, 193, 201–2, 206, 210–12, 239–41, 244–46, 251, 272–73
Aquinas on, 4–5, 16–17, 19, 26–31, 37–39, 68–69, 184–87, 188–90, 191–92, 193, 194, 199–203, 206–8, 212, 215, 225, 228, 238–41, 244–46, 251, 257–74, 290–91, 303n.42, 324n.67, 325n.82
Aristotle on, 4–5, 16–17, 18, 19
as considered absolutely, 26–31, 37–39, 191–92, 212, 238–41, 303n.42
Locke on nominal essences, 84, 93, 106–7, 149–51, 261, 311n.44

Locke on real essences, 84, 93, 149–51, 157, 261, 311n.44
as objects of knowledge, 199–203, 206–8, 225, 228, 239–41
Owen on, 27
Pieper on, 332n.55
Putnam on, 257–74, 289, 290
relationship to being, 28–30, 184–87, 188–90, 192, 193, 194
See also natural kinds

Fodor, Jerry, 11, 26, 156
on brain states, 128–29, 132
and British Empiricism, 113–14, 115–17, 125, 126, 132, 133
on broad content, 131–33, 144, 152–53, 155, 157, 247
on causal relation between mental representations and external world, 114, 116–17, 124–28, 129–33, 137, 155, 248, 317n.32, 318nn.37, 39
on causal relation between thoughts, 116–20, 121, 152–53
on computer models of the mind, 118–19, 124
on intentional attitudes, 119–20
on Language of Thought, 114, 115, 118, 119–24, 125–32, 137–38, 140, 143, 152–53, 316nn.2, 20, 318n.39
on mental representationalism, 6, 12, 113–33, 232–33
on narrow content, 131–33, 144, 152–53, 155, 157, 247
on natural languages, 122–23, 129, 140, 143, 317n.23
on private language, 115
Putnam on, 128, 137–38, 151–53
on resemblance, 114, 125, 126, 133
on science, 118–19, 318n.33
on Twin Earth argument, 152–53
on Wittgenstein, 114–15
formal identity
and intelligible characters, 37–39, 87–88, 237–49, 243, 251, 256, 257, 265
and intelligible species, 167, 170–71, 202, 212, 235
Frege, Gottlob, 6, 80, 113
vs. Aquinas, 303n.42
on concepts, 11, 168, 303n. 42
on definite descriptions, 204, 326n.12
on mental states, 11, 101–2

Frege, Gottlob (cont.)
  on objects, 204
  on proper names vs. predicates, 204
  on sense and reference, 102, 103, 136, 142, 143, 145, 313n.72, 314n.76

Geach, Peter, 221, 253–54, 285, 289, 331nn.26, 33
God, 185–86, 246, 281, 332n.55
  Aquinas on God's causality, 334n.32
  Aquinas on God's understanding, 200–201, 218
  Haldane on, 286, 288, 295, 334n.32
Goodman, Nelson, 105
Goris, Harm J.M.J., 331n. 27

Hacking, Ian, 9, 10
Haldane, John, 81, 82, 291
  on abstraction, 285–86
  on concepts, 166, 168–69, 285–89
  on *De ente et essentia*, 303n.34
  on God, 286, 288, 295, 334n.32
  on intellect, 285–86
  on linguistic convention, 36, 285
  on McDowell, 295–96, 333nn.24, 26
  on representation, 103
Hart, H.L.A., 152, 265
Honoré, A.M., 152, 265
Hume, David
  vs. Berkeley, 96, 97–98, 99
  on external objects, 96–97, 99, 312nn.57, 61
  on general terms, 97–99
  on ideas, 96–99, 99–100, 101, 104, 106, 160, 312n.57, 315n.78
  on language, 12
  vs. Locke, 96, 97–98, 99, 312n.52
  and mental representationalism, 81, 160
  Reid on, 81
  on signification, 97–98, 99
  Wittgenstein on, 12
Husserl, Edmund, 100–101, 107, 113, 313n.65

Ibn Rushd, 288
ideas
  Berkeley on, 92–96, 97–98, 99–100, 101, 112, 157, 278
  Descartes on, 90, 92–93, 160, 236, 288
  Dummett on, 6–7, 79–80, 82–83
  Hume on, 96–99, 99–100, 101, 104, 106, 160, 312n.57, 315n.78
  Locke on, 81, 83–93, 94–96, 98, 99–100, 101, 106, 157, 160, 161, 195, 261, 278, 310nn.16, 23, 313n.65
  Russell on, 81–83, 99, 101, 102, 112, 154
  See also concepts; meaning; mental representationalism; passions of the soul; words
imagination, 223
intellect, 3–5, 166–74, 278–79, 334n.45
  agent intellect, 166–68, 170–71, 177, 214–15, 218–24, 225, 278, 287–88, 295, 323n.57, 327n.25
  concepts as acts of, 31–32, 67–77, 88, 168–75, 176, 184, 189–90, 194, 200, 251–56, 257, 304n.43, 331n.27
  habits as structuring powers of, 174–75
  introspective self-knowledge of, 224–27
  obiecta of the, 82, 199–227, 228, 234–36, 241–44, 327n.25
  passive/possible intellect, 166–68, 170–71, 214–15, 218–19, 225, 281, 287–88, 295
  receptive character of, 237–38, 247–49, 278
  relationship to *entia rationis*, 190–94, 195, 325n.87
  relationship to intelligible species, 26–28, 30, 37, 166–68, 170–71, 173, 174–82, 199, 208–18, 219, 222, 224–25, 227–32, 235–36, 323n.52
  relationship to sense experience, 177–79, 206–8, 214–15, 216–17, 219–23, 224, 228–29, 327n.25, 328nn.36, 48
  See also abstraction; intellect, first act of; intellect, second act of; signification
intellect, first act of, 68–71, 188, 222–23, 236, 253–54, 331n.27
  essences (indivisibles) apprehended by, 4–5, 16–17, 19–21, 26–31, 68–69, 193, 201–2, 206, 210–12, 239–41, 244–46, 251, 272–73
  as immanent act, 169, 173–74
  as *modus significandi*, 19–22, 32, 38, 72, 73–74, 76, 87–88, 89
  relationship to intelligible characters, 26–28, 31–32, 36, 39
  relationship to intelligible species, 26–28, 30, 37, 166–68, 170–71, 173, 174–75, 176
  relationship to truth and falsity, 26, 67–68, 71, 244, 254, 263
intellect, second act of (composing or dividing), 68–71, 74, 166–68, 171, 188, 193, 222–23, 236, 253–54, 272–73
  Aristotle on, 4–5, 16–18, 19

as immanent act, 169, 173–74
and predication, 28–31
relationship to truth and falsity, 4, 17–18, 26, 67–68, 71, 167–68, 244–45, 250–51, 254
intelligible characters, 183–84, 205–6, 237–49, 267–74, 276, 324n.67
as considered absolutely, 26–31, 37–39, 191–92, 212, 238–41, 303n.42
and formal identity, 37–39, 87–88, 237–49, 243, 251, 256, 257, 265
and identity of concepts, 37–39, 251, 256, 257
as principles of knowing, 165
relationship to first act of intellect, 26–28, 31–32, 36, 39
relationship to words, 276
See also abstraction; essences; intelligible species
intelligible species, 165–82, 208–36
as accident of intellect, 165, 238
as efficient cause, 176–82, 232–35
as formal cause, 165, 176–82, 217, 232, 233–35, 247–49
and formal identity, 167, 170–71, 202, 212, 235
and God's understanding, 200–201
as likeness, 208, 214–15, 217–18, 225, 236, 248
as means of knowing/understanding, 214–18, 222, 232, 235–36
relationship to intellect, 26–28, 30, 37, 166–68, 170–71, 173, 174–82, 199, 208–18, 219, 222, 224–25, 227–32, 235–36, 323n.52
See also abstraction; essences; intelligible characters
Internalist Thesis, 13, 144, 153, 154, 155, 156–57, 226, 237, 241, 247, 264, 276–77
Introspectibility Thesis, 12–13, 156, 200–201, 203, 208–24, 225–26, 227, 231–32, 236, 239, 241, 264, 266

Jacobs, Jonathan, 327n.38
John of St. Thomas, 328n.46

Kant, Immanuel, 162
Kenny, Anthony, 285
Kretzmann, Norman
on Boethius, 12, 43–44, 46–47, 57–58, 60, 63, 80

on encoding/decoding symbols, 45, 47–48, 48, 49, 52–54, 56, 91
on Locke, 90, 311n.36
and πρώτως, 43, 44–56, 57–58, 60, 65, 80
on regular association, 48, 51
on signs (σημεῖα), 41–42, 43–55, 56, 57–58, 63, 69, 76, 77, 80, 138–40, 305n.25
on symbols (σύμβολα), 41–42, 43–55, 56, 57–58, 63, 69, 76, 77, 80, 138–40, 305n.25
on written vs. spoken words, 51–52, 306n.45
Kripke, Saul
vs. Aquinas, 34–35
on natural kinds, 34, 149–51
on objects, 205
on rigid designators, 150
on Wittgenstein, 105–6, 108, 109–11, 284

Lewis, C. S., 1
likeness. *See* similitude/likeness
lingua mentis, 2, 6, 137–40
Locke, John
on abstract ideas, 94, 98, 99, 106–7, 116, 311n.44
vs. Aquinas, 87–89, 91–92
vs. Berkeley, 92–96, 99
vs. Boethius, 64, 65, 77, 86, 89, 91
on complex ideas, 83–84, 86, 91, 254, 310n.23
on external objects, 83–84, 85–86, 91–92, 96, 310n.26
on general words, 87–92, 94, 98, 99
vs. Hume, 96, 97–98, 99, 312n.52
on ideas, 81, 83–93, 94–96, 98, 99–100, 101, 106, 157, 160, 161, 195, 261, 278, 310nn.16, 23, 313n.65
on language, 12, 53, 64, 87–92, 95–96, 195, 278
on mental representations, 53, 84–86, 89, 90–91, 160, 161
on nominal essences, 84, 93, 106–7, 149–51, 261, 311n.44
on primary and secondary qualities, 85–86, 116
Putnam on, 149–51
on real essences, 84, 93, 149–51, 157, 261, 311n.44
on signification, 64, 65, 77, 86–92, 94–95, 99, 278, 311n.36

Locke, John (*cont.*)
  on simple ideas, 83–84, 85, 86, 93, 310nn.16, 23
  on universality, 95
  Wittgenstein on, 12
Lonergan, Bernard, 55–56

MacIntyre, Alasdair, 291, 296
Magee, John, on *De interpretatione,* 45, 49, 55, 59, 61–62, 65, 70, 305n.25, 306nn.45, 47, 309n.81
Maritain, Jacques, 291
McDowell, John, 105, 111, 287, 313n.72
  on Aristotelian first nature, 161–63, 230, 279–81, 295–96
  on Aristotelian second nature, 109–10, 161–63, 230, 279–81, 286, 333n.24
  Haldane on, 295–96, 333nn.24, 26
  on inner vs. outer world, 160–62
  on mental representationalism, 106–10, 131–32, 133, 157
  on myth of the given, 162
  on Putnam, 236, 277
  on Wittgenstein, 106–10, 162–63
McGinn, Colin
  on externalism, 238, 255–56
  on Frege, 102
  on Putnam, 146, 157
  on Wittgenstein, 110–11, 284–85, 286, 288, 290
meaning, 5–13
  as extension, 141–43, 145–51, 152–53, 265, 267–68, 318n.41
  Frege on sense and reference, 102, 103, 136, 142, 143, 145, 313n.72, 314n.76
  as intension, 141–43, 145–51, 152–53, 265, 267–68, 318n.41
  as public, 5–6, 8–9, 24, 99
  Putnam on, 135–49, 152–55, 212
  Quine on, 142–43, 264, 288–89
  relationship to mind, 11–12
  Searle on, 11
  Searle on reference, 11
  Wittgenstein on, 2, 3, 6, 15, 102–11, 283–88
  See also ideas; signification; words
memory, 223
mental impressions. *See* passions of the soul
mental representationalism
  and Aristotle's semantic triangle, 5–7, 8, 39, 53, 64, 77, 79–81, 153–55
  criticisms of, 6–7, 8, 19, 39, 44–45, 99–112, 114, 125, 126, 136, 137, 140–49, 159–61, 212–13, 215–16, 225–26, 236, 239, 241, 257, 276–78
  definition of, 81
  Fodor on, 6, 12, 113–33, 232–33
  Husserl on, 100–101, 107, 113, 313n.65
  "in itself" character of mental representations, 97, 108–9, 130, 131–33, 144, 155, 156–57, 237, 241, 247, 264
  inner mental representations vs. external objects, 159–65
  McDowell's Master Thesis regarding, 106–10, 131–32, 133, 157
  mental representations as brain states, 128–29, 160–61
  Putnam on, 6, 8, 81, 135–57, 159–61, 212, 236, 237, 257, 276–78, 319n.25
  and resemblance, 85–86, 103–4, 114, 116, 133, 137, 138–39, 219, 328n.48
  Searle on mental representations, 160–61, 324n.66
  vs. Thomistic-Aristotelian tradition, 164–65
  Wittgenstein on, 101, 102–11, 112, 113, 114–15, 139, 140, 141, 160, 208, 228, 236, 277, 278
  See also concepts; ideas; Internalist Thesis; Introspectibility Thesis; passions of the soul; Third Thing Thesis
Methodological Solipsism. *See* Internalist Thesis
Mill, John Stuart, 6, 100, 135–36
Minio-Paluello, L., 54, 56–57, 58, 59, 70
Moerbeke, William of, 58–59, 60, 71, 306n.47, 309n.81

natural kinds, 34–35, 146–51, 256, 258–74
  See also essences
Nietzsche, Friedrich, 332n.11

objects
  as goals, 205, 206, 214
  of the intellect, 82, 199–27, 228, 234–36, 241–44, 327n.25
  See also things
Ockham, William of, 176, 177, 178, 179, 180, 181, 235, 323n.52
Oesterle, Jean, 44
Owen, Joseph, 27, 28

Pasnau, Robert, 175–82, 207, 323nn.52, 57
  on noncognitive apprehension, 232–36

Index 355

passions of the soul
  Aquinas on, 12, 16, 19–20, 21–22, 33, 36–37, 55–56, 65–67, 68–71, 72–76, 208–13, 224–32, 237–49, 281, 297
  Aristotle on, 5–6, 15–16, 19–20, 31, 36–37, 39, 65–67, 68, 138–40, 276
  as concepts, 19–20, 21–22, 31–32, 36–37, 60–61, 62–63, 66, 73–76, 88, 89, 165, 208, 217, 227–32, 236, 237–49, 250–51, 276
  as likenesses of things, 5–6, 16, 19, 23–26, 31, 32, 33, 36–37, 38, 42, 47, 50, 52–53, 55, 80, 89, 138–39, 165, 173, 208, 227–32, 236, 238–39, 250, 257, 276, 308n.80
  as logically ordered, 5–6
  as means of signifying *res extra animam*, 25, 31–32, 38, 74, 89, 214
  relationship to signs (σημεῖα), 41–42, 43–55, 56, 57–58, 63, 69, 76, 77, 80, 138–40, 305n.25
  relationship to symbols (σύμβολα), 41–42, 43–55, 56, 57–58, 63, 69, 76, 77, 80, 138–40, 205n.25
  relationship to things, 5–6, 16, 19, 23–26, 31, 32, 33, 36–37, 38, 41–77, 80, 87–88, 89, 138–39, 165, 208, 214, 227–32, 236, 238–39, 250, 257, 276–78, 308n.80
  relationship to universality, 25, 32, 87, 302n.21
  relationship to words, 5–6, 15–16, 19–20, 31–36, 37–38, 41–77, 55–56, 80, 87–88, 138–41, 156–57, 236, 276–78, 308n.80
  as same for all, 16, 36, 37, 62, 66, 68, 70, 245, 251
  See also concepts; ideas; intellect; soul
Percy, Walker, 23, 160
Pieper, Joseph, 273–74, 332n.55
Plato, 218, 318n.37
  Aquinas on Plato, 20, 22, 73, 83, 167, 199–200, 201, 210–11, 212–13, 215, 265, 266, 268, 269, 291, 296
  vs. Aristotle, 20, 22, 73, 83, 154, 167, 199–200, 201, 210–11, 256
  Cratylus, 2, 5, 10, 41
private language
  Fodor on, 115
  Wittgenstein on, 89–90, 115, 311n.36
proper names, 11

Protagoras, 209
psychologism, 193, 325n.88
Putnam, Hilary
  on Aquinas, 34–35
  on Aristotelian form, 159–60, 257–58
  on Aristotelians vs. Platonists, 200, 201–2, 203
  on Aristotle's semantic triangle, 6, 8, 12, 135–36, 140–41, 145–46, 239, 246, 277
  brain-in-a-vat thought experiment of, 161
  on concepts, 194–98, 200, 237, 257–72
  on conceptual schemes, 194–98, 258, 262–63
  on division of linguistic labor, 256
  on essences, 257–74, 289, 290
  on Fodor, 128, 137–38, 151–53
  on formal identity, 237, 240
  on language as public, 8–9
  on Locke, 149–51
  McDowell on, 236, 277
  McGinn on, 146, 157
  on meaning, 135–49, 152–55, 212
  on meaning as extension, 141–43, 145–51, 152–53, 212
  on meaning as intension, 141–43, 145–51, 152–53
  on mental attention to concepts, 302n.19
  on mental/psychological states, 143–44, 147, 319n.25
  on mental representationalism, 6, 8, 81, 135–57, 159–61, 212, 236, 237, 257, 276–78, 319n.25
  on metaphysical realism, 194–98, 259
  on methodological solipsism, 144, 153, 154, 155, 157, 237, 241, 276–77
  on mind/soul, 159
  and narrow-broad content distinction, 132
  on natural kinds, 34
  on objects, 203, 205
  on science, 258–72, 280
  Twin Earth argument, 132, 141, 143, 146–49, 152–55, 265, 318n.39, 320n.28
  on Wittgenstein, 159, 257–58

Quine, W. V. O., 82, 142–43, 204–5, 264, 288–89

reference
  Frege on sense and, 102, 103, 136, 142, 143, 145, 313n.72, 314n.76

reference (cont.)
  Searle on, 11
  See also meaning
Reid, Thomas, 81
  on Berkeley, 99–100
  on Descartes, 99–100
  on Locke, 99–100
Rorty, Richard, 7–9, 10
Russell, Bertrand, 6, 81–83, 99, 101, 102, 112, 136, 154–55, 160, 326n.12

Schmidt, Robert W., 304n.43
science
  Aquinas on, 34–36, 39, 161–63, 210–13, 221–22, 255, 266, 267–68, 271–74, 281, 291
  Aristotle on, 34–36, 39, 161–63, 210–11, 250, 281, 330nn.21, 22
  Fodor on, 118–19, 318n.33
  Putnam on, 258–72, 280
Searle, John, 6, 26, 109, 157
  on intentional states, 324n.66
  on mental representations, 160–61, 324n.66
  on reference, 11
sense experience, Aquinas on, 177–79, 206–8, 214–15, 216–17, 219–23, 224, 228–29, 327n.25, 328nn.36, 48
signification
  Aquinas on, 12, 19–21, 22–26, 31–36, 65–67, 70–77, 88, 276, 301n.14
  Berkeley on, 94–96, 99, 278, 312n.50
  direct vs. indirect, 43, 54–56, 62–65, 66, 73, 75–76, 80
  Hume on, 97–98, 99
  immediate vs. mediated, 61, 63–64, 70, 72–76, 77, 88–89, 91–92, 95, 99
  Locke on, 64, 65, 77, 86–92, 94–95, 99, 278, 311n.36
  modus significandi of general words, 19–22, 32, 72, 73–74, 76, 87–88
  primary vs. secondary, 42–43, 44–52, 54–67, 69–71, 73–76, 77, 80, 88–89, 95, 99, 236
  signification₁, 22–26, 31, 32, 33–34, 35–36, 37–39, 76, 88, 247, 253–54, 256
  signification₂, 22–26, 31, 32, 33–34, 35–36, 38, 76, 88
  See also intellect; meaning; words
similitude/likeness
  Aquinas on, 25, 89, 165, 173, 208, 214, 217–18, 225, 227–32, 236, 244, 246, 248, 257, 276, 278

Fodor on resemblance, 114, 125, 126, 133
intelligible species as likenesses, 208, 214–15, 217–18, 225, 236, 248
in mental representations, 85–86, 103–4, 114, 116, 133, 137, 138–39, 219, 328n.48
passions of the soul as likenesses of things, 5–6, 16, 19, 23–26, 31, 32, 33, 36–37, 38, 42, 47, 50, 52–53, 55, 80, 89, 138–39, 165, 173, 208, 227–32, 236, 238–39, 250, 257, 276, 308n.80
and relationship between cause and effect, 228–29
and simple concepts, 250, 251
and transient vs. immanent activity, 214–15
skepticism
  and Aquinas, 213, 241–42, 258, 273
  epistemological, 7, 8–9, 83–84, 92, 93–94, 96, 99, 102, 112, 213, 241, 258, 273
  semantic, 9, 10–11, 105–6, 108, 109–11, 313n.72
Smith, Barry, 204
Sokolowski, Robert, 81, 82, 101, 166, 182
soul
  Aquinas on, 205–6, 242, 267–69, 289–91, 334n.45
  Aristotle on, 205–6, 237, 267–68
  Putnam on, 159
  Wittgenstein on, 290
  See also intellect; passions of the soul
Spade, Paul Vincent, 301n.14

things
  relationship to passions of the soul, 5–6, 16, 19, 23–26, 31, 32, 33, 36–37, 38, 41–77, 80, 87–88, 89, 138–39, 165, 208, 214, 227–32, 236, 238–39, 250, 257, 276–78, 308n.80
  relationship to words, 1–3, 5–6, 15–16, 22–26, 31–32, 41–77, 80, 87–88, 236, 276–78
  res vs. *ens*, 182–98, 325n.85
  res vs. *obiecta*, 205–6
  as same for all, 15, 42–43
  See also objects
Third Thing Thesis, 12–13, 155, 165–82, 189, 194, 199, 200, 226, 251, 264
Thomistic-Aristotelian tradition, 3–13
  criticisms of, 3, 6–8, 12, 19, 24, 77, 159–60, 212, 215–16, 257–72, 273, 274, 275, 277–79, 287–88

and epistemological skepticism, 7, 8–9
vs. *Linguistic Turn* tradition, 7–12, 24, 77, 136, 171–72, 197–98, 203–5, 207–8, 212–13, 266, 275, 277–78, 283–90, 298
vs. mental representationalism, 164–65
See also Aquinas, St. Thomas; Aristotle; Internalist Thesis; Introspectibility Thesis; Third Thing Thesis
truth and falsity
Aquinas on, 4, 26, 67–68, 71, 167–68, 244–45, 250–51, 254, 263, 275, 297
Aristotle on, 17–18, 26, 67–68, 70, 71, 300n.2
relationship to first act of intellect, 26, 67–68, 71, 244, 254, 263
relationship to second act of intellect, 4, 17–18, 26, 67–68, 71, 167–68, 244–45, 250–51, 254
Wittgenstein's picture theory of truth, 102–3, 314n.75
Twin Earth argument, 132, 141, 143, 146–49, 152–55, 265, 318n.39, 320n.28

understanding. *See* intellect
universality
and abstraction, 302n.21
Aquinas on, 32, 39, 87–88, 202–3, 217–18, 221–24, 238, 243, 252–53, 272–73, 302n.21
Berkeley on, 95, 97–98, 312n.50
Locke on, 95
relationship to passions of the soul, 25, 32, 87, 302n.21

Verbeke, Gerald, 58, 59
verbum mentis, 12, 300n.16
verificationism, 115

Wiggins, David, 303n.42
Willard, Dallas, 325n.88
Wittgenstein, Ludwig
on Berkeley, 12
Brown Book, 106, 107–8
on grammatical vs. logical form, 171–72
on Hume, 12
on interpretation, 3, 10, 103–6, 107–8, 109–11, 140, 284
on Locke, 12
McDowell on, 106–10, 162–63
McGinn on, 110–11, 284–85, 286, 288, 290
on meaning, 2, 3, 6, 15, 102–11, 283–88

on mental representationalism, 101, 102–11, 112, 113, 114–15, 139, 140, 141, 160, 208, 228, 236, 277, 278
on objects as goals, 205
on our natural history, 162–63
Philosophical Investigations, 2, 3, 107–8
on picture theory of truth, 102–3, 314n.75
on pointing, 3
on private language, 89–90, 115, 311n.36
vs. Putnam, 148
Putnam on, 159, 257–58
on souls, 290
on substantives, 169, 174
Tractatus Logico-Philosophicus, 102–3, 159, 257–58, 314n.75
on words as pictures, 2
words
Aristotle on, 5–6, 15–16, 17–18, 19–20, 238, 268, 276, 289
and behavioral dispositions, 2
conventions regarding, 5, 33–36, 37, 42, 47, 49–51, 138
differences between, 9–11
identity of, 37–39
meaning of, 5–13, 24
modus significandi of general words, 19–22, 32, 38, 72, 73–74, 76, 87–88, 89
and natural kinds, 34–35
nominal definitions, 250, 251–52, 255–56, 281
public meaning of, 5–6, 8–9, 24, 99
relationship to concepts, 31–36
relationship to passions of the soul (*passiones animae*), 5–6, 15–16, 19–20, 22–26, 31–36, 37–39, 41–77, 80, 87–88, 138–41, 156–57, 236, 276–78, 308n.80
relationship to things, 1–3, 5–6, 15–16, 22–26, 31–32, 41–77, 80, 87–88, 236, 276–78
as signs (σημεῖα), 41–42, 43–55, 56, 57–58, 63, 69, 76, 77, 80, 138–40, 305n.25
as symbols (σύμβολα), 41–42, 43–55, 56, 57–58, 63, 69, 76, 77, 80, 138–40, 305n.25
See also ideas; meaning; mental representationalism; signification; written language
written language
Aquinas on, 292–93
Aristotle on, 5, 16
Kretzmann on, 51–52, 306n.45

Zeis, John, 327n.38

JOHN P. O'CALLAGHAN
is assistant professor of philosophy at Creighton University.
He is co-editor of *Recovering Nature: Essays in Natural Philosophy, Ethics
and Metaphysics in Honor of Ralph McInerny,*
also published by the University of Notre Dame Press.

www.ingramcontent.com/pod-product-compliance
Lightning Source LLC
Chambersburg PA
CBHW021116300426
44113CB00006B/166